ZEITSCHRIFT FÜR GEOMORPHOLOGIE

Annals of Geomorphology – Annales de Géomorphologie

Supplementband 92

Some Contributions to the Study of Landforms and Geomorphic Processes

edited on behalf of
Deutscher Arbeitskreis für Geomorphologie

by Dietrich Barsch and Roland Mäusbacher

with 7 photos, 91 figures and 19 tables

1993

GEBRÜDER BORNTRAEGER · BERLIN · STUTTGART

ISBN 3-443-21092-9 / ISSN 0044-2798

Printed in Germany by K. Triltsch, Würzburg

Contents

Z. Geomorph. N.F.	Suppl.-Bd. 92	V–VII	Berlin · Stuttgart	Mai 1993

Introduction

by

DIETRICH BARSCH and ROLAND MÄUSBACHER, Heidelberg

In 1989, the German geomorphologists had the possibility to host the Second International Conference on Geomorphology in Frankfurt/Main. For this event, a number of Geomorphological Societies presented volumes reporting part of the activities of their members. This was greatly acknowledged by all participants, because the results of recurrent research are often not published in easy accessible journals. To enhance international cooperation and an international feedback, this Supplementband of the Zeitschrift für Geomorphologie presents some papers written by German speaking authors. They are dedicated to the participants of the Third International Conference on Geomorphology held in Hamilton (Canada), where this volume will be distributed to the international scientific community in geomorphology. Perhaps, this book will contribute to the success of the conference and to its main goal which is to demonstrate the importance of our discipline in the modern world.

The papers presented form by no means a complete cross-section of the recent activities of members of the Arbeitskreis. They display some of the most recent results. A great number of joint projects carried out by other colleagues are not named because their results have already been published. Thus, information on the priority program "Fluvial Dynamics during the Younger Quaternary" which is financed by the Deutsche Forschungsgemeinschaft are given by HAGEDORN & PÖRTGE (1989). Others are just starting, e.g. the priority programm on "Landscape Development during Late Pleistocene and Holocene" which will be coordinated by ANDRES (Frankfurt/Main). The groups of geomorphologists gathered in the Academy of Sciences in München (chairman: H. HAGEDORN) or in Göttingen (chairmen: H. MENSCHING and J. HAGEDORN) are publishing their own reports. Thus, the first one is centered on "Relief – Boden – Paläoklima", the second on "Geomorphologische Prozesse, Prozeßkombinationen und Naturkatastrophen in den Landschaftszonen und Höhenstufen der Erde".

Most of the papers in this book discuss Polar and Periglacial Geomorphology. This may be caused by the deep impression the Canadian North has made on German geomorphologists. Polar Geomorphology plays an important role in the research scheme of the German expeditions to Spitsbergen 1990–92, which are mainly centered on a better understanding of sediment fluxes from land to sea (c.f. BLÜMEL 1992). It is a joint project of several German and Swiss universities. WOLF-DIETER BLÜMEL (Stuttgart) as the coordinator of the program reports some of the first geomorphological results. Part of the material published in the following two papers

0044-2798/93/0092-0001 $ 0.75

was collected during this expedition. KING & HELL demonstrate the importance of geometric information on push moraines. They show that the push moraines in northern Spitsbergen are inactive in contrast to the active ones of the Queen Elisabeth Islands (N.W.T., Canada). BARSCH, GUDE, MÄUSBACHER, SCHUKRAFT, SCHULTE & STRAUCH discuss the problem of slush streams (Sulzströme) which are triggered in the thalwegs during beginning snowmelt. According to velocity and transport rates, there are low energy slush flows (Sulzfließen) and high energy slush torrents (Sulzmuren). Slush streams form an important transport system during the short transition from Arctic winter to Arctic spring when water starts to flow again.

Periglacial denudation in former unglaciated polar areas is a controversial problem. FRIED, HEINRICH, NAGEL & SEMMEL demonstrate that periglacial pediments in front of the Richardson mountains (northern Canada) are not actively formed today. They believe that they date back to older periods of the Holocene during which sheet flows have been possible. Changes of geomorphic activity in the sub-nival belt of the eastern Tyrolean Alps form also the central part in the paper by VEIT & HÖFNER.

In recent years, another controversial issue concerns the question whether the last Pleistocene glaciation caused inland ice on the Tibetean Plateau. HÖVERMANN, LEHMKUHL & PÖRTGE demonstrate that this is unlikely at the present state of knowledge, only isolated glacierized areas existed in mountain areas at the eastern rim of the Tibetan Plateau. Pleistocene environments are discussed in the next paper. Syngenetic polar permafrost in the Alpine Foreland in Bavaria influenced the bedding of outwash gravel during the maximum of the last glaciation. This has consequences for the paleoclimatic interpretation of the structures of glaciofluvial sediments. Periglacial solifluction sheets form wide-spread slope deposits in all upland areas ("Mittelgebirge") of Germany outside the Alps. There exists an old debate whether these strata can be differentiated according to the geoecological conditions during the time of their formation. VÖLKEL describes about the differences between the strata and the influence of these contrasts on soil formation in the Bavarian Forest.

The Cenozoic relief development in the area around the German Continental Deep Drill Site has been clarified by the use of fission-track analysis by BISCHOFF, SEMMEL & WAGNER. The authors accept LOUIS's view of differential horst-like uplift of fault blocks as correct rather than BÜDEL's homogeneous geomorphic development. They also successfully demonstrate the possibilities of the fission-track method for resolving geomorphic questions.

The reunification of Germany has made it possible to report experiences and information on areas of Asia which have been not accessible to Western-German geomorphologists. OPP & H. BARSCH report on an expedition to the mountain forest steppe of Northern Mongolia. The main focus of this project is centered on a better use of the natural potential of the area studied. The main evaluation was done by remote sensing and a ground check using geomorphic methods. A similar aim has the study by MÄCKEL & WALTHER who are monitoring geomorphological processes on test sites to evaluate a sustainable range management in semi-arid tropical Kenya.

Working in the Great Basin, KLEBER tried to combine methods used in Germany for the differentiation of periglacial deposits with the pedo-stratigraphic concept developed in the area of Lake Bonneville. He demonstrates that the influence of the periglacial parent material is often overlooked, but is of decisive importance on the formation of the fossil soils.

Tropical geomorphology has a long tradition in Germany. Thus, in BÜDEL's system the formation of tropical planation surfaces is one of the main paradigmas. It is, therefore, of great interest that HANNA BREMER reviews etchplanation in the tropics. BÜDEL's model is accepted as valid, because it provides the basic principles for an explanation of the major landforms in the tropical zone. The development of tropical lateritic crusts in Togo has been used to discuss again BÜDEL's Rumpfflächenhypothesis in northern Togo. RUNGE displays that the lateritic crusts are very different in origin and depend heavily on local petrography; therefore, these crusts can not be used as indicators of old erosion surfaces.

Soil erosion presents heavy problems in many agricultural areas of the world because it effects not only soil fertility, but also causes water pollution. BAADE, D. BARSCH, MÄUSBACHER & SCHUKRAFT describe field experiments which demonstrate ways of getting a better control on linear and diffuse soil erosion in small agricultural catchments in south-west Germany, without hampering agricultural techniques too much.

Relief should form today an important part in all models centered on fluxes in our environment. Unfortunately, the development of computer based methods is still not adequate to meet the developing demand. DIKAU reports possibilities of geomorphographic and geomorphometric approaches to relief which can be incorporated in Geographical Information Systems. The new techniques now allow us to discuss questions which were originally asked already by ALBRECHT PENCK (1894).

The papers presented comprise a wide variety of themes; they are published to commemorate the occasion of the Third International Conference on Geomorphology held in Hamilton/Canada. Three is already a tradition! The international community of geomorphologists has reached one of its main goals. Others are still to be gained.

References

BLÜMEL, W.-D. (1992): Geowissenschaftliche Spitzbergen-Expedition 1990 und 1991: „Stofftransporte Land – Meer in polaren Geosystemen". – Stuttg. Geogr. Arb. **117**: 416 pp.

HAGEDORN, J. & K.-H. PÖRTGE (1989, ed.): Beiträge zur aktuellen fluvialen Geomorphodynamik. – Gött. Geogr. Abh. **86**: 143 pp.

PENCK, A. (1894): Morphologie der Erdoberfläche. – Bd. 1 & 2: 471 & 696 pp; Stuttgart.

Address of the author: Geographisches Institut der Universität Heidelberg, Labor für Geomorphologie und Geoökologie, Im Neuenheimer Feld 348, D-6900 Heidelberg.

Z. Geomorph. N.F.	Suppl.-Bd. 92	1–19	Berlin · Stuttgart	Mai 1993

Contributions to Polar Geomorphology by the German Spitsbergen Expeditions 1990–1992

by

WOLF DIETER BLÜMEL, Stuttgart

with 10 figures and 1 table

Summary. Concept, structure and dominant aims of the "Geoscientific Expedition to Spitsbergen 1990–1992" (SPE 90–92) of the German Working Group for Polar Research (coordinated by W. D. Blümel) are presented. This multi- and interdisciplinary project is connected with the more enlarged program PONAM of the European Science Foundation and has the frametheme "Land to sea sediment transports and material fluxes in polar geosystems". Seventeen geo- and bioscientific institutes from Germany, Switzerland and Norway take part in a three-years investigation program, which is working from a basecamp the surroundings of the Liefde- and Bockfjord in NW-Spitsbergen.

Preliminary results from some working groups concerning geomorphological or paleoclimatic aspects are presented in this short paper. Following themes are mentioned: Geomorphic forms and processes; weathering and soil formation; remarks on glacial history; recent glacier settings and actual fluvial processes and transports. To all participating sections preliminary results have been published under: W. D. BLÜMEL (ed.): Geowissenschaftliche Spitzbergen-Expeditionen 1990 und 1991 "Stofftransporte Land–Meer in polaren Geosystemen" – Zwischenbericht –. Stuttgarter Geographische Studien Bd. 117, 1992, 416 Seiten (ISBN 3-88028-117-3; ISSN 0343-7906).

Zusammenfassung. Es werden das Konzept, die Struktur und die wichtigsten Zielsetzungen der „Geowissenschaftlichen Spitzbergen-Expedition 1990–1992" (SPE 90–92) des Deutschen Arbeitskreises für Polargeographie (Koordinator: W. D. BLÜMEL) vorgestellt. Das fächer- und disziplinübergreifende Vorhaben ist angelehnt an das Projekt PONAM der European Science Foundation und trägt das Rahmenthema ‚Stofftransporte Land–Meer in polaren Geosystemen'. Es sind geo- und biowissenschaftliche Institute aus Deutschland, der Schweiz und Norwegen an einer dreijährigen Geländekampagne beteiligt, die von einem Basislager aus das Gebiet um den Liefde- und Bockfiord in NW-Spitzbergen bearbeiten.

Aus der Zahl von insgesamt 17 Teilprojekten werden erste Ergebnisse zu geomorphologischen und paläoklimatischen Untersuchungen vorgestellt bzw. zusammengefaßt. Angesprochen werden die Themenbereiche: Geomorphologische Formen und Prozesse; Bodenbildung; Permafrost und Klimaveränderungen; Vereisungsgeschichte; rezente Vergletscherung sowie aktuelle fluviale Prozesse. Zu allen Teilprojekten wurden Zwischenergebnisse publiziert unter: W. D. BLÜMEL (Hrsg.): Geowissenschaftliche Spitzbergen-Expeditionen 1990 und 1991 „Stofftransporte Land–Meer in polaren Geosystemen" – Zwischenbericht –. Stuttgarter Geowissenschaftliche Studien, Band 117, 1992, 416 Seiten (ISBN 3-88028-117-3; ISSN 0343-7906).

0044-2798/93/0092-0001 $ 4.75

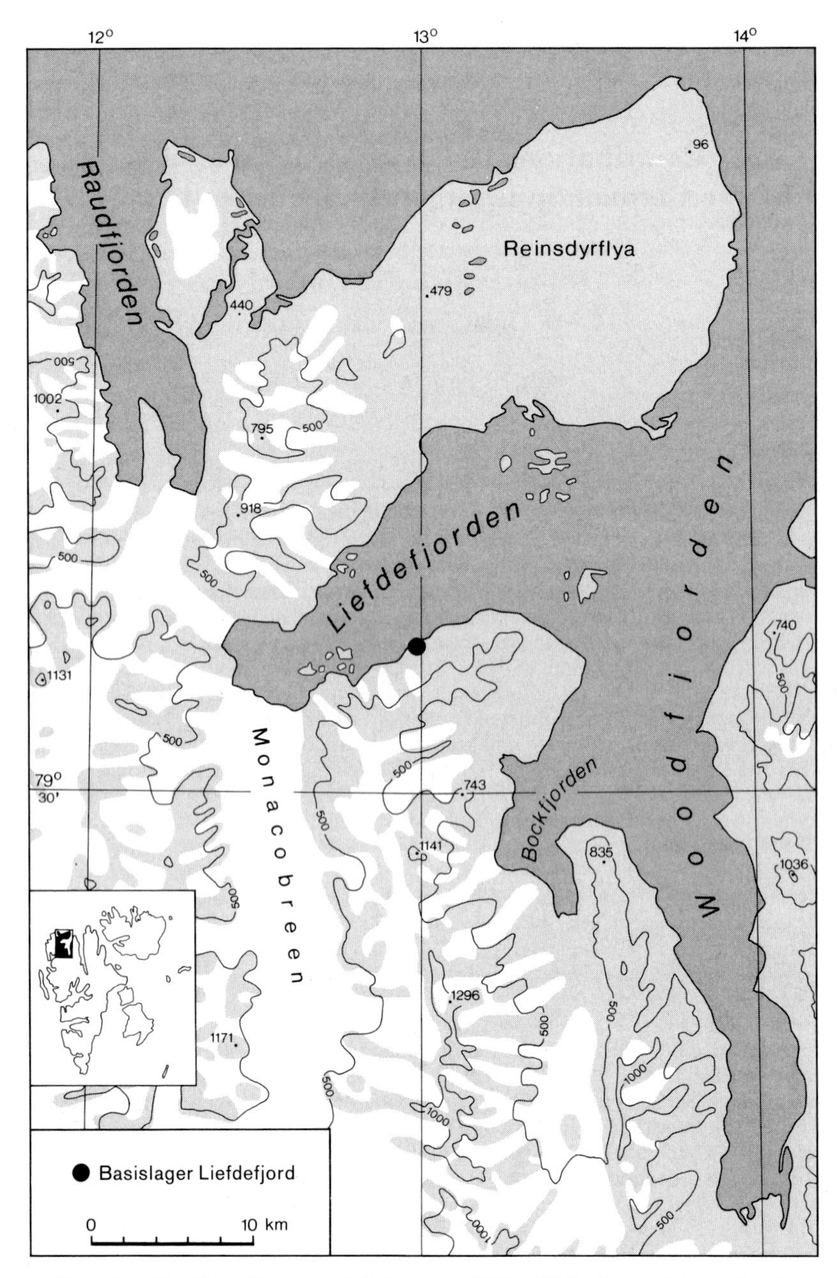

Fig. 1. Geoscientific Expedition to Spitsbergen 1990–1992: Area of investigation around Liefdefiord and Bockfiord (NW-Spitsbergen).

1 *Geoscientific Expedition to Spitsbergen 1990–1992 (SPE 90–92): Conception and goals*

The Geoscientific Expedition to Spitsbergen (1990–1992) continues a more than 100 years old tradition of German Geographical investigations in polar regions. More than 120 scientists are involved in a three years running program which was drafted by the 'German Working Group for Polar Research' (see LESER, BLÜMEL & STÄBLEIN 1988). Working field is the surrounding of the Liefde- and Bock-Fiord in NW-Spitsbergen (Fig. 1). A basecamp consisting of six huts was installed in 1989 on the south rim of the Liefdefiord. It is logistic, technical, experimental and social focus of different geographical, geological and biological activities. Crossings by rubber dinghy lead to distant mobile field camps or itinerant investigations.

In contrast to former projects, there exists a comprehensive, inter- and multidisciplinary, dominantly geographical research program, sponsored by the German Science Foundation (DFG) and the Swiss National Fund (SNF). The program has no singular main focus (like geomorphology, climatology or geobotany), but is separated into sections (Fig. 2) which contribute to a common framework: Goal of the expedition is to research in detail polar cryogenic geoecological systems in a relatively undisturbed arctic environment of high latitude. The connecting frametheme is "Land to sea sediment transports and material fluxes in polar geosystems". All topics are in connection with more enlarged programs of the European Science Foundation (ESF). SPE90 (meanwhile extended for two additional years) is a preliminary project of ASDEX (Arctic Sediment Dynamics Experiment) within the PONAM program (Polar North Atlantic Margins).

SPE is an international expedition which studies during the polar summers 1990–1992 within the active season the natural ecological processes in high-arctic landscapes. The investigations take place in catchments, in adjacent marine 'forelands' and surroundings of the Liefdefiord, which are observed representatively under time-space and eco-functional aspects. The scientific conception comes from a geoecosystem-model which is valid for actual and previous ecosystems.

This integral concept with four general sections share 17 working groups of Physical Geography with their specific disciplines from Germany, Switzerland and Norway (see Fig. 2). Connected are geological, geodetical/photogrammetrical/cartographical and biological projects. The integral structure and internal relations between sections, projects or topics shows Fig. 3.

An integrated recording of a landscape or ecosystem is only possible if the actual morphodynamic processes including weathering and soil formation are known. Additional: Only if morphogenesis (in the sense of landscape evolution) and its inherited relief forms, substrata and relictic/fossil soils as an historic-genetic inventory are included, the understanding for the actual ecological conditions and correlations can be developed. That is the reason why the cycle of materials and uncovering of the geomorphic and landscape history are additional central ideas of SPE 90-92: Mobilization, transport or accumulation at present as well as in the past are integrating, regulating and functional parts of the recent ecosystems. From this point of view the frametheme 'Land to sea sediment transport and material fluxes in polar geoecosystems' is to be seen and the scientific structure of SPE to be understood (see Fig. 3).

Section: GEOECOLOGY
- weathering and soil formation (BLÜMEL / Stuttgart)
- material fluxes in geoecosystems (LESER / Basel)
- vegetation / geobotany (THANNHEISER / Hamburg)
- organic trace elements and ion reactions (HERRMANN / Bayreuth)
- climatic ecology and remote sensing (PARLOW / Basel)
- bioecology of arctic marine and inshore waters (HARTMANN / Hamburg)

Section: FLUVIAL AND MARINE GEOMORPHODYNAMIC
- fluvial dynamics and processes (BARSCH / Heidelberg)
- off-shore sedimentation (fiord) (MÄUSBACHER / Heidelberg)
- litoral periglacial processes and erosion measuring (PRIESNITZ / Göttingen)
- coastal geomorphology (BRÜCKNER / Düsseldorf)

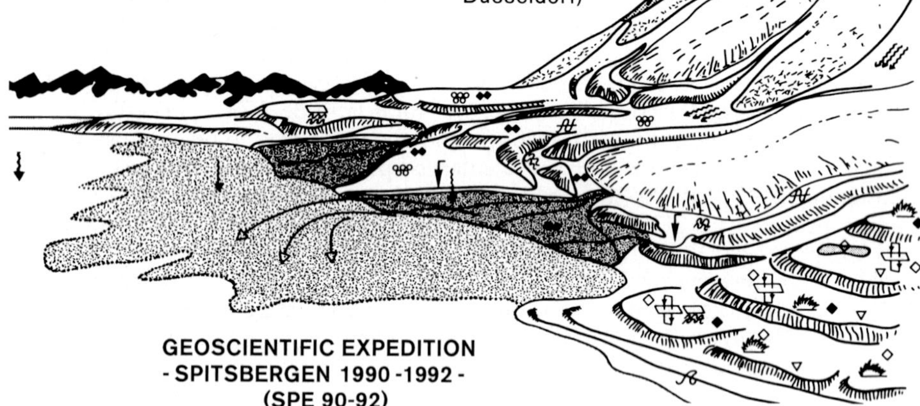

Section: GLACIAL AND PERIGLACIAL GEOMORPHODYNAMIC
- glaciation development and moraine dating (FURRER / Zürich)
- glacial geomorphology and ecology (KING / Giessen)
- glaciers and their forefront (SOLLID / Oslo)
- periglacial surface layers and abluation processes (LIEDTKE / Bochum)
- permafrost and relief development (STÄBLEIN / Bremen

ADDITIONAL PROJECTS
- geological mapping (THIEDIG / Münster)
- geodesy and photogrammetry (HELL / Karlsruhe)
- thematical cartography (BRUNNER / München)
- paleobotany (SCHWEITZER / Bonn)

Fig. 2. Investigation area (schematic diagram); sections; working groups/projects and scientists in charge.

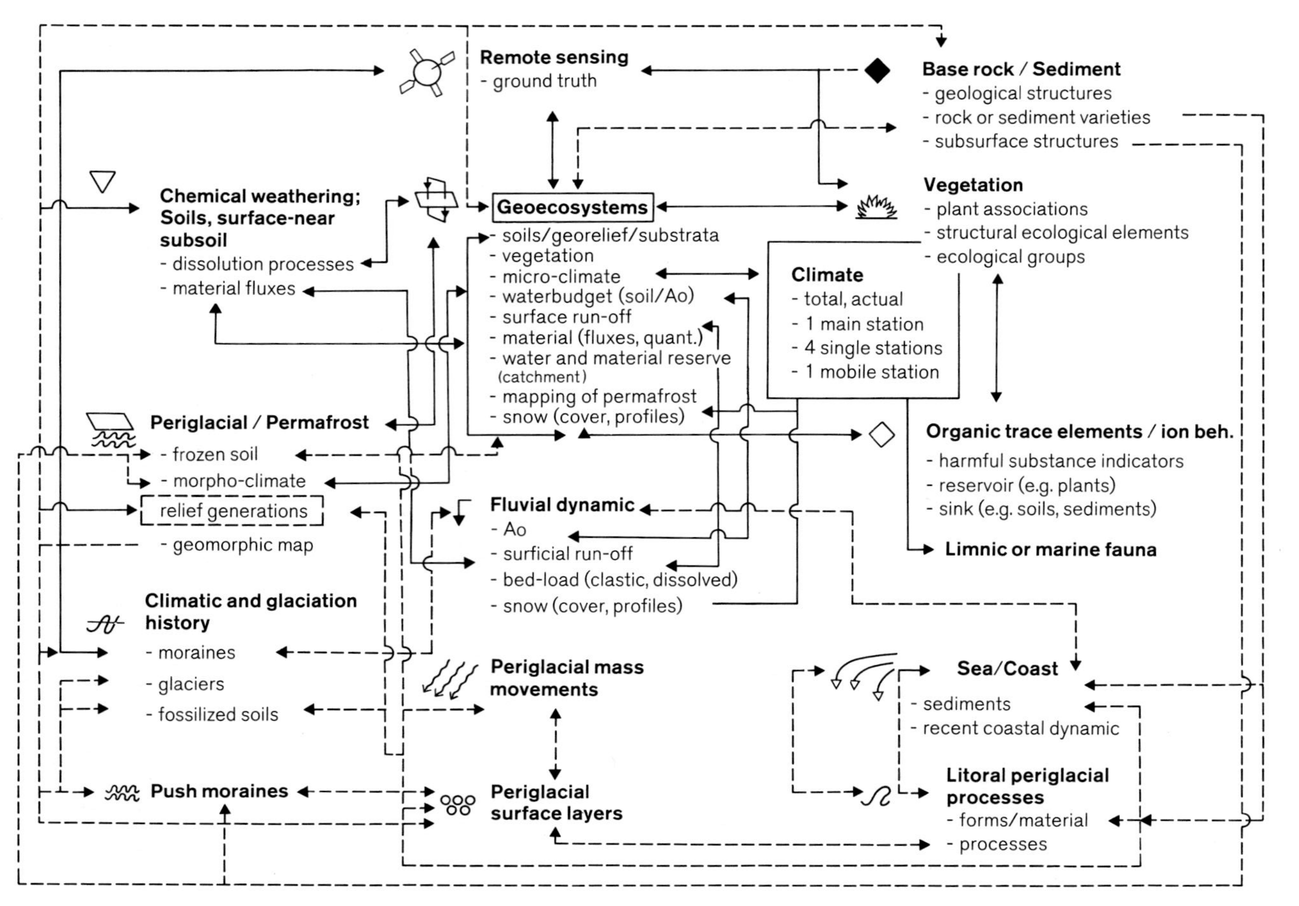

Fig. 3. Structure and sections of SPE 90–92 arctic investigations (transl. from H. LESER 1992).

To realize these integral concept, a geomorphic and geological configuration as a common test field was selected in NW-Spitsbergen (79,5°N/13°E; see Fig. 1). This area should have ready a multiple landscape inventory with different origin and dating. An outline of the expedition aims and the selection criteria of the area may be cited out of the scientific programm (see LESER, BLÜMEL & STÄBLEIN 1988):

1. Summary of the scientific framework

The project aims to describe not only the state or conditions of arctic geosystems from spatial, temporal or functional aspects, but also to work out the steering factors of the marine, litoral and terrestrial systems in polar landscapes as a fundamental correlation in extended areas. This formulation or concept corresponds to the geo-ecological environmental reality.
(. . .)
The area under study is situated in NW-Spitsbergen around the Liefdefiord and more distant surroundings. (. . .) In contrast to other areas of Spitsbergen (and other polar coasts of the northern hemisphere), NW-Spitsbergen represents a dominantly natural/virgin ecologic state. Therefore this area has a basic significance as a northern-hemispheric geoscientific-geoecological reference and comparison field.
(. . .)

2. Survey over the entire project

.. The plan in discussion since few years has the aim to investigate
* Land to sea sediment/material transports in polar geosystems. Basic idea is the research of a connected system between terrestrial geoecosystems, litoral and sea. The focal points are transport processes, geoecological and sedimentological accumulators. Their functional interaction shall be investigated spatially and (as far as possible and necessary) chronologically. (. . .)
The investigation area Germania-Peninsula (Fig. 1) lies south of Liefdefiord. Here are situated several catchments in parallel configuration with or without glaciers, which drain and deposit directly into the Liefdefiord. Nearly all varieties of relief formation and transport processes between terrestrial and marine systems are developed. The most remarkable features of the area are:

1. specific key position for glacial history, regional development of deglaciation and isostasy;
2. longer existing non-glaciation in postglacial periods;
3. undisturbed ecosystems, mostly still free of harmful substances;
4. one of the northermost areas with tundra vegetation; the high-latitude position enables long-distance comparison in zonal-climatic changes;
5. narrow-spaced geoecological variety;
6. example of an individual climate without direct gulfstream-anomalies;
7. leeward position to the Westspitsbergen-Stream;
8. variously equipped catchments;
9. easily comprehensible areas for investigations which can be instrumented by low means;

10. close meshed land-sea-systems;
11. material transports land-sea can be studied in transects on narrow space;
12. prolongation of transport-processes into deepwater possible;
13. Research subjects for all sections/projects in a common area can be carried out. This results in substantial and logistic connections.
14. up to now scarcely touched by scientific investigations; import results can be expected;
15. logistic connections to Ny Alesund are possible."

(end of the translated quotation)

2 *Preliminary results of geomorphological and paleoclimatic investigations*

In the following notes, some aspects and selected preliminary results especially of geomorphological projects shall be mentioned. They derive from a symposium held in Stuttgart at the beginning of the last year. Different papers from all groups involved in SPE (1990 and 1991) have been prepared. They do not cover all investigations or topics and cannot claim to fulfil the general framework of the expedition. The latter will be tried in a comprehensive volume after the evaluation of the third (and last) campaign in 1992.

The proceedings of the symposium mentioned above will be published under: W. D. Blümel (Hrsg.): Geowissenschaftliche Spitzbergen-Expeditionen 1990 und 1991 "Stofftransporte Land–Meer in polaren Geosystemen" – Zwischenbericht –. Stuttgarter Geographische Studien, Band 117, 1992, 416 Seiten (ISBN 3-88028-117-3; ISSN 0343-7906).

All authors quoted in the following extracts have published their preliminary results in this volume.

Projects or working groups with main emphasis on geoecology, botany, zoology, geology, meteorology, paleobotany or geodesy cannot be taken into consideration in this paper. Results and aspects of affiliated investigations are to be found in Blümel [ed.] (1992).

2.1 *Geomorphic forms and processes*

Geomorphic indications characterize the area of Germania-Peninsula/Liefdefiord as a long-time glacial as well as periglacial shaped landscape. "The constant cryogenic processes effected a form-process-accordance" (Stäblein 1992): A mature wealth of forms resulted by recent and subrecent periglacial processes in the investigation area (Fig. 1). Typical forms of the periglacial relief like sorted circles, palsas or others are rather seldom. Only non-sorted circles and stripes (mudpits and mudpit stripes) besides scarcely developed polygones and debris stripes occur more frequently. As Stäblein mentiones (and was observed by other geomorphologists, too), those cryogenic forms are often covered by tundra vegetation and represent the nowadays widespread inactivity of cryoturbation and solifuction in deeper situated relief positions.

Slopes are formed by amorphic, free or partly impeded solifluction ('cryofluction'). Slopes are dominantly smoothed out. Dissections occur occasionally, caused by periodical snow spots. On gentle slopes, especially on the parts of the lower slopes, in places cryoabluation and filter drainage is to be found within the active layer. Stäblein concludes that the recent periglacial denudation is only light. Because of the correlation between the recent periglacial geomorphic forms and the actual processes, a longer existing non-glacial period within holocene times can be postulated.

Abluation is the dominant recent denudation process on lower slopes. This term combines mainly denudative, in places accumulative processes caused by surficial or surface-near water runoff. (These dynamic was widespread active in the sand-rich elder glacial landscapes ‹Saalian glaciation› of Northern Germany or on the northern borders of the German Mittelgebirge during Middle-Weichselian times ‹65,000–25,000 yBP›). Liedtke & Glatthaar (1992) carried out intensive studies and detailed mapping on different testfields to assess the mechanism and quantitative efficiency of abluation processes. Abluation has only weak shaping effects and leaves only coarse-grained residual sediments on the surfaces, but is a significant part of material transfer in polar regions. Details of mapping and results see Liedtke & Glatthaar (1992).

2.2 *Remarks on soil formation*

The material transport by surficial water runoff or surface-near rinsing (ablution) is an important factor of the soil formation (eluvial or colluvial profiles) and influences the vegetation cover. The main steering factor for the efficiency of abluation is (besides grainsize and structure) the availability of water. Below snow-niches which produce meltwater all over the summer, this dynamic is to be observed impressively. The main source of ablual material is the 'Frostschuttzone' in the sense of Büdel – the zone above tundra-formations. The continuous supply/accumulation of fresh sediment leads to disturbances in soil-profiles and partly to the degradation of the vegetation (far more than by eolian deposits). Soils with up to four buried initial Ah-horizons and restitutions of humiferous layers could be observed (Fig. 4; see Eberle & Blümel 1992).

In the investigated area, parent material is dominated by series of Devonian sand-, silt- and mudstones. Soil associations, soil formation and weathering were observed by the Stuttgart working group and extensively mapped in its remarkable variety and heterogenety: The development of landscape controlled and effected by glacial, periglacial and marine geomorphological processes, causes a different and complex composition of parent material in various relief positions. Owing to this it effects soil formation.

Analogue to arid or semiarid regions with extreme climatic features, petrography, geomorphic position and processes (catenary relationships), vegetation, recent and former climatic conditions play a deciding role in soil formation and geoecology. Soils in northern Spitsbergen cannot be combined in narrow simplifications or monocausative factors like 'climate'. Soils represent individual features and express the complex local integration of geo- and biofactors.

Fig. 4. Arctic Brown Soil (gelic Cambisol) on marine terrace sediments: On top of the profile, fine-grained layers (each with initial humiferous horizons) have been deposited by abluation processes (photo BLÜMEL, July 1990).

Soil thinsections combined with chemical and mineralogical analysis give new informations about the intensity of in situ-weathering. Well developed brown soils (gelic Cambisols) were only found on coarse textured and well drained locations with geomorphological stability since a long time (EBERLE & BLÜMEL 1992). Soils with finer 'loamy' texture are subjected to cryoturbation and abluation. Therefore horizontation is perturbed. Typical soils in such positions are gelic Regosols and gelic Leptosols. Impeded drainage restrains chemical processes and weathering; well-drained locations with unsaturated soil solutions show the most intensive pedological development and alteration of minerals.

The differentiation between lithological and pedological features (e.g. iron oxides, clay minerals) is of great significance relative to the correct estimation of weathering rates and soil formation: The clay mineral composition in Bv-horizons of arctic brown soils does not differ from the lithogene spectrum of the Devonian sedimentary rocks. A pedogene transformation cannot be proved. Consequently, decarbonization and oxidation (pedogenic free iron oxides; see Weber & Blümel 1992) are the dominant weathering and pedogenic features.

The Bale working group on geoecology applies sophisticated methods and measurements on different test fields (tesserae). Some first results have been published by Leser et al. (1990) and in several papers in the proceedings volume mentioned above. Additional geomorphological and geoecological investigations are persued by the Giessen team (see King et al. 1992).

Fig. 5. "Cryostatic block" still covered with fresh subterranean material within a stabilized periglacial milieu (tundra vegetation) hints to young elevation and deeper thawing of the active layer. The probably reason is the recent warming-up in high latitudes (photo Blümel, July 1990).

Fig. 6. Spontaneously moving masses ('sliding tongues'): This middle part of the slope was rather inactive and covered dominantly by moss-tundra. The actual deeper thawing leads to a renaissance of periglacial mass-movements by temporary oversaturation (photo BLÜMEL, July 1990).

2.3 *Permafrost and climatic variations*

Geoelectric measurements (STÄBLEIN) to investigate the permafrost layer under consideration of soil temperatures, snow cover and winterly snow temperatures come to the result that the thickness lies between 30 m (coastal near) and up to 100 m in more elevated sections of the forelands.

As recorded by other working groups, too (Basel, Stuttgart), the active layer normally reaches deeper than 1 meter (1.50–2 m). Ice segments or pure ice structures are seldom in the upper parts of the continuing permafrost. An outstanding phenomenon represent the "cryostatic blocks" (Fig. 5). These moraine boulders are recently elevated out of the ground by a deeper thawing layer and make visible – together with frequent young mass-movements ('sliding tongues'; see Fig. 6) – an actual climatic variation towards upwarming. The same effect shows Fig. 7: The depth of the soil-formation (andosolization) is correlated with the former active layer. Now the active layer reaches much more deeper.

Aspects of holocene glacier variations are mentioned in chapter (2.5). The youngest push of terrestrial and outlet glaciers – culminating in the last century at the end of the 'Little Ice Age' – left behind prominent ice-core moraines. Thermoerosion is proceeding now. A lot of material is transported by reinforced thawing

Fig. 7. Hints to recent upwarming in a soil profile with a relictic ice-wedge on volcanic terrace sediments (Bockfiord). The basis of the soil documents the former depth of the summerly active layer (about 1 m). Now the thawing reaches 2,5 m (end of the rule; photo EBERLE, July 1991).

of the ice-cores (slumping of lateral moraines and the inner parts of terminal moraines).

The escalating retreat can be observed by aerial photographs: F.e. the large Monacobreen (Fig. 1) at the end of the Liefdefiord has lost nearly 2 km since 1966. All larger terrestrial glaciers possess young steep moraine walls. Up to now, they have retreated with mostly three stops about several hundred meters. Thermoerosion and melting back of glaciers are additional, more evident signals for nowadays climatic changes in polar regions.

2.5 *Remarks to glacial history*

In the opinion of the author, the extent system of deep and long fiords in Spitsbergen goes back into a maximum glaciation phase in the Younger Tertiary. The distribution of glacial and erratic boulders up to 700 m above sealevel shows a mighty, widespread glaciation elder than the last glaciation. The Weichselian (Wisconsin) ice-net only covered the lower parts of the fiords and the side of Reinsdyrflya. A correlation of moraine stages with glacial periods is tried (see Abb. 1 and Tab. 3 in STÄBLEIN 1992). The weak differences concerning state of weathering, soil formation or composition of moraines are not sufficient to differentiate in age because of the heterogeneous petrography of the catchments. It seems that the Woodfiord was only partly filled with ice in its southern and middle sections during last glaciation. A clear reconstruction by geomorphic features and field methods is very difficult.

Within the soil profiles, geomorphological forms, correlated processes and sediments – and especially moraine complexes – numerous hints to climatic fluctuations/variations are to be found but difficult to date. The working group of G. Furrer (Zürich) succeeded in dating and revealing the younger glacial resp. climatic history of northern Spitsbergen. The reconstruction of the holocene glaciers fluctuations bases mainly on the stratigraphic analysis of peats, bogs and fossil soil horizons connected with moraine deposits. The radiocarbon dating provided the chronological time scale. Paleobotanical aspects derive from the analysis of macrorests and pollen.

"Numerous fossil organic horizons (fAh) could be found by digging through the leeward side of morainic ridges, mostly frozen by permafrost, and than tracing the transition between rock basement and glacial deposits. These soils are interpreted to be overridden and buried by former glacial advances. The radiocarbon dating proves four different periods of distinct glacier advances (phase A, B, and D; see Fig. 8) occurring during the early and late stages of the Older Subatlantic (PZ IX: 2800–1000 y BP) and the early and middle phases of the Younger Subatlantic (PZ X: since 1000 y BP). The pollen analysis of the fossil soil layers suggest a strong similarity to the present vegetation cover close to the coastline. Therefore the climatic conditions during the soil and peat bog formation periods were at least and most likely not less favorable than they are today.

The results obtained by this project can be summarized as follows (see Fig. 8):

Lateglacial

Based on our field observations we suppose, that the region of the Liefdefiord was deglaciated prior to the beginning of the Alleröd (PZ II: 11,800–11,000 y BP). The glaciers then reached at least an extent similar to the recent position.

Younger Dryas

Evidence of late Weichselian (Würmian) or of early holocene advances could not be found in the studied area. The question concerning the glacier behaviour during the cold phase of the Younger Dryas (PZ III: 11,000–10,200 y BP) is therefore still unanswered.

Holocene

The glacial history of the Late Holocene can be divided into four different advance periods. Probably during the entire Holocene the glaciers have been oscillating always within the same limits surrounded by the moraines of the Subatlantic and the

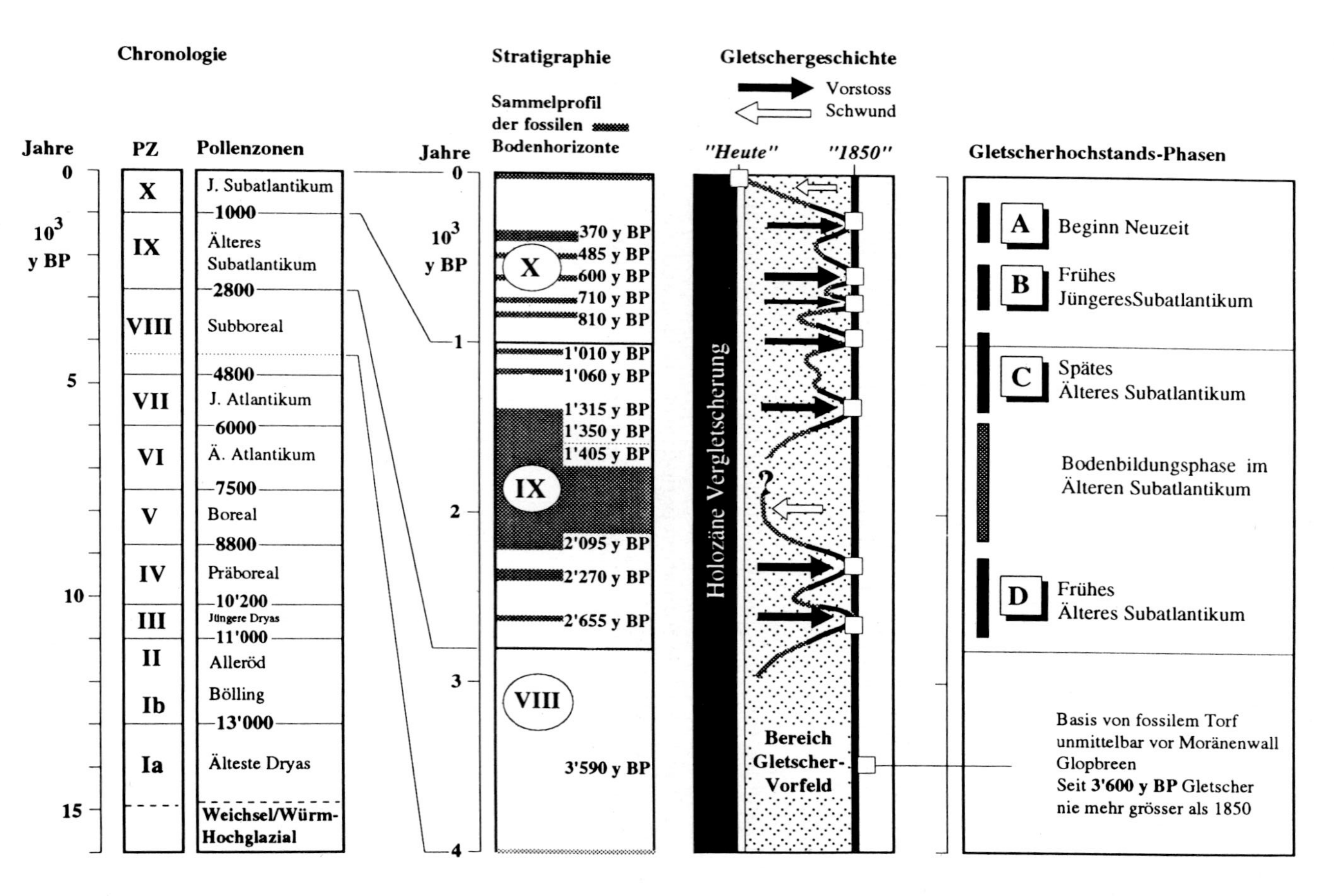

Fig. 8. Compilation of radiocarbon datings and subdivision of the glacier pushes into the four different maxima within the Younger Holocene (A, B, C and D), (FURRER 1992).

Little Ice Age period. This advance and retreat behaviour is particularly typical also for the holocene glacier fluctuations found in the Alps.

Isostatic uplift – sea level regression

An isostatic uplift leading to a sea level regression on the order of 1.60 m since 4000 y BP can be derived from data marine deposits found in a profile close to the Erikbreen." (FURRER 1992)

2.6 *Recent glacier settings*

Waterbudget and glaciofluvial sediment transfer of a Liefdefiord glacier (Erikbreen) are studied (beside additional aspects of glaciology) by a Norwegian working group from Oslo (see VATNE et al. 1992). They contribute to the framework of SPE 90–91 – material fluxes land to sea – by observing a larger glacial catchment. Suspended sediments and dissolved salts have been recorded (see Fig. 9).

Waterbudget and sediment transfer of Erikbreen were monitored in July and August 1990 and 1991. The total runoff was 8.3 million $m^3/49$ days (1990) and 3,4 million $m^3/44$ days (1991). In 1990 3747 tons were transferred as suspended sediments and 1836 tons as dissolved salts. For 1991 the transfer was only 553 tons

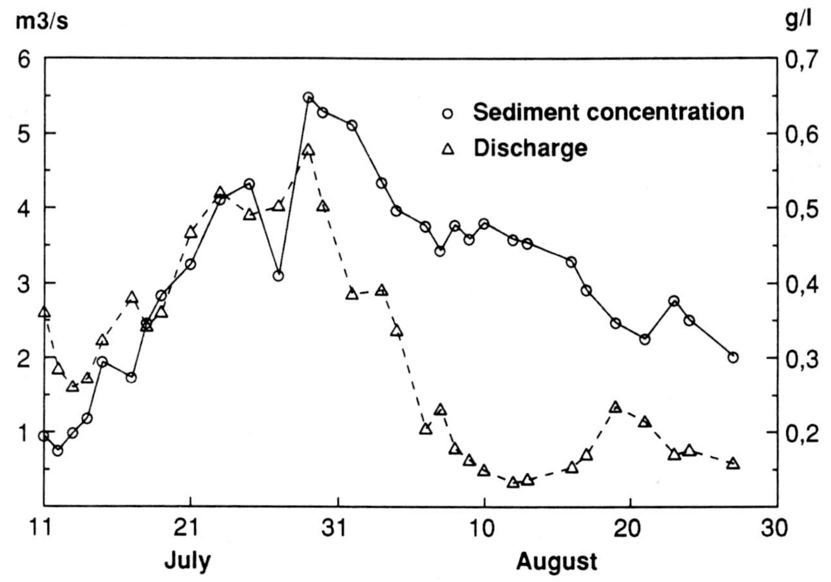

Fig. 9. Runoff and concentration of suspended sediments from the lake in front of the Erikbreen during 1990 (VATNE et al. 1992).

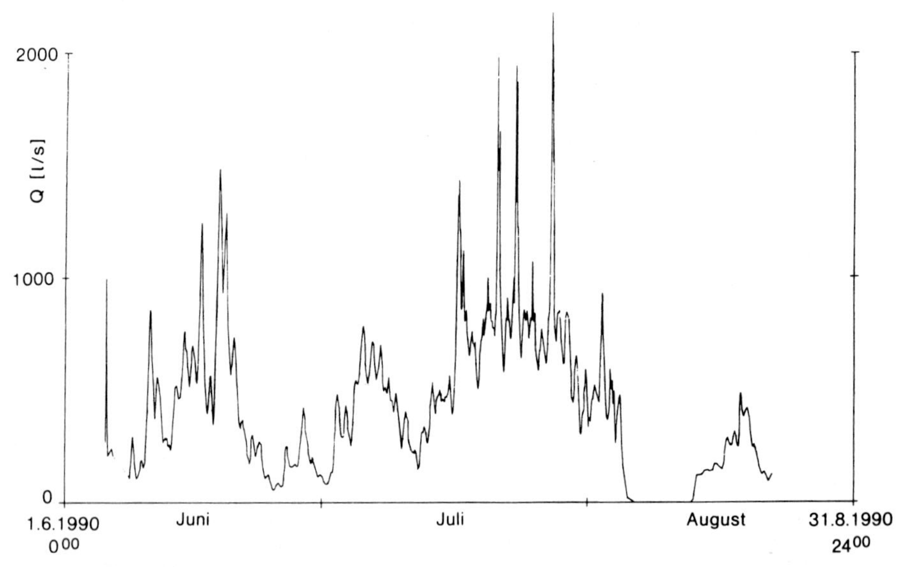

Fig. 10a. Discharge 1990 at the gauge Kvikkaa (BARSCH et al. 1992).

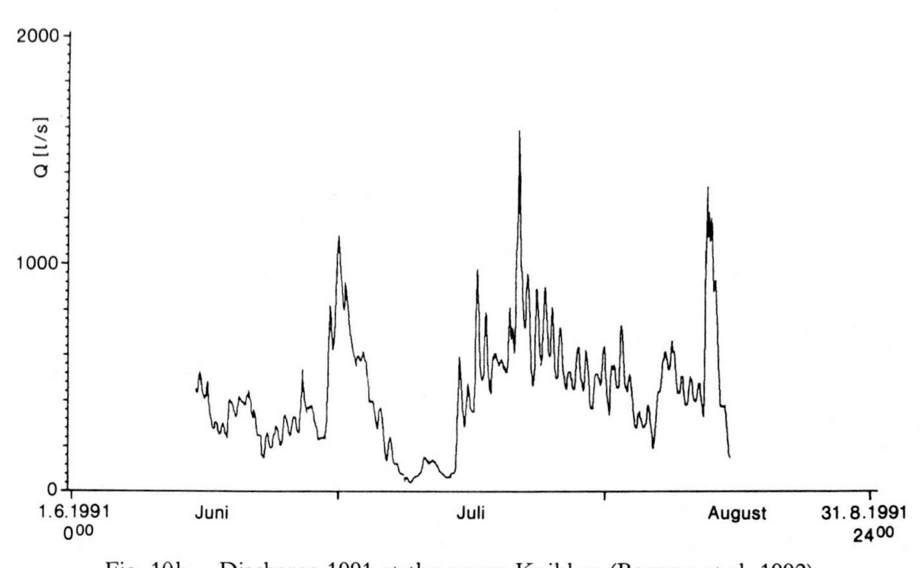

Fig. 10b. Discharge 1991 at the gauge Kvikkaa (BARSCH et al. 1992).

(suspension) and 275 tons (dissolution). Additional 3600 tons resp. 550 tons in total 1990 and 1991 deposited in a lake in front of the glacier.

These summarized short dates show a high variability of discharge and sediment transfer in this polar area. The concentration of suspended sediments and dissolved salts indicate extensive subglacial drainage as well as the presence of lodgement till beneath the glacier.

Temperature measurement and radio-echo soundings show that Erikbreen is a subpolar glacier with an upper cold layer and a temperate sole except along the margins.

Compared to glaciers of the same size in Spitsbergen, Erikbreen has high flow velocities. A maximum mean velocity of 44 m/year is measured in the central ablation area. This is more than a magnitude larger than observed on glaciers further southwest in the Kongsfiord area. (More details about methods and measurements applied at Erikbreen see VATNE et al. 1992).

2.7 *Actual fluvial processes and transports*

The Heidelberg working group studied fluvial processes and transports in three small catchments with dimensions between 4.6 and 5.1 qkm on the northwestern part of Germania-Peninsula (southern rim of Liefdefiord). Fluvial dynamic (geomorphodynamic) is the interface between the terrestrial and the marine sphere. Beneath the glaciers, the recent discharge and sediment dynamic is the final link of

Table 1 Comparison of fluvial transports from the Kvikkaa catchment during 1990 and 1991 (BARSCH et al. 1992).

		1990	1991
Water equivalent of the snow cover		1,2 mio qm	(*)
discharge:	total discharge (gauge)	2,8 mio qm	2,4 mio qm
	Q_{max}	2200 l/sec	1600 l/sec
	events >1000 l/sec	8	4
	start of discharge	June, 5th	June, 14th
suspension:	transport through gauge	237 tons	124 tons
	total suspension export at the mouth	177 tons	103 tons
	specific transport/erosion	46 g/qm	24 g/qm
	number of great events	3	0
bedload:	transport trough gauge (basket/sink)	5/– tons	2,2/3 tons
	total bedload export at the mouth (tub/sink)	226/– tons	60/90 tons
	max. transport (baskets)	112 kg/h	9 kg/h
solution:	total transport	89 tons	109 tons
	input of aerosols	16 tons	(*)
	specific transport	17,3 g/qm	21,8 g/qm

(*) not yet determined.
– in 1990 not installed.

a complex chain of mass and energy transfer within the catchment. Concerning the marine milieu, they are primary sizes for the sedimentation rates in the fiord. Sophisticated methods for measurement of discharge and fluvial transports have been applied. Some aspects of the results or trends shall be referred (see Barsch et al. 1992).

Both summers are similar regarding the total discharge of the Kvikkaa catchment:

1990: 2.8 million cubicmeters
1991: 2.4 million cubicmeters

The flood peak, however, was higher during 1990 by 40% (2200 liter/sec. versus 1600 liter/sec. in 1991; see Fig. 10a, b). Therefore, bedload transports were greater by a factor of 2.3 (5 tons versus 2.2 tons).

Table 1 summarizes the most important results of both campaigns, containing discharge, transport in suspension, bedload and transport in solution. A short conclusion is that transport in suspension is the most effective one, transport in solution and especially bedload are less important. About 18% of transport in solution has been related to aerosol input during the winter. (This input is thought to be partly responsible for the triggering of slush-flows. A model for the development of this snow melt and transport mechanism is presented in the paper mentioned above.) Discharges about 600 liter/sec. start bedload movements. More details, explanations and interpretations see Barsch et al. 1992.

2.8 *Investigations on sediments*

Two working groups will complete the geoscientific program in 1992: The Late Cenocoic – holocene sedimentation in Liefde- and Woodfiord will be investigated by R. Mäusbacher (Heidelberg). Echography and seismic methods shall record the mightiness of the Late Cenocoic up to recent times accumulation. Informations about dynamic and rates of sedimentation are expected from cores as well as hints to climatic changes and (de-)glaciation history. The same aims are pursued by R. Baumhauer (Trier) and U. Glaser (Würzburg): They take sediment cores in seasonal lakes and fluvioglacial deposits between moraines and fiords.

Acknowledgement

All authors and participants of the expedition are grateful to the German Science Foundation (DFG; Deutsche Forschungsgemeinschaft, Bonn) and the Swiss National Funds (SNF) for the financial support of this arctic project.

References

Barsch, D., M. Gude, R. Mäusbacher, G. Schukraft & A. Schulte (1992): Untersuchungen zur aktuellen fluvialen Dynamik im Bereich des Liefdefjords in NW-Spitzbergen. – In: Blümel [Hrsg.] (1992).

BLÜMEL, W. D. [Hrsg.] (1992): Geowissenschaftliche Spitzbergen-Expeditionen 1990 und 1991 „Stofftransport Land–Meer in polaren Geosystemen“ – Zwischenbericht –. – Stuttgarter Geographische Studien, **117**: 416 S.

EBERLE, J. & W. D. BLÜMEL (1992): Substratgenese und Bodenentwicklung im Bereich devonischer Sedimentgesteine des Liefde- und Bockfjordes (NW-Spitzbergen). – In: BLÜMEL [Hrsg.] (1992).

FURRER, G. (1992): Zur Gletschergeschichte des Liefdefjords. – In: BLÜMEL [Hrsg.] (1992).

KING, L., E. SCHMITT & D. WOLLESEN (1992): Untersuchungen zur Geomorphologie und Geoökologie eines glazial geprägten arktischen Küstengebietes. Liefdefjorden, Spitzbergen. – In: BLÜMEL [Hrsg.] (1992).

LESER, H. (1992): Methodische und ökologische Aspekte im SPE-Projekt: Idee und Wirklichkeit. – In: BLÜMEL [Hrsg.] (1992).

LESER, H., W. D. BLÜMEL & G. STÄBLEIN (1988): Wissenschaftliches Programm der Geowissenschaftlichen Spitzbergen-Expedition 1990 (SPE 90) „Stofftransporte Land–Meer in polaren Geosystemen“. – Materialien und Manuskripte 15, Bremen.

LESER, H., S. REBER & A. REMPFLER (1990): Geoökologische Forschungen in Nordwest-Spitzbergen. Erster Bericht über das Teilprojekt Geoökologie der Geowissenschaftlichen Spitzbergen-Expedition 1990 (SPE ’90) zum Liefdefjord. – In: Die Erde, **121**: 255–268.

LESER, H., K. DETTWILER & CH. DÖBELI (1992): Geoökosystemforschung in der Elementarlandschaft des Kvikkaa-Einzugsgebietes (Liefdefjorden, Nordwest Spitzbergen). – In: BLÜMEL [Hrsg.] (1992).

LIEDTKE, H. & D. GLATTHAAR (1992): Abluation auf den Gesteinen des Siktefjellet am Liefdefjord (Spitzbergen). – In: BLÜMEL [Hrsg.] (1992).

STÄBLEIN, G. (1992): Zur quartären Klima- und Permafrostentwicklung am Liefdefjorden/Nordwest-Spitzbergen. – In: BLÜMEL [Hrsg.] (1992).

STÄBLEIN, G. & H. KÖNIG (1992): Temperaturmessungen zum lokalen Bodenwärmehaushalt am Liefdefjorden / Nordwest-Spitzbergen. – In: BLÜMEL [Hrsg.] (1992).

VATNE, G., B. ETZELMÜLLER, R. ODEGARD & J. L. SOLLID (1992): Waterbudget and glaciofluvial sediment transfer of a subpolar glacier, Erikbreen, Svalbard. – In: BLÜMEL [Hrsg.] (1992).

WEBER, L. & W. D. BLÜMEL (1992): Methodische Probleme bei bodenchemischen Untersuchungen an Böden aus dem Liefdefjord/Bockfjord-Gebiet (NW-Spitzbergen). – In: BLÜMEL [Hrsg.] (1992).

Address of the author: Prof. Dr. W. D. BLÜMEL, Geographisches Institut der Universität Stuttgart, Silcherstraße 9, D-7000 Stuttgart 1.

Z. Geomorph. N.F.	Suppl.-Bd. 92	21–38	Berlin · Stuttgart	Mai 1993

Photogrammetry and Geomorphology of High Arctic Push Moraines, Examples from Ellesmere Island, Canadian Arctic, and Spitsbergen, Svalbard Archipelago

by

LORENZ KING, Giessen, and GÜNTER HELL, Karlsruhe

with 10 figures

Summary. Actual glacial geomorphological processes have been researched in two arctic areas: Liefdefjorden in Spitsbergen, Svalbard Archipelago, and Expedition Fiord and Oobloyah Bay in the Queen Elizabeth Island, Canadian Arctic Archipelago. The build-up and the transformation of moraines after deposition have been investigated with modern photogrammetric techniques using semi-metric cameras. The method is described detailedly and the accuracy obtained is discussed.

The semi-metric camera proved to be an important working tool on arctic expeditions, especially in view of their flexibility and their easy use for aerial photography during move-in or take-out when helicopters or planes are available anyhow. The following geomorphological results have been obtained mainly with semimetric cameras but also other field methods such as mapping or geoelectrical soundings.

The push moraine of Thompson Glacier, Expedition Fiord, has remained active for more than thirty years and shows a horizontal displacement of 400 m to 500 m. In spite of this considerable dislocation, the main structures of the large moraine complex have been preserved in the older parts and thermal erosion is relatively rare. Whereas the horizontal displacement of the moraine complex was more than 20 m per year in the sixties, the velocity decreased to an average of about 14 m per year in the last 20 years and to about 10 m per year in the eighties. The movement and the topography of the moraine are compared with the characteristics of push moraines in front of a stagnant glacier (Troll Glacier, Oobloyah Bay) and of a receding glacier (Hare Fiord).

In Spitsbergen, push moraines are often ice-cored and it is difficult to distinguish between frozen debris and debris covered dead glacier ice. Thermal erosion in neoglacial and older moraines is widespread, and once started, melting of debris covered ice and frozen debris in moraines is progressing rapidly with melt-down rates of more than 5 cm/year over large areas.

Zusammenfassung. In zwei arktischen Gebieten, dem Liefdefjord in Spitzbergen und am Expedition Fjord der Oobloyah Bay in den Queen Elizabeth Islands, kanadische Hocharktis, wurden aktuelle glazialmorphologische Prozesse untersucht. Das Interesse galt vor allem dem Aufbau und der Überformung von Moränen nach ihrer Ablagerung, wobei moderne photogrammetrische Arbeitsweisen zum Einsatz gelangten (Halbmeßkammer, semi-metric camera). Das Vorgehen und die erhaltene Genauigkeit werden ausführlich beschrieben und diskutiert.

0044-2798/93/0092-0021 $ 4.50

Infolge ihrer vielseitigen Verwendbarkeit und Handlichkeit hat sich die "semi-metric camera" als wichtiges Hilfsmittel sehr bewährt, insbesondere bei der Aufnahme von Luftbildern während des Fluges zum Basislager, da dabei kaum zusätzliche Kosten entstehen. Die photogrammetrischen Arbeitsmethoden wurden ergänzt durch geomorphologische Kartierungen und geoelektrische Sondierungen.

Die Stauchmoräne des Thompson Gletschers am Expedition Fiord ist seit nunmehr über dreißig Jahren aktiv und zeigte während dieses Zeitraums eine horizontale Verschiebung von 400 bis 500 Metern. Alle wichtigen geomorphologischen Elemente des großen Moränenkomplexes wurden dabei aber trotz des hohen Verschiebungsbetrages, insbesondere in den älteren Teilen der Moräne, erhalten. Bereiche mit Thermoerosion treten selbst bei der vorhandenen ausgeprägten Topographie nur selten auf. Über den rund dreißigjährigen Meßzeitraum hinweg nahmen die horizontalen Verschiebungsbeträge ab. In den sechziger Jahren lagen sie noch bei mehr als 20 m pro Jahr, im Mittel der letzten 20 Jahre bei 14 m und in den achtziger Jahren bei etwa 10 m pro Jahr. Die Bewegung und die Topographie dieser aktiven Moräne werden verglichen mit den Grundzügen von Stauchmoränen vor der Stirn eines stationären Gletschers (Troll-Gletscher, Oobloyah Bay) bzw. eines zurückschmelzenden Gletschers (Hare Fiord).

Im Unterschied zu den untersuchten Stauchmoränen in Kanada zeigen Moränen in Nordspitzbergen oft einen Kern aus massivem Gletschereis, wobei es meist schwierig ist, zwischen gefrorenem Schutt und schuttbedecktem Toteis eine Unterscheidung zu treffen. Thermoerosion ist in neuzeitlichen und auch älteren Moränen in Spitzbergen weit verbreitet, wobei nach ihrem Einsetzen vertikale Abschmelzbeträge von mehr als 5 cm pro Jahr gemessen wurden. Die geomorphologischen Aufnahmen zeigen, daß dieser über lange Zeiten sich hinziehende Vorgang durch mehrjährige Ruhephasen unterbrochen werden kann, indem Toteis durch abgetrocknete Schlammströme überdeckt wird.

1 *Introduction and aim of study*

High arctic glacial environments are quite widespread today, especially in the Canadian Arctic, Greenland and in the Svalbard Archipelago, but e.g. also on the Russian arctic islands. Many other regions may be included if high mountain areas in subarctic latitudes are added, especially in Canada and Alaska. Whereas the periglacial geomorphology is comparatively well known in these areas, the glacial processes are less studied mainly due to access problems and logistic problems arriving in the field (e.g. large meltwater streams). These areas, however, are of great interest, first due to the fact that vast regions have experienced glacial processes in arctic environments during the Quaternary glaciations that left impressive remains in today densely population areas, secondly and more recently within the discussion on climatic change and its geomorphological consequences.

One very impressive glacial feature, the push moraine, has attracted the interest of researchers for almost 100 years. They have mainly been described from areas of Quaternary ice margins in Europe, e.g. the Netherlands, North Germany, Poland, but also in North America (CHAMBERLAIN 1893, GRIPP 1929, CARLE 1938, LAMERSON et al. 1957, RUTTEN 1960, MACKAY et al. 1964). The interest and research of active push moraines has been intensified in the last few years (VAN DER MEER & BOULTON 1986, BOULTON & VAN DER MEER (eds.) 1989, RITZEBOS et al. 1986, and recently by VAN DER WATEREN 1992) and it started even in the Alps (HAEBERLI 1979, EYBERGEN et al. 1987).

Arctic push moraines are certainly as impressive as relict features. A classical and very thorough study on an active arctic push moraine is the thesis of KÄLIN (1971) about the Thompson Glacier push moraine, and although he showed that there are about 35 similar features in the Queen Elizabeth Islands, none of them has been studied since (cp. ADAMS 1987). We have chosen therefore primarily this classical Canadian region for our research. The aim of our study was to investigate geomorphological forms and processes in glacially influenced high arctic environments. Special interest was put then to push moraines and thermo-erosion features, and to the accurate measurement of longterm glacial and geomorphological changes. It was felt, that accurate long term data is often missing for the interpretation of geomorphological processes, thus forcing geomorphologists to unsafe suggestions. Besides the presentation of new geomorphological data and results the paper intends to show to the geomorphologist the far reaching possibilities that are offered by modern photogrammetry, and it tries to interest and guide geomorphologists as well as geodetic surveyors in using modern semi-metric cameras in arctic research.

2 *The natural environment of the investigation areas*

Both main investigation areas are located at a latitude of about 80°N. The Expedition Fiord area is on the west coast of Axel Heiberg Island in the Queen Elizabeth Islands, the northernmost triangle of the Canadian Arctic Archipelago. Liefdefjorden is at the northern coast of Spitsbergen, the main island of the Svalbard Archipelago (Fig. 1). In both areas medium-sized and large glaciers originate in glacial cirques or in vast icefields that reach up to more than 1000 m a.s.l.

The climates of the two research areas show clear differences. Whereas the mean annual air temperature (MAAT) at Expedition Fiord is about −15 °C, the MAAT at Liefdefjorden is probably in the order of −4 °C. However, both areas experience a similar summer climate with about +5 °C mean July temperatures, and display the rich flora, that is typical for these more protected "inner fiord areas" (cp. SCHMITT, 1992). Morphologically important and common to both areas is also their location within the continuous permafrost zone. As a result of the different mean annual air temperatures the average permafrost thickness in non-coastal areas is about 500 m in Axel Heiberg or Ellesmere Island (JUDGE et al. 1981) and between 100 and 200 m in Svalbard.

The main parts of our studies were focused on the snouts of Troll Glacier on Ellesmere Island, of Thompson Glacier on Axel Heiberg Island, and of Glopbreen at Liefdefjorden. Besides these main study subjects a larger number of comparative features have been researched in the vicinity in order to check the representativity of the investigated features. Maps from aerial photographs have been produced of a glacier at Hare Fiord, Queen Elizabeth Island. In Svalbard, several prominent glacier tongues of the Liefdefjorden and Bockfjorden catchment areas could be visited (e.g. Erikbreen, Börrebreen, Adolfbreen, Schjelderupbreen, Nigardbreen), and four maps 1:25,000 have been prepared within the context of the German Geoscientific Expeditions 1990 to 1992 (HELL & BRUNNER 1992).

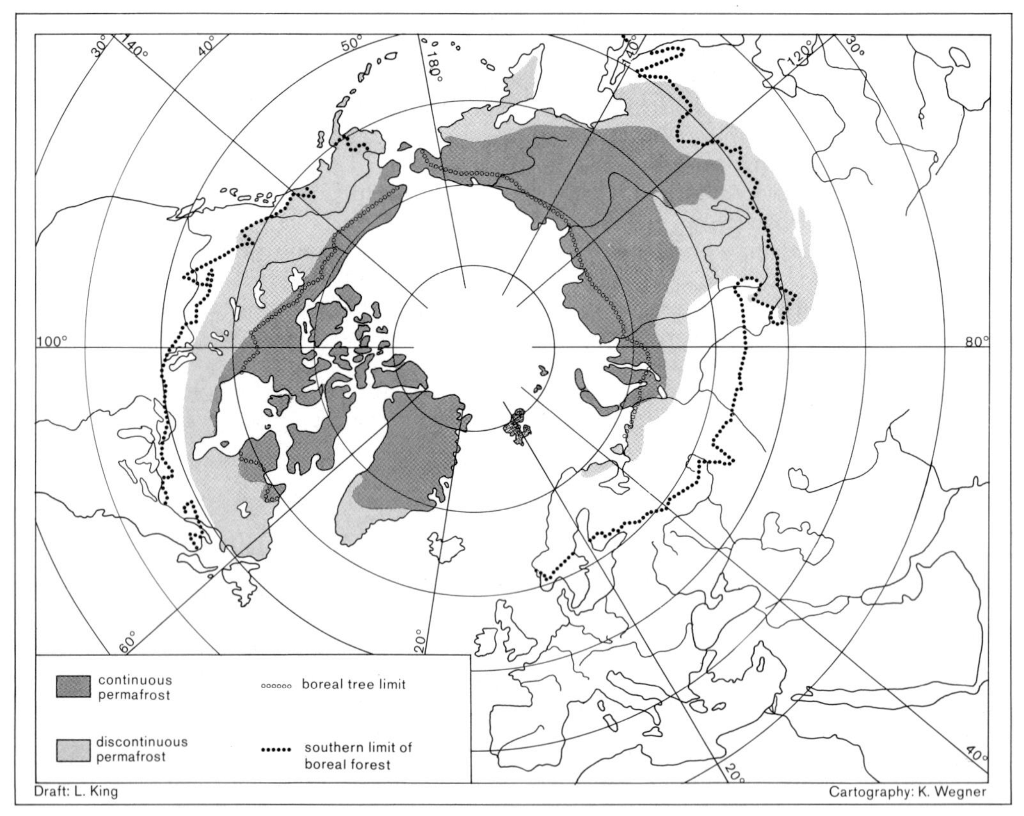

Fig. 1. Map of the northern hemisphere. The investigated areas in the Queen Elizabeth Islands (cp. Fig. 2) and in northern Spitsbergen are situated in the continuous permafrost zone.

3 *Modern photogrammetry as a working technique*

Photogrammetry allows to calculate accurately the topography of photographed areas, if at least some information (as e.g. coordinates, distances, altitudes) of a few field points are known. If photographs have been taken from the same topic but at different times, it is also possible to calculate the movement of prominent features, e.g. boulder, during this time interval. Photogrammetry can therefore be a very useful working tool for geomorphologists, e.g. for the measurement of solifluction (HELL 1981), rockglacier movement (HELL 1976) or even for the geodetic evaluation of glacier mass balances (REINHART & RENTSCH 1986).

The photogrammetric techniques have been greatly developed during the past years. This was made possible especially due to the application of modern computer software that allow the solution of very complicated mathematical problems. Better field instruments have also been developed, and here, today's photogrammetric practice can no longer be imagined without the so-called semi-metric camera, that can easily be used also by the amateur photographer.

These camera systems are developed from high quality amateur cameras and contain a reseauplate in the image plane. It is also possible to fix reproducably certain survey distances. It is obvious, that the light-weight construction of the semi-metric camera with its option to change objectives cannot maintain the inner orientation of the camera as accurately as a conventional metric camera with a weight of more than 10 kg for the equipment and a corresponding size and price. However, using the values of a former calibration the obtained minor accuracy of semi-metric cameras is usually sufficient for the desired geomorphological results.

Apart from the use of semi-metric cameras, however, the photogrammetric data collection and its evaluation have to be done by the professional geodetic surveyor. Today's standard equipment for the stereo compilation at photogrammetrical institutes consists of analytical plotters without limits concerning survey geometry (tilt, base-height ratio) and camera parameters (calibrated focal length, distortion). It is therefore possible to compile directly contourlines, even of an unusual survey geometry such as oblique photographs obtained with semi-metric cameras. Professional measuring tasks have thus become much more flexible and easier.

Our field camps have often been flown in with the help of helicopters or fix-winged planes. This offers the opportunity to use these flights for the photogrammetric data collection. Though only tilted pictures are obtained (instead of the classical vertical aerial metric camera imagery), these are sufficiently accurate for many photogrammetric purposes. In addition, aerial views although tilted avoid dead angles that often cannot be avoided with terrestrial photogrammetry. In the present study, these techniques have successfully been applied and combined with other photogrammetrical and geodetical methods by providing accurate data for the solution of geomorphological problems.

4 *Geodesy at push moraines in the Canadian Arctic*

4.1 *Troll Glacier, Oobloyah Bay, Ellesmere Island*

The locations of our research features are displayed on Fig. 2; most of them are near sea-level. The push moraine of Carl Troll Glacier is situated in the Borup Fiord area north of Greely Fiord, just 2 km away from the head of Oobloyah Bay. The push moraine has a semi-circular shape and surrounds the left half of the glacier front, where it reaches a relative height of about 25 m and a width of 200 to 300 m. The outwash plain in front of the moraine has an altitude of 25 m a.s.l. In its right part, the glacier front seems to ride over frozen, ice-rich till and the push moraine is missing there, but a smaller moraine consisting of shear and ablation material is present. The glacier history of the area is described in KING (1983).

The glacier was visited by the authors during two expeditions in 1978 (BARSCH & KING, eds., 1981) and in 1988, respectively. Photogrammetry was chosen in order to supply data to morphogenetic studies at the moraine. A triangulation network for the improvement of an orthophotomap 1:25,000 was established during the first visit

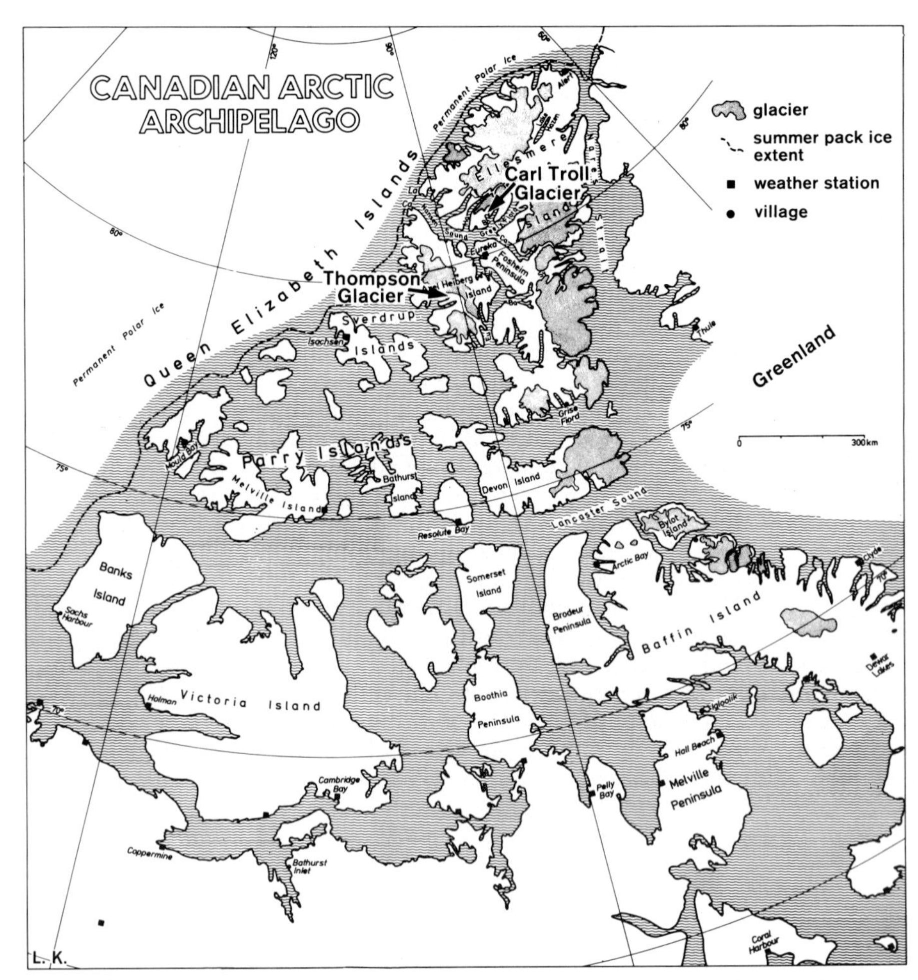

Fig. 2. The eastern Canadian Arctic with location of the investigated Troll Glacier and Thompson Glacier in the Queen Elizabeth Islands.

(HELL 1981) and some of these triangulation points could also be used as reference points for the 1988 photogrammetry. In addition the National Air Photo Library, Ottawa supplied aerial photographs taken in 1959/60 (scale: about 1:50,000).

At the end of the 1988 expedition a photosurvey was done using a Rollei metric 6006 camera with a 40 mm Distagon lens during move-out. Two strongly overlapping flight lines with 12 pictures each could be taken, the first line was taken over the glacier with view towards the frontal moraine and the fiord, the second line in front of the glacier with view towards the accumulation area (cp. Fig. 3). Both flight lines were taken at an altitude of about 750 m a.s.l. and with a nadir distance of about 80 gon.

Fig. 3. Rollei metric 6006 photograph of the frontal area of Troll Glacier. It was taken in 1988 from an altitude of about 750 m a.s.l. and used for the stereo compilation. The picture shows the investigated push moraine. On its right half (left margin of figure) the glacier overrides its own moraine, consisting mainly of sheared subglacial debris. The extent of the Aufeis in the central part varies from year to year.

The photogrammetric evaluation was done as follows: By means of the pictures taken in 1959/60 and 1978 and together with six terrestrially determined points of the 1978 expedition the control points for a bundle adjustment of the semi-metric camera imagery of 1988 were defined. The small picture scale limited the accuracy of the coordinates to ± 1 m. The 1988 topographical conditions had to be constructed then from the 1988 imagery. As the surface structures had changed quite considerably during the last 10 or even 30 years, the necessary control points could not be determined directly from the 1959/60 or the 1978 models. Nine photos out of the 24 taken in 1988 and covering well the investigated area were chosen for the bundle adjustment. About 80 natural points were picked as tie points and as future control points for the stereo compilations of the Rollei exposures, and their picture coordinates were measured with the Planicomp C100. The simultaneous determination of the coordinates and the parameters of the outer orientation of the photos was effected with a program for bundle adjustment (BINGO). The accuracy achieved for the tie points was about ± 2 m for the location and ± 1.5 m for the altitude ($\Sigma 0 = \pm 16$ µm), an accuracy judged to be sufficient for the task of mapping at a scale

Fig. 4. Push moraine in the Hare Fiord area. Whereas the glacier tongue retreated a distance of 600 m between 1959 and 1988, the push moraine itself shows hardly any changes. Photograph taken during the first Kanarktis expedition in 1978, NAPL Ottawa.

of 1:10,000. A higher accuracy could not be achieved for the following reasons: inaccuracy of control points, uncertainties of the inner orientation of the semi-metric camera and the use of natural (instead of marked) points as tie points.

The mapping of the 1988 situation at the scale of 1:10,000 was based on the now available control points and carried out with eight models of the Rollei exposures. Some additional models were applied in order to fill otherwise invisible areas. Because of the strongly tilted pictures (about 20 gon) and the distortion of the Rollei 6006 camera, it was indispensable to carry out the stereo compilation with an analytical plotter. The comparison of the pictures (1960/1979/1988) shows a significant dislocation of the glacier surface, but only hardly visible changes at its front and in large parts of the moraine. Some clear morainal deformations may especially be found in the right hand part of the moraine, probably due to melting of ice. In spite of an almost stagnant behaviour of the glacier front itself, glaciotectonic activities could be observed in the field: the outermost ridges of the moraine give the impression, that relatively thin frozen thrust plates have been pushed up recently, connected with a possible dislocation of the moraine complex for about 3 to 5 meters.

Three geoelectrical soundings have been done 500 to 800 m in front of the moraine on an moss-covered, dry accumulation plain that reaches only a few decimeters above ground water. Below the active layer of about 50 cm, the sounding graphs show a high-resistivity horizon with a thickness of 10 m, 6 m and 4 m, respectively, which may be interpreted as perennially frozen sediments. The resistivity relation (active layer/permafrost) is 1:100 for the first two soundings and 1:15 for the third one. Although a permafrost thickness of 550 m has been obtained in an instrumented drilling hole on nearby Neil Peninsula (JUDGE et al. 1981), our geoelectric graphs show again low apparent resistivities in greater depths and it is supposed that only relatively thin permafrost occurs in this near-shore area of young sediments and strong isostatic rebound. A weekness layer in a depth of only a few meters functioning as a sliding plane also corresponds to the morphological observations of thin thrust plates in the outermost parts of the push moraine. However, it must be stressed, that the total permafrost thickness may be considerably more than the mentioned 4 to 10 m, because the geoelectrical resistivity (and the shear strength) of frozen sediments is strongly dependent of temperature, and is relatively low in permafrost with near-zero temperatures.

4.2 *Hare Fiord, Ellesmere Island*

No terrestrial research could be done at the glaciers in the Hare Fiord area, but aerial photographs could be taken in 1978 and 1988 (Fig. 4). Glacier changes have already been evaluated earlier by comparing photographs taken in 1958 and 1978 (cp. differential map produced by HELL ed. 1986). A comparison with the 1988 photographs shows that this glacier has further melted back 600 meters since 1958 or 100–200 meters since 1978, respectively. The push moraine, however, has been very well preserved even in the parts situated in the fiord. The conclusion may be drawn, that moraines may well be preserved in a marine environment, at least in areas with low energy waves and isostatic rebound (KING 1985).

Fig. 5. Terrestrial view of the 700 m wide Thompson Glacier push moraine, taken in July 1988 (cp. Fig. 6 for scale).

4.3 *Thompson Glacier, Axel Heiberg Island*

Thompson Glacier is bulldozing up an impressive push moraine that is more than 2 km long an 700 m wide (Fig. 5). It rises generally about 45 m and in places even more than 100 m above the outwash plain. Extensive research has been done at the Thompson Glacier as early as 1960 and some excellent maps based on aerial photographs and geodetic determination of control points have been produced in 1963 by HAUMANN in a scale of 1:5,000. Some geomorphological work has been done by King in 1975 and the area was again visited by the authors in 1988 and aerial photos were taken in a similar manner as at Troll Glacier.

The photogrammetrical work was done as follows: The existing maps served both, as references for the surface changes within 28 years as well as a base for establishing the necessary control points for the stereo compilation of the 1988 photographs. A bundle adjustment (43 tiepoints on 6 photos) with 7 control points and 5 height points referring to the coordinate system of the 1960 maps yielded a $\Sigma 0$ value of ± 12 µm. The planimetric accuracy of the tie points – used in the following stereo compilation of the Rollei exposures as control points – is about ± 2.5 m, the

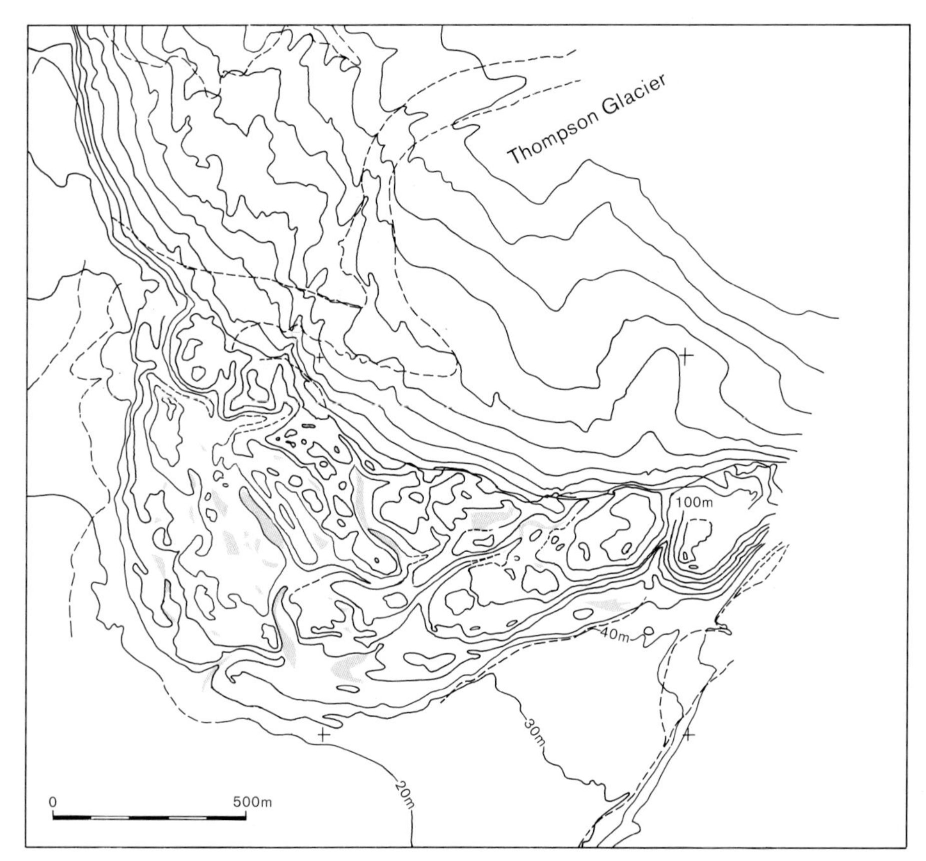

Fig. 6. Graphic map of the Thompson Glacier push moraine, compilated from 6 main models of the 1988 photographs. The push-moraine reaches up to more than 120 m a.s.l. (contour interval = 10 m).

accuaracy in height is about ± 1.5 m. Even these values are tolerable for a scale of 1:10,000. Besides the same influences as mentioned in the first example, the worse geometry of the tie points added to the inaccuracy, as only the pictures of one flight line could be used for the adjustment due to a very small overlap of only about 30%. The stereo compilation with 6 main models and again several supplements from other models allowed the compilation graphic map of the glacier tongue and the moraine in the scale 1:10,000 with 10 m contour lines (Fig. 6).

The geodetical and photogrammetrical evaluation showed that the moraine complex was moved as a whole for up to 500 meters in front of the advancing Thompson Glacier (Fig. 7). The main structures have been preserved surprisingly well within the moraine complex, e.g. the relative location of sediment blocks and lakes. Most changes occurred in the outermost parts, where relatively thin new

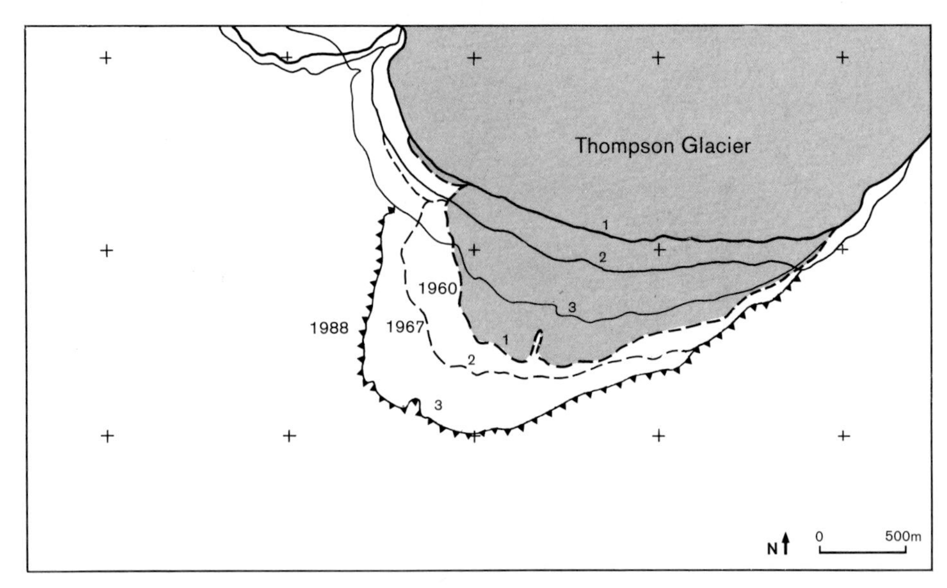

Fig. 7. The dislocation since 1960 is shown with dashed lines and reaches up to 550 m without much change in the main topographical features (cp. Fig. 7).

sediment plates were integrated into the push moraine complex. The observations led to the suggestion, that structurally weak layers serving as shear planes might exist in a few meters depth in the frozen sediments and the geoelectrical subsurface resistivity was measured and interpreted therefore. They are difficult to interprete probably due to local thermal anomalies (warm springs) and low resistivity bedrock (e.g. gypsum) in the surroundings of the moraine.

Clear signs of activity could be observed in the push moraine in June 1988. The outermost, youngest ridges consisting of frozen sediment plates must have recently been lifted 4 to 5 m above the valley floor and fresh radial cracks with a surficial width of about 1 to 4 m ran into the outwash plain. The width of cracks was difficult to measure, because thermokarst enlarged the surficial width considerably but the width of the crack in its lower parts was 30 cm at least. Further and smaller cracks extended to the outwash plain in the eastern part of the moraine; their width was only a few cm to dm, there. On its western part, the Thompson Glacier showed a steep ice cliff in 1988 with water falls; at the glacier base large ice blocks seemed to be overridden by the advancing glacier front. Very similar fresh radial cracks and overridden ice blocks have been observed by the author already in June 1975.

→

Fig. 8. Liefdefjorden and Bockfjorden (northern Spitsbergen) with investigated glaciers described in the text. In addition, four map sheets 1:25,000 have served as a topographical base for a large number of geomorphological studies during the German Spitsbergen Expeditions SPE '90–92.

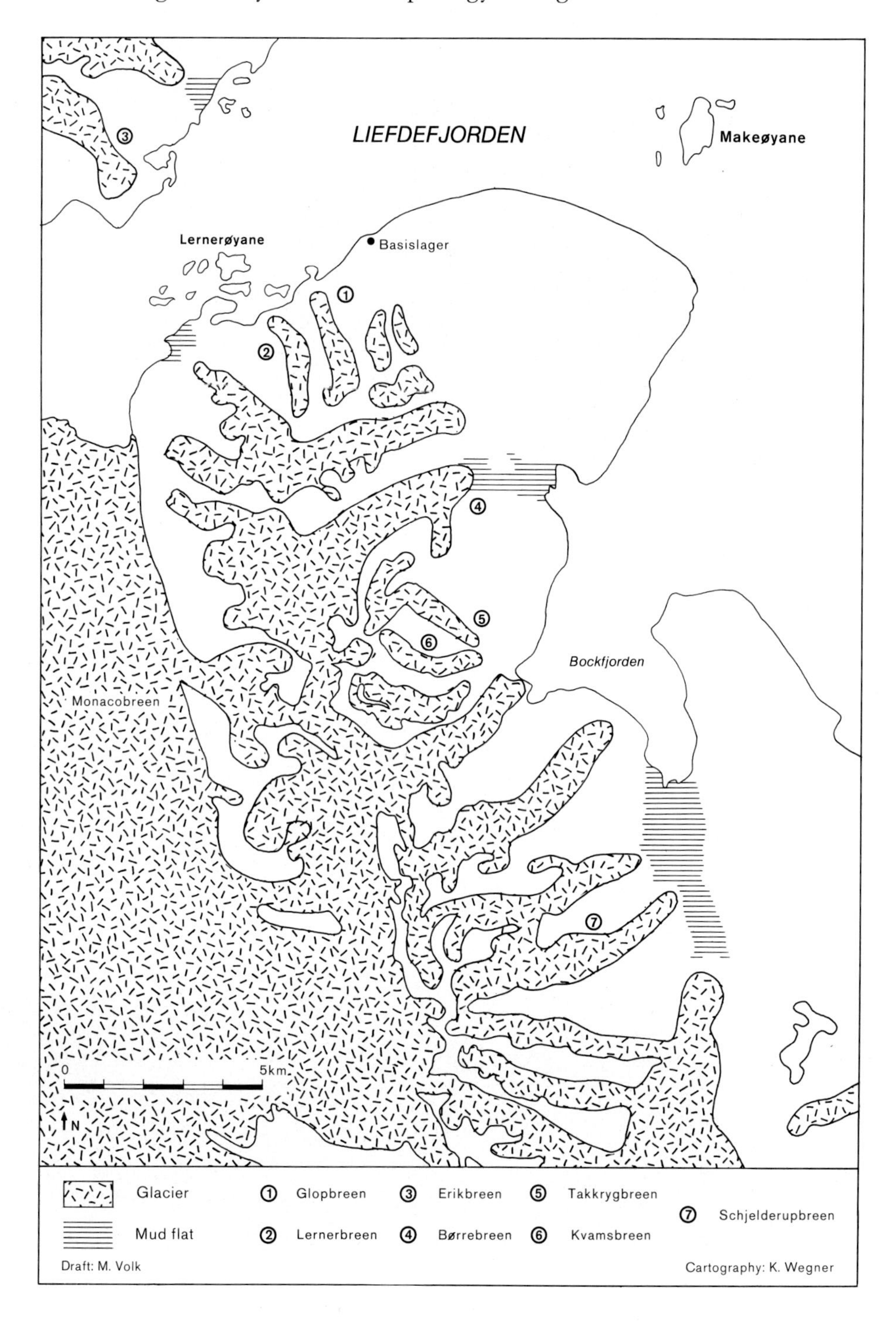

LIEFDEFJORDEN
Makeøyane
Lernerøyane
Basislager
Bockfjorden
Monacobreen
0
5km
N
Glacier
Mud flat
① Glopbreen
② Lernerbreen
③ Erikbreen
④ Børrebreen
⑤ Takkrygbreen
⑥ Kvamsbreen
⑦ Schjelderupbreen
Draft: M. Volk
Cartography: K. Wegner

The eastern and highest part of the moraine gave a more inactive impression during June 1988 than in 1975. The frozen, stratified sediment plates were also lifted and tilted, but showed signs of surficial melting and mass wasting. Although the push moraine is much higher here, it is smaller in its width, and the glacier seems even to run onto the moraine. Ground ice could be observed in some of the lakes within the push moraine. The geomorphology of the moraine is interpreted in LEHMANN (1989, 1992).

5 *Geomorphological and geodetical studies in Spitsbergen*

Terrestrial techniques with the described semi-metric cameras have also been applied during the expeditions to the Liefdefjorden area, northern Spitsbergen (Fig. 8) in the years 1990, 1991 and 1992 (KING et al. 1992a and 1992b). The geomorphology of the push moraines of Erikbreen and Glopbreen has been detailedly researched by VOLK (1992) and geodetic and photogrammetric work has been done there in addition and will be evaluated at the Institute for Geodesy and Cartography in Karlsruhe.

The morainic ridges themselves are usually very impressive and the outer slopes often show push moraine features with strongly tilted sediment layers. Dating usually at the foot of the outer slopes has been done by FURRER et al. (1991). The rockglacier-like movement of the moraine at Glopbreen has been measured by HELL et al.; similar movements may be expected at many other morainic features if the deposit has sufficient thickness and slope for permafrost creep.

In contrast to the Canadian glaciers visited, the glaciers at Liefdefjorden are strongly receding, leaving vast debris covered areas that easily experience thermal erosion. At Glopbreen (Fig. 9) the surficial movement as well as the responsible meteorological parameters have been detailedly investigated, but similar or even larger thermo-erosive areas exist at many other glaciers, e.g. Erikbreen (KING & SCHMITT 1992). Semi-metric camera techniques and geodetical surveying were combined in the inner parts of this same moraine and allowed to measure repeatedly the rates of thermal erosion between July 1990 and August 1992. The results are displayed in Fig. 10 and show an average movement of 100 cm for the summer period of 1990. The vertical thermo-erosive displacement of the moraine surface is often more than 5 cm per year. Similar values were obtained during the summer periods of 1991 and 1992.

Besides illustrating quantitatively the process of thermal erosion, the geomorphological conclusion is, that many of the investigated push moraines in Svalbard consist of a solid core of glacier ice, and to a lesser amount of glacial debris, compared with the investigated Canadian moraines. Soon after deposition thermal erosion of the moraine will start at the inner parts of the moraines, where the landscape will be transformed to a great extent. The outer parts of the moraines, although often ice-cored, usually have a sufficient thick debris cover to protect its core from melting.

Fig. 9. Terrestrial view of the intra-morainal area at Glopbreen with strong thermo-erosive activity; a weather station registers the meso- and microclimatic parameters. Surface movements were measured in the lower half and in the left part of the picture.

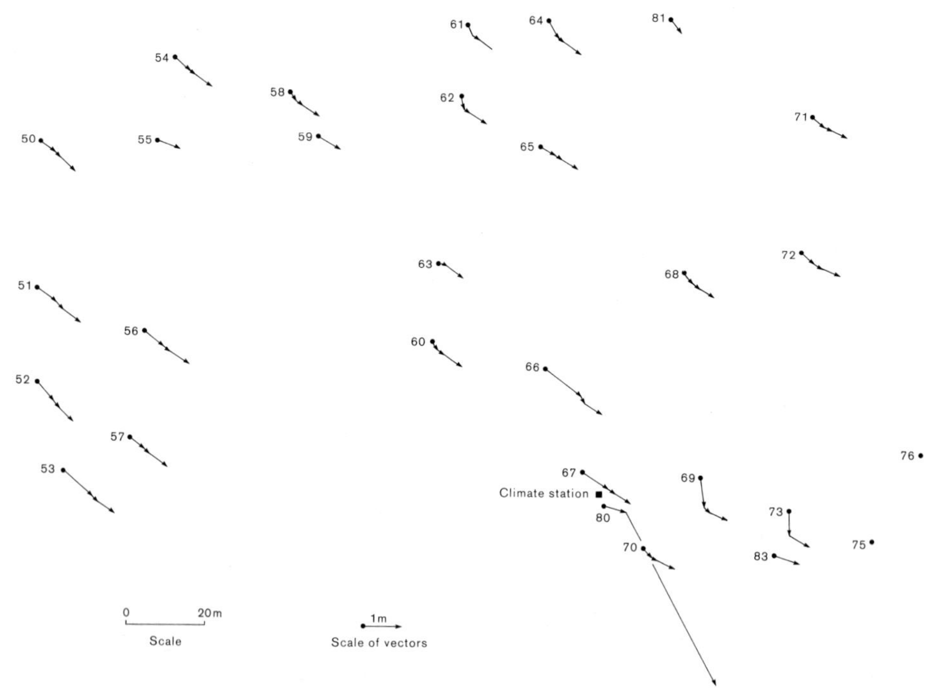

Fig. 10. Vectors showing the horizontal displacement of boulders in the investigated thermo-erosive area at Glopbreen between July 1990 and August 1991.

6 *Conclusions*

The aerial as well as the terrestrial photogrammetry supply valuable techniques for the study of geomorphological changes over time. The semi-metric camera is an especially helpful working tool due to its easy use for terrestrial photography in difficult terrain or for aerial photography.

In the Canadian Arctic active push moraines exist as very prominent features, usually in front of advancing or more or less stagnant large valley glaciers. They consist of pushed-up plates of frozen, often marine sediments. In Svalbard, similar push moraine features exist as relict forms probably originating from the Little Ice Age advances. Active features are comparatively rare. Due to the usually smaller size of the valley glaciers and often a steeper topography, their morphology consists mainly of a more or less complex rim system that surrounds the area of the former glacier tongue now often replaced by a proglacial lake. Whereas the usually very impressive moraine ridges consist to a large extent by dead glacier ice, covered with thick morainic debris (usually rocks) that insulates the ice from melting, thermal erosion is concentrated to the inner slopes of the moraines and may lead to extensive landscape transformation.

Acknowledgements

Sincere thanks are given to our colleagues in the field and their cooperative attitude: WOLFGANG ZICK as surveyor and RAINER LEHMANN as geomorphologist during the 1988-Kanarktis expedition, and MATHIAS GERBER, PETER HARTMANN, SUSANNE RESSING, PAUL RAWIEL and MARTIN VOLK during the Svalbard expeditions 1990 to 1992, respectively. All expeditions were mainly financed by the Deutsche Forschungsgemeinschaft, Bonn, the Kanarktis expeditions got additional logistical support from the Polar Continental Shelf Project, Ottawa. Many friends and colleagues abroad readily gave oral and written comments and advice, especially PETER ADAMS, HEINZ BLATTER, MARTIN JEFFRIES, FRITZ KOERNER, SIMON OMMANNEY and MARIUS VAN WIJK. For the preparation of the 1988-expedition we could use the facilities of the Geography Department of McGill University and got readily advice and help from THOMAS MOORE and DANIEL DESROCHERS, and in the field we could use the camp facilities of McGill University on Axel Heiberg Island.

References

ADAMS, W. P. (ed.) (1987): Field research of Axel Heiberg Island, N.W.T., Canada. – Bibliographies and Data Reports, McGill University, Montreal: 207 pp.

BARSCH, D. & L. KING (eds.) (1981): Results of the Heidelberg Ellesmere Island Expedition 1978. – Heidelb. Geogr. Arb. **69:** 573 pp.

BOULTON, G. S. & J. J. M. VAN DER MEER (eds.) (1989): Preliminary report on a expedition to Spitsbergen in 1984 to study glaciotectonic phenomena (Glacitecs '84). – Rapport No. 37, Fysisch Geografisch en Bodemkundig Laboratorium, Universiteit van Amsterdam: 195 pp.

CARLE, W. (1983): Das innere Gefüge der Stauch-Endmoränen und seine Bedeutung für die Gliederung des Altmoränengebietes. – Geol. Rundsch. **29**: 27–51.

CHAMBERLAIN, T. C. (1894): Proposed genetic classification of Pleistocene glacial formations. – J. Geol. **2**: 517–538.

EYBERGEN, F. A. (1987): Glacier snout dynamics and contemporary push moraine formation at the Turtmannglacier, Wallis, Switzerland. – In: VAN DER MEER, J. J. M. (ed.): Tills and Glaciotectonics. – Rotterdam: 217–231.

FURRER, G., A. STAPFER & U. GLASER (1991): Zur nacheiszeitlichen Gletschergeschichte des Liefdefjords (Spitzbergen), Ergebnisse der Geowissenschaftlichen Spitzbergen-Expedition 1990. – Geogr. Helvet. **46**, 4: 147–155.

GRIPP, K. (1929): Glaciologische und geologische Ergebnisse der Hamburgischen Spitzbergen-Expedition 1927. Abh.-Naturw. Ver. Hamburg, **XXII**: 147–249.

HAEBERLI, W. (1979): Holocene push-moraines in alpine permafrost. – Geogr. Ann., **61A**: 43–48.

HELL, G. (1976): Photogrammetrische Bewegungsmessungen am Blockgletscher Murtèl 1, Oberengadin, Schweizer Alpen. – Z. Gletscherkd. Glazialgeol. **XI**, 2: 111–141.

– (1981): Geodätische und photogrammetrische Arbeiten an der Oobloyah Bay, N-Ellesmere Island, N.W.T., im Rahmen der Heidelberg Ellesmere Island Expedition. – Heidelb. Geogr. Arb. **69**: 35–46, and map 1:25,000 (48 cm × 103 cm).

– (ed.) (1986): Veränderungen der Zunge eines Gletschers am Harefjord (NWT, Kanada) zwischen 1958 and 1978. – (Maps 1:20,000), Karlsruhe.

HELL, G. & K. BRUNNER (1992): Zur Arbeitsausgabe der Orthophotokarten der Spitzbergen-Expedition 1990, + Karten 1:25,000. – Bremen.

Judge, A. S., A. E. Taylor, M. Burgess & V. S. Allen (1981): Canadian Geothermal Data Collection. Northern Wells 1978–1980. – EMR, Earth Physics Branch. Geothermal Service of Canada, Geothermal Series 12, Ottawa.

Kälin, M. (1971): The active Push Moraine of the Thompson Glacier, Axel Heiberg Island, Canadian Arctic Archipelago. – Axel Heiberg Island Research Reports, McGill University, Montreal. Glaciology **4**: 68 pp.

King, L. (1983): Contribution to the glacial history of the Borup Fiord area, northern Ellesmere Island, N. W. T., Canada. – In: Schroeder-Lanz, H. (ed.): Late- and post-glacial oscillations of glaciers: glacial and periglacial forms. – Rotterdam: 305–323.

– (1985): Land-Sea Interactions in an arctic marine low energy environment, northern Ellesmere Island, N. W. T., Canada. – 14th Arctic Workshop, Arctic Land-sea interaction, Nov. 6–8, Dartmouth N.S., Canada: 144.

King, L., E. Schmitt & D. Wollesen (1992): Untersuchungen zur Geomorphologie und Geoökologie eines glazial geprägten arktischen Küstengebietes, Liefdefjorden, Spitzbergen: Forschungskonzept und erste Ergebnisse. – Stuttg. Geogr. Stud. **117**: 123–140.

King, L. et al. (1992): Geomorphologische und geoökologische Daten der Gießener Arbeitsgruppe der SPE '90-Expedition zum Liefdefjorden, Spitzbergen. – „Basler Datenband", im Druck.

King, L. & E. Schmitt (1992): Solifluction and thermal erosion in high Arctic ecosystems, Northern Spitsbergen: Processes and effects. – In: B. Frenzel (ed.): Palaeoclimate Research, Vol. 7. Special Issue: ESF Project "European Palaeoclimate and Man". European Science Foundation, Strasbourg.

Lamerson, P. P. & L. F. Dellwig (1957): Deformation by ice push of lithified sediments in south-central Iowa. – J. Geol. **65**, 5: 546–550.

Lehmann, R. (1989): Geomorphological mapping of an arctic push moraine – Thompson Glacier, Axel Heiberg Island, N. W. T., Canada. – In: Materialien und Manuskripte, Heft 17, Universität Bremen – Studiengang Geographie: 25–28.

– (1992): Arctic Push Moraines, a case study of the Thompson Glacier Moraine, Axel Heiberg Island, N.W.T., Canada. – Z. Geomorph. N.F., Suppl.-Bd. **86**: 161–171.

Mackay, J. R. & W. H. Matthews (1964): The role of permafrost in ice-thrusting. – J. Geol. **72**: 378–380.

Riezebos, P. A., G. S. Boulton, J. J. M. van der Meer et al. (1986): Products and effects of modern eolian activity on a nineteenth-century glacier-pushed ridge in West Spitsbergen, Svalbard. – Arc. Alp. Res **18**: 389–396.

Rutten, M. G. (1960): Ice pushed ridges, permafrost and drainage. – Am. J. Sci. **258**, 4: 293–297.

Schmitt, E. (1992): Ökologische Untersuchungen zur Vegetationsdifferenzierung und -dynamik im Liefdefjord. – In: L. King, E. Schmitt & D. Wollesen (1992): 133–140.

van der Meer & G. S. Boulton (1986): Hernieuwde belangstelling voor onderzoek van stuwwallen. Eerste resultaten van de Glacitecs 1984 Expeditie naar Spitsbergen. – In: K.N.A.G. Geografisch Tijdschrift **XX**, 3: 236–244.

van der Wateren, F. M. (1992): Structural Geology and Sedimentology of Push Moraines. – Diss. Universität Amsterdam.

Volk, M. (1992): Beiträge zur Geomorphologie glazial geprägter Gebiete, Liefdefjorden und Bockfjorden, Nordspitzbergen (in prep.).

Addresses of the authors: Prof. Dr. L. King, Geographisches Institut, Justus-Liebig-Universität, Senckenbergstraße 1, D-6300 Giessen. Prof. Dr. G. Hell, FB Vermessungswesen und Kartographie, Fachhochschule Karlsruhe, Moltkestraße 4, D-7500 Karlsruhe.

Z. Geomorph. N.F.	Suppl.-Bd. 92	39–53	Berlin · Stuttgart	Mai 1993

Slush Stream Phenomena – Process and Geomorphic Impact

by

DIETRICH BARSCH, MARTIN GUDE, ROLAND MÄUSBACHER, GERD SCHUKRAFT, ACHIM SCHULTE, and DIRK STRAUCH, Heidelberg

with 6 figures and 2 tables

Summary. In two catchments in the high arctic area of NW-Spitsbergen, slush stream phenomena were studied in three proceeding years (1990–1992) during a project concentrated on fluvial transport from land to sea. Slush streams are caused by an abrupt mobilisation of oversaturated snow in thalwegs at the beginning of the snow melt period. Slush flows belong to the class of slush stream phenomena. The catchments with a size of approx. 5 km^2 each, show an altitudinal difference of 750 m between the outer rim of the catchment and the fjord. The studies reveal the large variability in space and time of slush stream processes, concerning its physical properties and its geomorphic impact. The triggering of the events is related to a rapid melt of snow in early summer. Small slush flows with low velocities (max. 3 m/s) occur as quite a frequent process, often from year to year, as channel break-up in the rivers; its magnitude, however, remains low as far as certain thresholds in snow mass and meltwater release are not exceeded. They are confined to thalwegs. Transport of valley bottom sediments can be caused by this process. In contrast, large events run down the valleys in some years with high velocities of at least 20 m/s. Their geomorphic impact is tremendous, it is similar to that of avalanches or debris flows. The process can be evaluated by means of equations adopted from other mass movement analysis. These processes are named "slush torrent". Although the geomorphic impact in a single year exceeds the fluvial activity by far, the low frequency is a limiting factor.

Zusammenfassung. In zwei hocharktisch geprägten Einzugsgebieten im NW von Spitzbergen wurden im Rahmen der Untersuchung der fluvialen Transporte vom Land zum Meer in drei aufeinanderfolgenden Jahren (1990–1992) Sulzstrom-Phänomene studiert. Wir verstehen darunter die plötzliche Mobilisierung von stark durchfeuchteten Schneemassen in den Tiefenlinien zu Beginn des sommerlichen Auftauens. Zu dieser Erscheinung gehört vor allem das Sulzfließen (slush flow). Die Einzugsgebiete umfassen jeweils ca. 5 km^2; sie weisen einen Höhenunterschied vom Gipfel bis zur Mündung in den Fjord von 750 m auf. Die Untersuchungen verdeutlichen die große räumliche und zeitliche Variabilität von Sulzstrom-Prozessen bezüglich der physikalischen Eigenschaften als auch der geomorphodynamischen Auswirkungen. Die Initialisierung dieser Phänomene steht im Zusammenhang mit schneller Schneeschmelze im Frühsommer. Kleine Ereignisse mit geringen Geschwindigkeiten (max. 3 m/s) erscheinen als ein mehr regelmäßiger Prozeß des Aufreißens der Gerinne. Ihre Größenordnung ist allerdings beschränkt, solange bestimmte Schwellenwerte in bezug auf Schneemassen und Schmelzwassermobilisierung nicht überschritten werden. Da das Sulzfließen auf den fluvial aktiven Bereich eingeschränkt ist, werden vorherrschend Talbodensedimente transportiert. Im Gegensatz dazu stehen die großen Ereignisse, die mit Geschwindigkeiten von über 20 m/s durch die Täler schießen. Ihre geomorphodynamische Wirksamkeit ist enorm, sie

0044-2798/93/0092-0039 $ 3.75

ist vergleichbar mit Wirkungen von Lawinen und Muren. Der Abflußprozeß kann durch Gleichungen beschrieben werden, die aus der Analyse anderer Massenbewegungen stammen. Daher wird die Bezeichnung Sulzmure (slush torrent) für diese Prozesse vorgeschlagen. Obwohl die geomorphodynamische Wirksamkeit dieser Prozesse die fluviale Aktivität in einzelnen Jahren bei weitem übertrifft, wirkt sich die meist geringe Wiederkehrfrequenz limitierend aus.

1 *Introduction*

The significance of slush stream phenomena (slush flows, slush avalanches, etc.) has been subject of discussion for more than three decades (WASHBURN & GOLDTHWAIT 1958, RAPP 1960). The process probably is restricted to arctic areas. Although information about slush stream dynamics is rare, it seems that with all of them, the beginning of the process is the same: during snow melt, a sudden collapse of snow stability results in a more or less rapid movement of slush down a valley. Slush stream phenomena may reach catastrophic magnitudes, therefore they can have great effect, both on man and natural environments (ONESTI 1985, HESTNES 1985).

Due to the fact that the process is shortlasting and highly variable in time and space, the investigation of the geomorphic impact normally is restricted to a discussion of the forms and sediments left by the process: erosion along the path and – more obvious – deposition in the areas, where path gradients decrease or the valleys widen and the flows run out.

A first and still valid definition of slush stream phenomena was published by WASHBURN & GOLDTHWAIT (1958), who already reflect on the problems connected with process studies. Since their definition implies phenomena with a wide range of speed, internal properties, and deposition characteristics, the process description is far from being clear. This leads to uncertainties concerning the definition of slush flows and slush avalanches in the following investigations published by different authors. RAPP (1960) therefore suggests the new terminus technicus "torrent avalanche", because he is not sure, whether his observations fit the definition given by WASHBURN & GOLDTHWAIT (1958). CAINE (1969) uses the term slush avalanche, although the process described by CAINE, and more recent by GARDNER (1983), implies similarities to wet snow avalanches, which are characteristic for steeper slopes. The same reference to steeper slopes is given by RAPP (1985). On the other hand, NYBERG (1989) suggests that the term slush avalanche should be applied only to major events. Slush flows are indicative for the more regular break-up of the channels, at least in some arctic rivers (WOO & SAURIOL 1980).

To complete the old definition, a number of authors suggest the addition of relief parameters that may determine the process. Thus, CLARK & SEPPÄLÄ (1988) distinguish slope slush avalanches and valley bottom slush flows, the former with reference to CAINE's (1969) model of slope development. The valley confined slush flows are described by numerous authors (e.g. WASHBURN & GOLDTHWAIT 1958, RAPP 1960, JAHN 1967, WOO & SAURIOL 1980), which lead to a common use of the term in the scientific community. HESTNES (1985) fixes, as a result of empirical investigations on several slush flows in Norway, the average inclination from the crown to the accumulation end at 12.5°, in no case exceeding 20°. Many so called slush avalanches, therefore, seem to be dirty wet snow avalanches, because they are not confined to valleys (in contrast to real slush avalanches) and occur on slopes steeper than 20°. NYBERG (1989) recognizes the problem in demanding a distinction of channel confined wet snow avalanches, slush flows and other stream break-up phenomena. As slush avalanches on steeper slopes can not be differentiated from wet snow avalanches, the term "slush avalanche", should be avoided. It is suggested to call the general phenomena "slush stream" ("Sulzstrom"). Smaller events ($v < 3$ m/s) are termed "slush flow" ("Sulzfließen"), high energy events "slush torrents" ("Sulzmure").

The triggering of slush streams has been discussed by several authors. It is governed by a number of principal factors; of which at least the following must be fulfilled (e.g. RAPP 1960, JAHN 1967, CLARK & SEPPÄLÄ 1988, NYBERG 1989, BARSCH et al. 1992):

- a sudden and high input of energy during the beginning of snow melt;
- a relief that encourages the storage of water in the snow cover in wider parts of valleys;
- a high meltwater input from the adjacent slopes into the snow deposit in the valley.

In addition, BARSCH et al. (1992) demonstrate in NW Spitsbergen that the depression of the freezing point by the input of airborne ions is a further parameter. This affects also the stratigraphy, the transmissivity, and the stability of the snow cover.

The supersaturation of the snow and, in consequence, the rising waterlevel in the snow results in an abrupt failure of the snow pack in the valley bottom.

In general, the factors influencing the release of slush streams are known. They can be reconstructed even after the event, since relief features as well as snow cover in adjacent areas are preserved. In contrast, the beginning of the flow process is difficult to reconstruct. This is due to the fact that along the flow path erosional and depositional features are often rare, only snow levees are piled up at the margins of the stream. A study limited to the analysis of depositional features in the runout-zone does not lead to a clear definition of the flow process, even if, in addition, relief parameters are taken into account. Process studies are nessecary to combine the common knowledge with data of flow properties, such as velocity, discharge, and material composition.

2 *Investigation area and methods*

The investigation of fluvial transports in the area of Liefdefjorden (NW-Spitsbergen, cf. Fig. 1) is part of the geoscientific project "Land-Sea Sediment Transport in Polar Geoecosystems" (cf. BLÜMEL 1992). From 1990 to 1992, three catchments with approx. 5 km² each were investigated, regarding fluvial dynamics, sediment, and solute transport in the rivers. The highest points of the catchments are located at about 750 m a.s.l. Two of the rivers, the Kvikkåa and the Beinbekken (Fig. 2), revealed activity in slush streaming during the break-up periods in early summer. During the three summer months, the rivers are fed by snow and permafrost meltwater and, in the case of the Kvikkåa, by meltwater of two small glaciers. The gradients of the valley slopes adjacent to the channels are gentle at the Beinbekken, whereas the Kvikkåa runs partly in a steep canyon with slope heights of 10–30 m

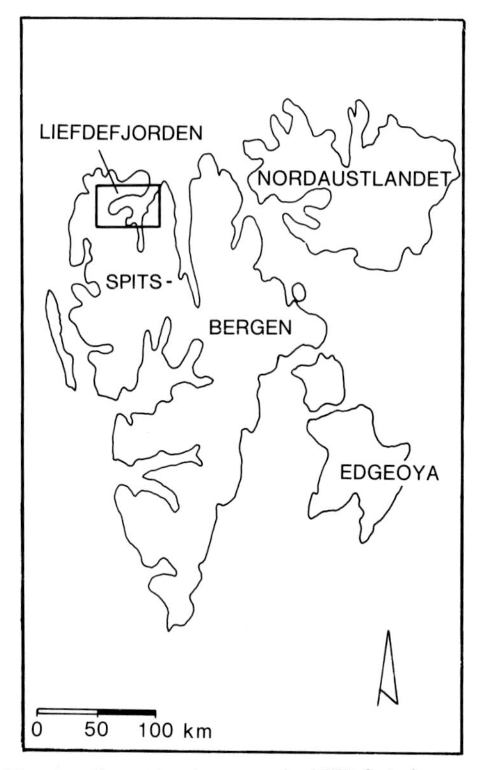

Fig. 1. Investigation area in NW-Spitsbergen.

and a valley floor of 5–10 m width. Both channels have approx. 8°–10° slope in the upper part and approx. 6° in the lower, where the Kvikkåa forms an alluvial fan with a length of approx. 500 m. The active alluvial fan of the Beinbekken is much smaller.

As the slush stream phenomena are related to the fluvial system – both in regard to the space affected and to the flow properties – they represent part of the fluvial research in the catchments. The observations and measurements of slush stream dynamics and mass properties in two catchments for three years offer new data for the distinction of slush stream phenomena. Two different forms related to rapid release of water-saturated snow in the valleys can be distinguished. Slush flows are the more or less regular form of break-up in the valley bottoms, they occurred in 1990 and 1991 with limited kinetic energy. Slush streams with a high energy impact took place in 1992. The term slush stream is suggested for the general event in the thalwegs of arctic regions. It comprises low energy slush flows (and slush avalanches) as described e.g. by CLARK & SEPPÄLÄ (1988). It also comprises high energy events which are called slush torrents. The term stresses a certain processual overlap with avalanches and debris flows. A general similarity to other mass movements was stressed by NOBLES (1966).

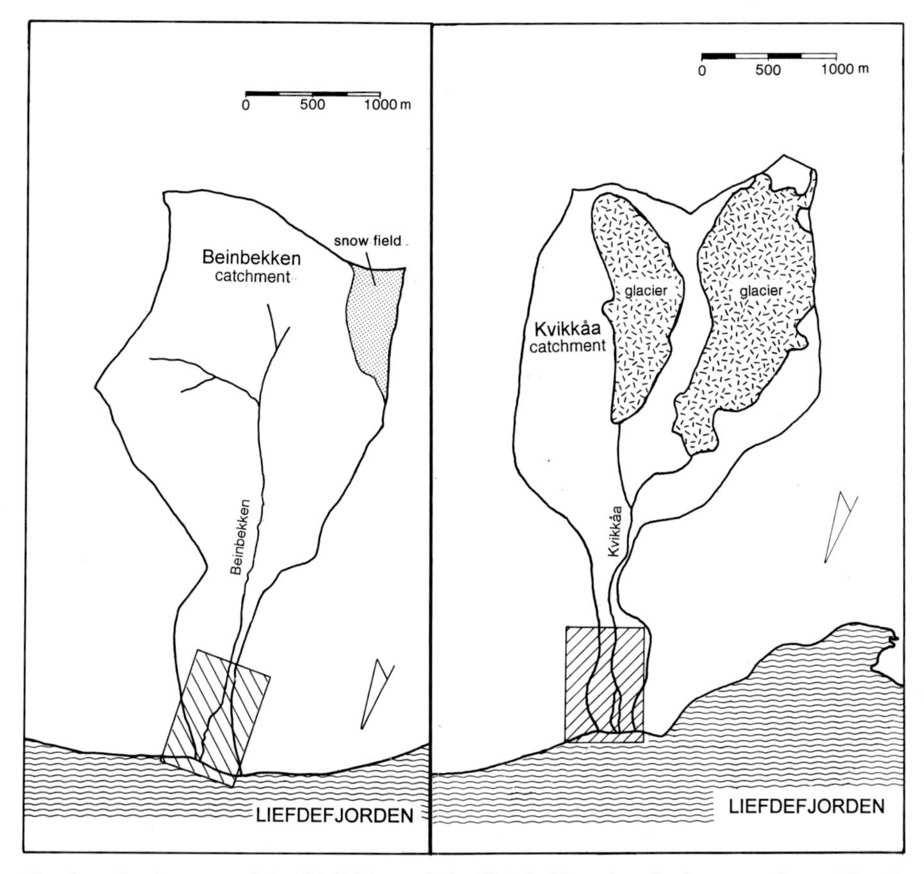

Fig. 2. Catchments of the Kvikkåa and the Beinbekken (marked areas refer to Fig. 4).

The investigations on slush stream phenomena were carried out by a combination of different methods related to both the erosional and depositional features and to the flow process. As far as the latter is concerned, the velocity, the discharge, and the flow mass properties are of predominant interest. Although slush flow phenomena are confined to fluvial channels and adjacent areas, no direct measurement techniques are applicable, particularly in the case of slush torrents. The high potential energy, the great variability of flow paths, as well as the restricted technical facilities in this arctic area require primarily indirect measurements and observation of the flow processes.

Measurements at the front section of the flows on definite reaches lead to estimates about the velocities. In addition, calculations based on physical parameters were adapted. The Froude equation, applicable in flowing water as well as in avalanches, marks a threshold of subcritical versus supercritical or slug flow

(SCHAERER & SALWAY 1980). The superelevation at the outer bends in curved valleys allow estimates of the velocity. Strictly, this is only valid for steady, subcritical flows, primarily because the wave velocity is not able to produce maximum elevation levels in supercritical and/or unsteady flows. If the deflection of a flow in the upper part of a bend is identified as a disturbance of the flow, the reaction will be the development of a wave. Thus, the calculated velocities reflect minimum values. The adoption of an equation developed for avalanche motion modelling stresses the similarity and differences to other mass movements (KÖRNER 1976, PERLA et al. 1980, RICKENMANN 1990).

A survey of the channel and valley relief provides important data about the flow process. The product of wetted cross-section area and velocity results in discharge. The total amount of discharge can be estimated either by the duration of the flow and the related discharge measurements, or by adding the snow losses due to the flow erosion along the path.

Slow moving slush flows allow to take samples during the flow. The samples can be separated into snow, water, and sediment content. Concerning the slush torrents, the study of mass properties is restricted to analysis of the depositions. As the free water leaves the deposited material rapidly after the flow runs out, no data about the water content of the flow in motion are available. The survey consequently is limited to snow and sediment in the depositions. Additionally, erosion along the flow path provides information about the sediment origin. This is supported by analysis of the flow depositions: valley slope material is angular, due to the lack of smoothing transport, whereas valley floor material is dominated by pebbles. Sediment features, therefore, facilitate the distinction of the predominance of lateral or downward erosion along the path, even if erosional features are rare. The survey of the spatial distribution of the depositions leads to estimations about the total sediment yield.

3 *Process description and analysis*

The rivers which have been examinated are characterized by a seasonal discharge typical for arctic areas: during a long winter period of approx. 9 months the channels are choked with snow and fluvial activity is suppressed due to low temperatures. In general, the break-up of rivers in early summer occurs rapidly, due to the input of energy (radiation etc.) into the catchments. In June 1992, the temperatures increased from approx. 2 °C up to 14 °C during one day. This initiated a rapid snowmelt which lead to the saturation of the snow cover, especially where water percolates from the valley side slopes into the snow on the valley bottom. This occurs either within the snow or on the snow base. Normally, impermeable ice layers, formed by refreezing of meltwater, act as an interflow base (WOO & HERON 1981). This results in rapid partial oversaturation of the snow in the valleys. Supported by the existence of ice layers, which can act as shear planes, a sudden failure of the snow masses initiates the slush flows and torrents, respectively.

Slush flows and slush torrents run down the valley at different velocities (from some cm/s to a few m/s and at approx. 20 m/s, respectively). Determined primarily by discharge and velocity, the erosion and the deposition are also variable.

3.1 *Slush stream phenomena in the Kvikkåa*

In 1990 and 1991, slush flows initiate the discharge period in the Kvikkåa catchment. They run down the valley as a mixture of snow and meltwater with velocities of not more than 3 m/s but in most parts of the thalweg the velocity is much smaller. The flows incorporate parts of the water-saturated snow on the valley floors. The flow path is characterized by levees with a maximum height of 1 m, limiting a track of a few meters width. In this track, the snow is partly eroded down to the ground-ice or to the gravels. The velocity in the cross profile is fastest in central part. The decrease of the gradient and the lack of steep valley slopes sometimes causes a branching on the alluvial fan.

In 1992, a slush stream with much higher velocities and sediment contents initiates the discharge. The release zone remains the same as in previous years, but the flow mass enlarges rapidly while running down the valley by entrainment of supersaturated snow and, therefore, gains a tremendous kinetic energy. This results in an important erosional impact on the valley and, consequently, in a change of flow on the alluvial fan. The possibility to spread out leads to the deposition of large amounts of sediment and snow. The first wave of the slush torrent with a duration of approx. 3 seconds is followed by two smaller ones (30 min and 180 min later, respectively; cf. Table 1). They comprise large quantities of water which seems to be derived from temporary water storages in the canyon.

In 1990 and 1991, no indication for a high energy release is apparent in the initial zone. The slush flows probably begin as a rupture of snow packs or as a snow dam decay, corresponding to that described by Xia & Woo (1992) and the flows move down the valleys at low velocities. In contrast, the slush torrent in 1992 reached a high velocity even in the initial zone with a length of approx. 50 m. It is caused by the release of a much larger mass of snow compared to the previous years.

The differences in the quantity of water-saturated snow released in the years 1990/91 and 1992 has an effect on the type of flow in the canyon. The slush flows of the two previous years entrain only a part of the water-saturated snow along their path. Due to the limited kinetic energy, the flow permanently deposits snow at its outer margins (levees). The wetted cross profile hardly exceeds the thickness of the

Table 1 The phases of the slush torrent at the Kvikkåa on June 11, 1992.

	starting time/duration	estimated mean discharge (m^3/s)	amount (m^3)
1. slush flow	ca. 12.00/ca. 3 h	0.3	ca. 3,000
2. slush torrent	ca. 15.00/ca. 3 s	6,000	ca. 20,000
3. slush torrent (flood wave 1)	ca. 15.30/ca. 5 s	200	ca. 1,000
4. normal discharge	ca. 15.40–17.50	1	ca. 8,000
5. slush torrent (flood wave 2)	ca. 18.00/ca. 5 s	1,200	ca. 6,000

snow choking the canyon floor, which had a maximum thickness of 3 m. Thus, the flow is constantly slowed down by the snow cover and the channel roughness. In 1992, the slush torrent rapidly grows while running down the canyon and exceeds the snow thickness by far. In doing this, the kinetic energy increases with the increasing mass. In this case, the main limiting factors for the velocity are the inherent friction and the course (sinuosity) of the canyon.

3.2 *Physical properties of the slush stream phenomena*

The motion of slush flows is transitional: steady and unsteady flow alternate, depending on the amount of meltwater and the resisting forces, such as snow density and thickness, as well as canyon floor roughness and gradient. The velocity, therefore, changes continuously and ranges between 0.3 and 3 m/s. As the drag is determined by the snow entrainment along the path, no velocity equations for water flow in open channels are applicable. Even the computation of the Froude index, indicating whether a flow is subcritical or supercritical, is not adoptable, because the ploughing produces a larger height compared to water flow. The limited possibilities of numerical evaluation of slush flows is counterbalanced by the fact that direct measurement techniques are more easily applicable than in the case of slush torrents.

The velocity of the slush torrent in 1992 was studied indirectly: observations and rapidly taken photographs by different witnesses were analysed independently. They lead to the estimation of a velocity of at least 20 m/s for the front wave at the mouth of the canyon. The velocity, the extreme unsteady flow comparable to a slug flow, and the fact that the flow grows rapidly by incorporating wet snow along the path, encourages the adoption of equations applicated in avalanche and debris flow analysis. To evaluate the flow characteristics, the computation of the Froude number is done by the following formula (MELLOR 1978, SCHAERER & SALWAY 1980):

$$(1) \qquad Fr = \frac{v}{\sqrt{g R}}$$

where v is the velocity, g the acceleration due to gravity and R the hydraulic radius. If the minimum velocity of 20 m/s and a R of 4.1 m (Fig. 3a) is assumed, the Froude number equals 3.2; a value that exceeds the critical limit by a factor of 3. This indicates a shooting flow. The computation for the cross profile in the middle reach of the canyon (Fig. 3b) equals a Froude number of 2.9. The slush torrent can not only be termed as supercritical or shooting, but as well as a slug flow, according to SCHAERER & SALWAY (1980). This represents an appropriate description of the features observed in situ: a single, thick and fast front wave is followed by a stream of slush and water with decreasing thickness and velocity. The flow dynamic reveals a similarity to avalanches or debris flows. Particularly the latter are characterized by a pronounced unsteady flow (HAEBERLI et al. 1991).

A computation of the velocity was also carried out by accounting to the superelevation at outer bends in curved reaches of the canyon. The equation

$$(2) \qquad v = \sqrt{\frac{dHg}{\ln (r_o/r_i)}}$$

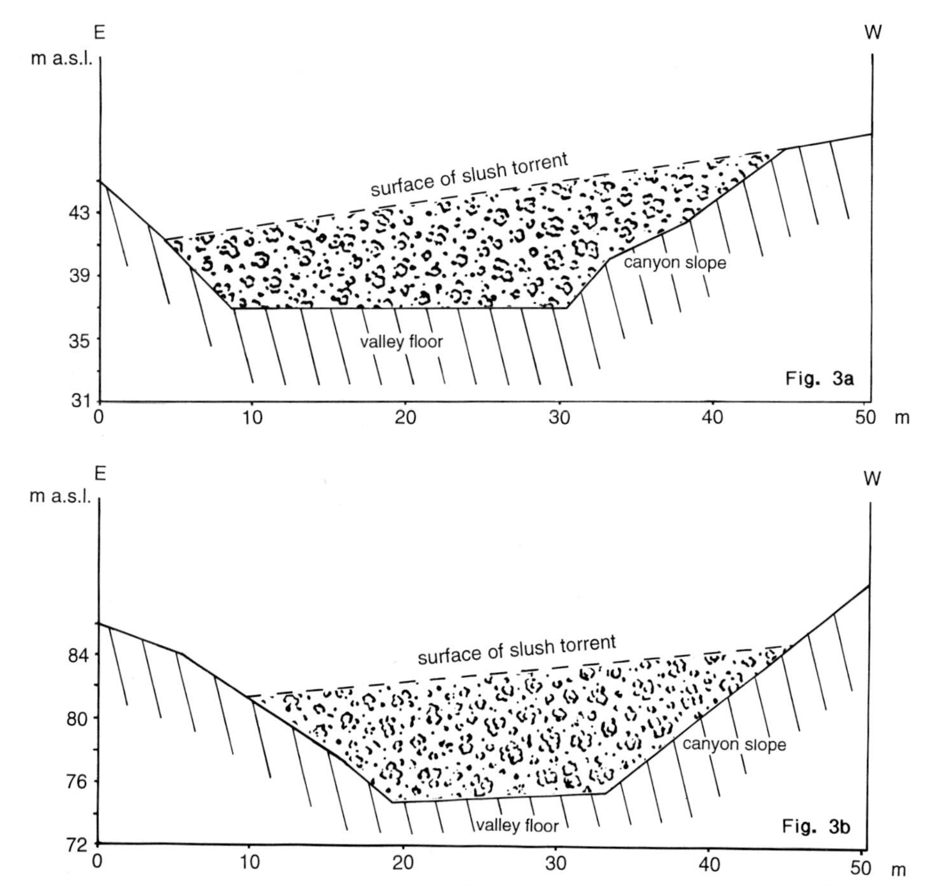

Fig. 3. Cross profiles of the slush torrent in 1992 in the canyon of the Kvikkåa: a) in a curved section in the lower part, b) in the middle section.

(where dH is the difference in elevation between outer radius r_o and inner radius r_i of the bend) is developed for water flow in open channels (MANGELSDORF et al. 1990). Nevertheless, satisfying results are also obtained, calculating debris flow velocities (HAEBERLI et al. 1991). Inserting the values documented for the cross profile in the lower part of the canyon (Fig. 3 a), where the outer radius is approx. 120 m, the velocity adds up to approx. 12 m/s. This must be assumed as a minimum value, as already mentioned above.

Another calculation of flow velocity derives from an equation developed by VOELLMY (1955) and modified by PERLA et al. (1980). The terminal velocity at a section of constant gradient is

$$v = \left[\frac{Mg}{D} (\sin\theta - \mu\cos\theta) \right]^{1/2} \quad (3)$$

where M is the mass at the point of consideration, D is an assumption of dynamic drag, θ is the gradient and μ the sliding friction coefficient. PERLA et al. change the VOELLMY-equation, in order to take into account the entrainment of snow along the flow path, which is represented in the term D. It is for straight channels:

$$(4) \qquad D = \frac{dM}{dS} + k$$

where dM and dS are the mass and distance changes, respectively, and K is a drag and ploughing term. The term k is estimated for avalanches to range between 10^3 and 10^4 kg/m (PERLA et al.). Considering the range of k, maximum velocities between 15 and 45 m/s are obtained for the slush torrents, which agrees with the field observations. The more interesting result of the application of this equation is that for the low canyon floor gradient of about 6° an infinitesimal sliding friction coefficient is needed to allow flow. Assuming the equation is applicable to this case, it reveals another indication for the flow process: the acceleration of the mass primarily derives from the confinement to the canyon and not from the gradient alone. Furthermore, the flow must have been a supercritical one, which is not significantly decelerated by canyon floor roughness. This is supported by the field evidence: along the flow path, the erosional features are much more obvious at the canyon walls than at the floor.

3.3 *Geomorphic impact*

Where parts of the canyon are cleared from snow, the protection against erosion is removed. Only the basal ice in the channel and the frozen active layer limit erosion. The erosional activity of the slush streams appear to be different in the observed years. In the years 1990/91, the slush flows affected primarily the canyon floor, because the shear stress at the slopes remained low, due to the moderate flow velocity. The mobilization of sediment, therefore, was restricted to the gravels that form the canyon floor. Thus, only a small amount of angular slope material was found in the sediments of the flows. The high velocity of the slush torrent in 1992 results in a predominant erosion at the canyon walls. This is demonstrated by the superelevation on the outside bends, which indicates a strong impact on these parts.

The deposition of the slush streams in 1990/91 and 1992 illustrates the variability in the size of the area affected (Fig. 4). As the velocity of the slush flows in 1990/91 on the upper part of the fan is low, no considerable sedimentation occurs. Only the Beinbekken slush flow in 1990 spread out when it left the mountain valley. Apart from this, the deposition areas are clearly distinct between slush flows and slush torrents. The first runs out on the ice covered fjord, leaving its sediments on top of the ice. Part of the sediment is incorporated in the levees lining the flow path, but the total amount of sediment transport remains low if compared with slush torrents. Nevertheless, the sediment output from the catchments by slush flows is significant in relation to the fluvial transport (BARSCH et al. 1992). The Beinbekken slush flow in 1990, which was the largest observed during the three years, yields a total

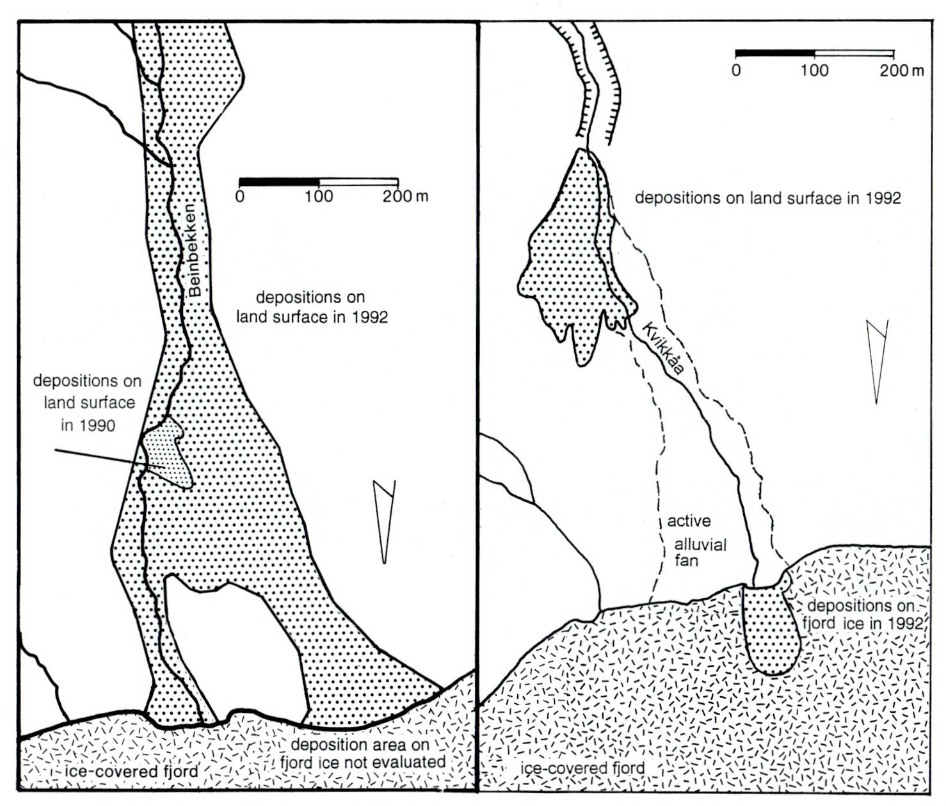

Fig. 4. Deposition areas of the slush torrents in 1992 at the Kvikkåa and the Beinbekken, the latter including the slush flow depositions of 1990 (marked areas refer to Fig. 2).

sediment load of approx. 50 t (on land surface and on fjord ice). The other slush flows deposited only a fraction of this.

The accumulations by the slush torrents are also partly situated on the fjord ice, but the main part rests in the run-out zone (Figs. 4 and 5). As apparent in Fig. 4, this deposition zone is situated mainly at the outer parts of the fluvial fan, where lichens, mosses, and vascular plants grow. The deposits partly covered this plant covered surface near to 100% (Fig. 6). The accumulation consists of a nonsorted mixture of grain sizes, ranging from silt up to single boulders with a diameter of c. 1 m. In general, boulders of 40–60 cm are deposited. The water waves, following the slush torrent at the Kvikkåa, had only minor effects on the deposition (Fig. 4) due to the predominance of finer grains in these waves. The sediment, mobilized by the slush torrents, primarily consists of angular material, which derives from the canyon walls.

Total sediment yields deposited on land surface by the slush torrents in 1992 amount to c. 1200 tons for the Kvikkåa, and to approx. 5000–6000 tons for the

Fig. 5. Photograph of the alluvial fan of the Kvikkåa, on the left side the active area, the light area on the lower right represents the area covered with sediments of the slush torrent in 1992 (July 1992).

Beinbekken (the latter is based on preliminary calculations). The comparison of sediment yields of the slush flows in the previous years emphasizes the necessity for the distinction of the slush stream phenomena. Even the largest slush flow observed (Beinbekken 1990), adds only 35 tons of the sediment accumulation on land surface. The relation to fluvial mobilized sediments stresses the geomorphic importance of slush torrents: total fluvial sediment yields in the Kvikkåa added up to 240 tons in 1990, and 130 tons in 1991 (BARSCH et al. 1993). The probably low recurrence interval of slush torrents limits the geomorphic impact. On the other hand, natural cuts in the fan of Beinbekken contain deposits of several metres thickness, which can only be explained as slush stream and probably as slush torrent deposits.

4 *Discussion on flow properties and geomorphic impacts of slush stream phenomena*

Summarizing the physical properties of the slush flows in 1990/91 and the slush torrent in 1992 at the Kvikkåa, the necessity for the distinction between the two types is obvious. Velocities, as well as other flow characteristics reveal clear differences (Table 2).

Slush flows are characterized by a mixture of water and snow, sometimes with sediment, and by relative low velocities. The resistance forces of the snow choking

Fig. 6. Photograph of the slush torrent sediment cover on the land surface beneath the active alluvial fan of the Kvikkåa (July 1992).

Table 2 Comparison of physical flow properties of slush stream phenomena in 1990/91 and in 1992 in the Kvikkåa catchment.

	slush flows of 1990/91	slush torrent of 1992
velocity	0.3– 3 m/s	>20 m/s
cross profile width	2 – 5 m	20–40 m
cross profile depth	0.5– 1.5 m	5–15 m
discharge (maxima)	0.3–20 m³/s	2,000–12,000 m³/s

the channel are high, and the amount of meltwater in the valley floor is limited. As the velocity is restricted, due to ploughing, drag, and friction, the geomorphic effect remains modest. Although some properties of slush flows are similar to open channel flow, the application of related equations is not possible.

Slush torrents develop flow characteristics that allow a comparison with avalanches or with other high energy mass movements, such as debris flows. Since slush torrents are confined to valleys with moderate gradients, they are well distinguishable from avalanches. The calculation of flow properties by the above cited equations leads to a satisfying description. The high Froude numbers support the suggestion to call this feature a slush torrent. Additionally, it provides information means to decide whether slush flows or slush torrents are the process under consideration.

The erosional and depositional features of both types represent differences in geomorphic impact: erosion tracks of slush flows are hardly recognizable, because they are in general confined to the channels and the active flood plain. Slush torrents, on the other hand, also effect areas remote of the channels. Their geomorphic impact can therefore influence valley slope development.

5 *Conclusions*

It is stressed that slush stream phenomena in arctic areas appear as processes with a wide range, regarding flow properties as well as geomorphic impacts. Whereas the geomorphic impact can be studied by mapping the erosional and depositional features, the analysis of the flow properties has to be deducted from empirical equations.

The geomorphic impact of slush flows and slush torrents reflects the difference in the energy involved. Sediment transport by slush flows remains low, nevertheless it adds up to a significant amount, compared to fluvial activity in some years. In contrast, the sediment mobilization of slush torrents is tremendous and comparable to other mass movements.

A clear distinction between the two types of slush stream phenomena is sincerely required. The less frequent – and consequently less often observed – process of slush torrent should be differentiated from slush flows, as well as from avalanches.

Acknowledgements

We have to thank many friends and colleagues for their help, but the most important task is to remember

ANDREAS FIEBER (13.09.1964–11.06.1992),

who died in a slush torrent in Kvikkåa. He was a friend with whom we have been working in the Elsenz catchment near Heidelberg, in Argentina, and on Spitsbergen.

Our thanks are due to THOMAS GLADE, HARTMUT GÜNDRA and BEATE SANDLER (Heidelberg) who worked with us during the summers of 1991/92 on Spitsbergen, to our friends from Basel and Giessen and from other universities, who joined the work on Liefdefjorden.

Thanks are due to WOLF-DIETER BLÜMEL, the coordinator of the program and to the Deutsche Forschungsgemeinschaft (Bonn) who financed the expedition.

References

BARSCH, D., M. GUDE, R. MÄUSBACHER, G. SCHUKRAFT & A. SCHULTE (1992): Untersuchungen zur aktuellen fluvialen Dynamik im Bereich des Liefdefjorden in NW-Spitzbergen. – Stuttg. Geogr. Stud. **117**: 217–252.

––––– (1993, in press): Sediment transport and discharge in a high Arctic catchment (Liefdefjorden, NW-Spitsbergen). – Lecture Notes (Springer, Heidelberg).

Blümel, W. D. (ed.) (1992): Geowissenschaftliche Spitzbergenexpedition 1990 und 1991 „Stofftransporte Land–Meer in polaren Geosystemen", Zwischenbericht. – Stuttg. Geogr. Stud. **117**: 416 pp.

Caine, N. (1969): A Model for Alpine Talus Slope Development by Slush Avalanching. – Journ. Glaciol. **77**: 92–100.

Clark, M. J. & M. Seppälä (1988): Slushflows in a subarctic Environment, Kilpisjärvi, Finnish Lapland. – Arctic and Alpine Research **20**, 1: 97–105.

Gardner, J.-S. (1983): Observations on Erosion by Wet Snow avalanches, Mount Rae Area, Alberta, Canada. – Arctic and Alpine Research **15**, 2: 271–274.

Haeberli, W., D. Rickenmann, M. Zimmermann & U. Rösli (1991): Murgänge. – Mitt. Bundesamt Wasserwirtsch. **4**: 77–88, Bern.

Hestnes, E. (1985): A Contribution to the Prediction of Slush Avalanches. Ann. Glaciol. **6**: 1–4.

Jahn, A. (1967): Some Features of Mass Movements on Spitsbergen Slopes. – Geogr. Ann. **49 A** (2–4): 213–225.

Körner, H. J. (1976): Reichweite und Geschwindigkeit von Bergstürzen und Fließschneelawinen. – Rock Mechanics **8/4**: 225–256.

Mangelsdorf, J., K. Scheurmann & F.-H. Weiss (1990): River Morphology. – 235 pp.; Springer, Heidelberg, New York, Tokio.

Mellor, M. (1978): Dynamics of Snow Avalanches. – In: Voight, B. (Ed.): Rockslides and Avalanches, **1**: 753–792.

Nobles, L.-H. (1966): Slush Avalanches in Northern Greenland and the Classification of Rapid Mass Movements. – Ass. Int. Hydrol. Scient. Publ. **69**: 267–272.

Nyberg, R. (1989): Observations of slushflows and their geomorphological effects in the Swedish mountain area. – Geogr. Ann. **71 A** (3–4): 185–198.

Onesti, L. J. (1985): Meterological Conditions that Initiate Slushflows in the Central Brooks Range, Alaska. – Ann. Glaciol. **6**: 23–25.

Perla, R., T. T. Cheng & D. M. Mc Clung (1980): A Two-Parameter Model of Snow-Avalanche Motion. – Journ. Glaciol. **26**, 94: 197–207.

Rapp, A. (1960): Recent Development of Mountain Slopes in Kärkevagge and Surroundings, Northern Scandinavia. – Geogr. Ann. **17** (2–3): 71–200.

– (1985): Extreme Rainfall and Rapid Snowmelt as Causes of Mass Movements in High Latitude Mountains. – In: Church, M. & O. Slaymaker (eds.): Field and Theory. Lectures in Geocryology. – Vancouver (University of British Columbia Press): 36–56.

Rickenmann, D. (1990): Debris flows 1987 in Switzerland: modelling and fluvial sediment transport. – Hydrology in Mountanious Regions; II. Artificial Reservoirs, Water and Slopes (Proceedings of 2nd Lausanne Symposia 1990), IAHS Publ. **19**, 4: 371–379.

Schaerer, P. A. & A. A. Salway (1980): Seismic and Impact-Pressure Monitoring of Flowing Avalanches. – Journ. Glaciol. **26**, 94: 179–187.

Voellmy, A. (1955): Über die Zerstörungskraft von Lawinen. – Schweiz. Bauzeitung Jg. **73**, 12: 159–162, 15: 212–217, 17: 246–249, 19: 280–285.

Washburn, A. L. & R. P. Goldthwait (1958): Slushflows. – Geol. Soc. Am., Bull., **69:** 1657–1658 (Abstract).

Woo, M.-K. & R. Heron (1981): Occurrence of Ice Layers at the Base of High Arctic Snowpacks. – Arctic and Alpine Research **13**, 2: 225–230.

Woo, M.-K. & J. Sauriol (1980): Channel Development in Snow-Filled Valleys, Resolute, N.W.T., Canada. – Geogr. Ann. **62 A** (1–2): 37–56.

Xia, Z. & M.-K. Woo (1992): Theoretical analysis of snow-dam decay. – Journ. Glaciol. **38**, 128: 191–199.

Address of the authors: Geographisches Institut, Universität Heidelberg, Im Neuenheimer Feld 348, D-6900 Heidelberg.

Z. Geomorph. N.F.	Suppl.-Bd. 92	55–69	Berlin · Stuttgart	Mai 1993

Periglacial Denudation in Formerly Unglaciated Areas of the Richardson Mountains (NW-Canada)

by

G. FRIED, J. HEINRICH, G. NAGEL and A. SEMMEL, Frankfurt/M.

with 5 figures and 1 table

Summary. Contrary to other publications, this study did not find evidence for the prevalence of active pedimentation under the present conditions. The well-developed pediments, which are now densely covered by tundra vegetation, are explained as remnants of colder climatic periods from the Pleistocene. Decisive for this opinion is the connection of pediment debris with former ice wedges. Patterned ground, even on steeper slopes, is indicating the predominance of geomorphic stability. Based on numerous ^{14}C dates, we conclude that the number of cryoturbations has increased during the last 5000 years.

Zusammenfassung. Im Gegensatz zu den Darstellungen in anderen Publikationen wurden im Untersuchungsgebiet keine Anzeichen für rezente Pedimentation gefunden. Die gut entwickelten Pedimente, die gegenwärtig mit dichter Tundrenvegetation bedeckt sind, werden als unter kälteren Bedingungen während pleistozäner Kaltzeiten entstanden angesehen. Zu dieser Zeit waren im Gegensatz zu heute Eiskeile im Permafrost ausgebildet. Frostmusterböden, auch auf stärker geneigten Hängen, zeigen die gegenwärtige weitgehende Formungsruhe zusätzlich an. Zahlreiche ^{14}C-Datierungen zwingen zu der Annahme, daß die meisten Kryoturbationen an der heutigen Oberfläche erst in den letzten 5000 Jahren entstanden, für das ältere Holozän also mit milderem Klima zu rechnen ist.

1 *Introduction*

Geomorphologists still discuss the problem, which type of landscape will develop under continuous periglacial conditions. In Russian (e.g. DEDKOV 1965) and US-American publications (e.g. PÉWÉ 1975) the opinion prevails that the typical periglacial landforms are denudational surfaces. In German publications, above all BÜDEL (e.g. 1977) expressed the opposite view: periglacial areas are dominated by fluvial dissection. This exceptional position of a substantial number of German publications was already mentioned by KARRASCH (1972) (cf. also WASHBURN et al. 1979: 78). PRIESNITZ (1981; see also 1988) studied this problem again by analysing a section of the Richardson Mountains in north-west Canada, which had never been glaciated, and which has now become relatively easily accessible after the construction of the Dempster Highway. We also visited this area in order to study the prevailing landforms and geomorphic processes. The field trip was sponsored by the

0044-2798/93/0092-0055 $ 3.75

Deutsche Forschungsgemeinschaft. An important aspect of our studies dealt with the following question: are the prevailing landforms developing under present conditions, or were they formed under a different climate? The latter seems more likely, if one considers that even an area, which had always belonged to the periglacial zone in the course of the Quaternary, must have been subject to climatic changes (see also TARNOCAI & VALENTINE 1989). Such changes may have caused, for instance, a deterioration or destruction of the vegetation cover, variations in the thickness of the active layer, or even a complete disappearance of the permafrost in certain periods. It is obvious that this will influence the intensity of denudation and erosion.

Therefore, our studies were not only concerned with a precise inventory of landforms and recent geomorphic processes, but also with the dating of these landforms. More than 30 ^{14}C samples were analyzed by the Labor für Geochronologie of the Niedersächsisches Landesamt für Bodenforschung in Hannover. A comparable number of pollen analyses was carried out in order to draw conclusions about the development of the vegetation. A preliminary report of the field results and first laboratory data were published by SEMMEL (1987) (cf. also HEINRICH 1990).

2 *Description of the research area and reasons for its selection*

Our research was centred on that part of the Richardson Mountains, where the Dempster Highway crosses the watershed between the Yukon and Mackenzie rivers (Rat River Pass). This coincides with the boundary between the Yukon and Northwest Territories (cf. Fig. 1). The area is situated at 67° northern latitude (Mount Cronin and Mount Sittichinli sheets of the Topographical Map of Canada 1:50,000). The landform is basically determined by a north-south running hogback, which is built up by resistant Jurassic and Cretaceous sandstones. The rocks are usually dipping to the west. The hogback is underlain by soft Devonian shale and capped by Cretaceous shale, which are also soft (Map 1519 A Geology Bell River and Map 1523 A Geology Eagle River 1:250,000). The bedding conditions are outlined in Fig. 2.

On both sides of the hogback pediments have developed. The pediments of the hogback front, i.e. on the east-facing slope, are more strongly dissected than the ones on the backslope. This is most likely related to glacifluvial influence, which caused a stronger erosion impulse on the lower reaches of the streams of the hogback front. There is no such influence on the backslope (cf. HARRIS et al. 1983: 94).

The hogback reaches an elevation of 1200 m, whereas the pediments are situated between 550 and 880 m above sea level. According to data, which were kindly provided by H. E. WAHL of the Yukon Weather Office in Whitehorse, the area has a mean annual temperature of approximately – 7.5 °C (1959–1978). The January mean temperature is ca. – 30 °C, the July mean temperature is ca. 10 °C. The mean annual precipitation is estimated to be about 400 mm (January mean ca. 10 mm, July mean ca. 50 mm).

The area is situated immediately above the tree-line. Based on its vegetation cover, it can be roughly subdivided into a tundra and a debris zone in the sense of

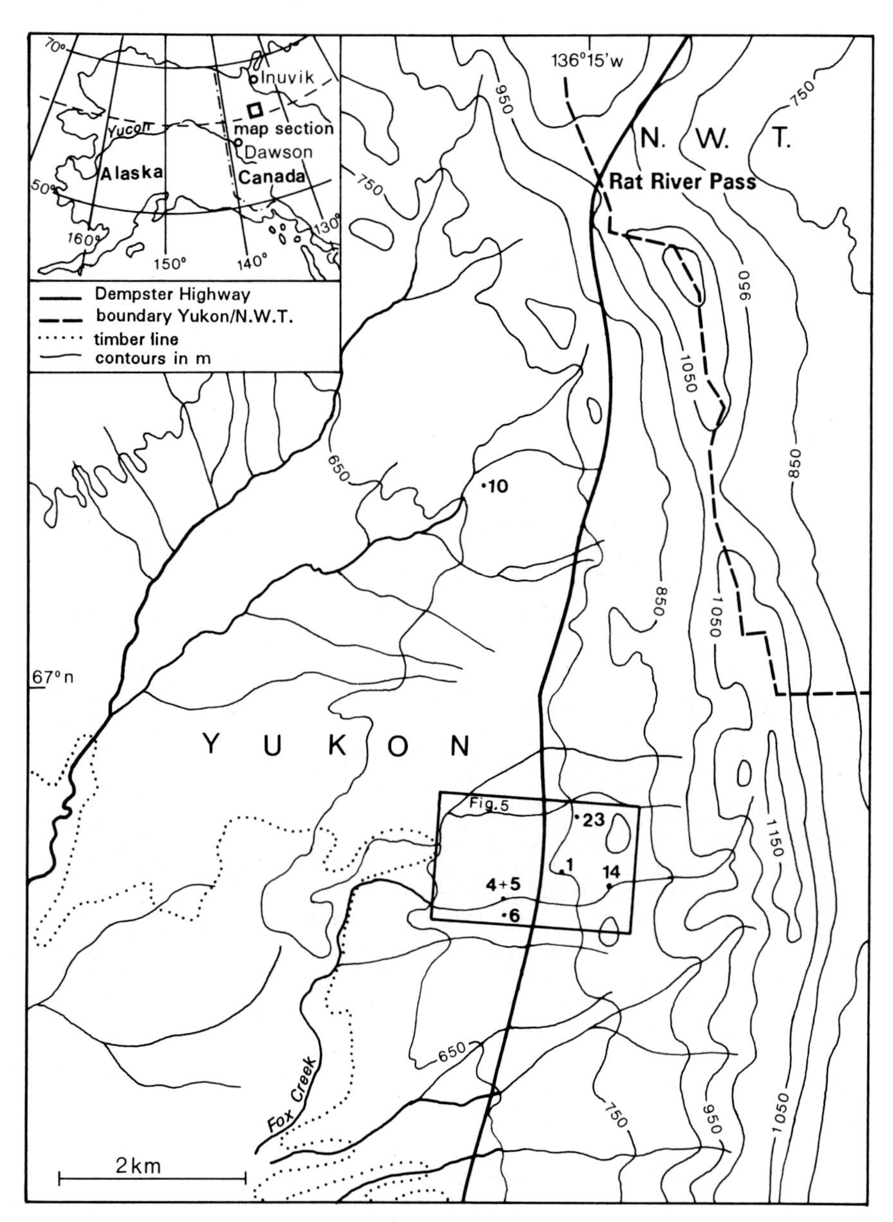

Fig. 1. Location of the research area.

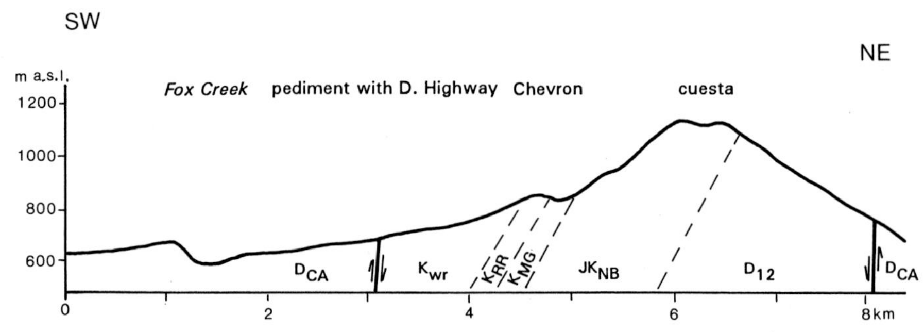

Fig. 2. Profile across the Richardson Mountains south of the Rat River Pass.

Fig. 3. Pediments with four chevrons. On the left (east) side of the photo begins the main hogback. Source: Copy of a postcard.

BÜDEL (1948) – without, however, sharing BÜDEL's concept of geomorphic process-combinations for these zones. The boundary between the two zones coincides with the boundary between pediments and hogback. Chevrons, which rise above the pediments, are related to outcrops of more resistant strata (Fig. 3). They are usually free of vegetation, and are therefore grouped into the debris zone. The reason for the lack of vegetation is not only the extreme exposition, but above all the substratum. The hard bedrock disintegrates into coarse boulders and slabs. The resulting edaphic conditions are very dry. As soon as shales reach the surface, the edaphic dryness disappears, and the ground is covered by the usual grass tundra. The preconditions for the development of the debris zone are similar to those described by SCHWARZBACH (1963: 92) from comparable areas in Iceland, which are basically edaphic deserts.

The central research area described above was selected, because here the hogback landscape with connected pediments is developed most beautifully. Even a quick comparison with geologically differently structured neighbouring areas shows that geology plays a decisive role also for the development of pediments. As soon as the bedding situation becomes more complicated, and as the lithological changes from hard to soft rocks occur within short distances, the pediments are missing or only weakly developed. The region south of the research area may serve as an example for these conditions. Near Mount Cronin, the Cambrian rocks possess respective lithologies (Geological Map 1:250,000, Map 1523 A Eagle River).

In the following, at first the steeper slopes will be described, which are generally free of vegetation, or only sparsely covered by vegetation. Then the results from the pediments will be presented.

3 *Steeper slopes*

The largest part of the hogback has strongly inclined slopes. The concordant back-slopes are covered by coarse quartzitic debris. The highest parts of the front slope, however, are free of debris. The bedrock is usually covered completely by lichens, which make everything look grey. Only very few places show beige colours, indicating fresh exposures of the bedrock. Elsewhere, a situation of prevailing geomorphic stability is dominating the scene. Shallow, rill-shaped depressions between block fields are densely covered by grass. They are filled with fine-grained solifluction deposits. These vegetated areas are also indicating geomorphic stability. Such places occur in the highest part of the hogback. Only in few localities, which are extremely exposed to the wind, there are outcrops of unvegetated or sparsely vegetated shale. Therefore, the absence of tundra vegetation is usually caused by edaphic reasons.

Despite its high content of fine material, the debris in the vegetation-covered rills does not move downslope. One may assume that the neighbouring and underlying blocks are very well drained and thus prevent the saturation of the material, which is necessary for a mobilization of the debris. In addition, the blocky material thaws quickly and to great depths. The top of the permafrost lies usually deeper than 2 m. The drainage of the active layer is therefore not impeded.

But also in areas, where the permafrost comes close to the surface, there is no downslope movement of material observable. The chevrons west of the hogback provide examples for this situation. In general, they are not as much covered with coarse blocks as the hogback itself. Their substratum consists to a great part of shale, with some thin quartzitic beds. As far as we could ascertain, the shale of all chevrons was indurated additionally through contact metamorphism by the intrusion of basalt. Because of these properties, rocks from the chevrons were used as construction material for the banks of the Dempster Highway. The indurated shales concentrate within the amorphous solifluction layers, which cover the chevrons. These solifluction layers help to smooth the ruggedness of the terrain, which results from the lithologic differences between harder and softer bedrock. Over resistant bedrock, the thickness of the debris diminishes. In some places it is totally absent. By means of these processes beautifully smooth slopes have developed across unconformities in our research area, as well as on other hogbacks. Similar forms did not only develop in the Richardson Mountains. They can be found in all present periglacial areas of north-west Canada and Alaska. Especially impressive forms can be seen in the Ogilvie-Mountains, where unvegetated, smooth slopes pass into stands of firs. There, as in our more northern research area, the unvegetated debris layers of the smooth slopes do not extend down to the local base levels. In the climatically sheltered, lower areas, a closed vegetation cover on top of the heads of the solifluction layers testifies their immobility.

The surfaces of some smooth slopes seem to contradict this result, as they are covered by numerous nonsorted stripes. Since PENCK (1912: 244), they were always interpreted as a convincing proof for movement within the substratum. But as in many other localities of the recent periglacial areas (cf. SEMMEL 1989: 9ff.), the patterns do no extend deeper into the soil than a few centimetres. At present, solifluction does not lead to any considerable mass transport. The intensity of denudation can be considered to be of minor importance.

The patterns on the surface are related to the growth of some vegetation in the shallow depressions, which are confining the stripes with unvegetated stone pavements. Both are generally underlain by an amorphous solifluction layer. The lower part of this bipartite solifluction layer is usually missing over resistant bedrock. Exposures over softer shale show a dark and clayey, basal debris layer of 50 to 100 cm thickness. It contains fragments of shale, which are furnished by the fracturing and downhill bending of the strata. The basis of the annual active layer lies at a depth of approximately 1 m. Above the dark, clayey debris follows a brown, sandy to silty debris layer, which covers almost completely the entire slope area. A similar subdivision into a lower, more "autochthonous" debris and an upper, more "allochthonous" debris layer has been recognized in other periglacial areas (SEMMEL 1969; 42ff.; BIBUS et al. 1976; 32ff.). Heavy mineral analyses indicate a higher garnet-content in the upper debris layer. The extremely low heavy mineral content (0.02–0.04%), however, provides only a very limited absolute number of minerals. The figures are therefore inadequate for the common statistical requirements. The presence of an aeolian component in the upper debris layer cannot be confirmed. One should, on the other hand, expect to find some traces of the "Old Crow tephra" (WESTGATE et al. 1983), whose age was determined to be 60,000 to 120,000 years. The differences between the two debris layers can thus best be explained as follows: the lower one is characterized by a higher content of shale, which is indicated by its high content of clay and middle-size silt. The upper layer contains distinctly more quartzite and sandstone, which results in higher quantities of sand and less clay in the substratum. This sandy component stems most likely from quartzitic sandstone beds, which are found further upslope. This coincides well with the increasing amount of coarse-grained debris immediately below the outcrop of these strata. The differences in grain size cannot, by any means, be explained as the result of clay illuviation. If this were the case, then clay eluviation should be observable even in those places, were the lower, clayey debris is missing.

The subdivision of the solifluction layers as described above shows parallels to the subdivision into "Deckschutt" (cover debris) and "Basisschutt" (basal debris) found in Central Europe (SEMMEL 1964). Here, however, the top layer always contains a distinct loess component. It could also be shown in many places that both layers did not develop simultaneously. In the Richardson Mountains, the different ages of the two layers could be demonstrated in one exposure. Here, the lower, clayey layer was separated from the hanging, sandy debris by a peat layer. In this case, one can also exclude the assumption that two different layers were moving downslope synchronously, however, at a different speed. Such situations were frequently described (e.g. SEMMEL 1985; 10 ff.). They can be grouped to those processes, which were named "episodic" or "periodic" solifluction by BÜDEL (1959) (cf. also SPÄTH 1986: 18).

The statement that debris layers at present evidently do not reach the local base level is, however, only valid, if one excludes all places, were streams are undercutting the slopes. In these places, the debris migrates, or rather falls into the streams, comparable to the bedrock above.

As soon as the vegetation cover becomes denser in more sheltered locations, forms of vegetation-banked solifluction ("gebundene Solifluktion" in the sense of TROLL 1944) will develop, above all solifluction lobes. As mentioned before, these forms are often joined to nonsorted stripes further upslope. The front of the solifluction lobes will generally move downslope over older superficial humus horizons. This phenomenon was observed in many periglacial areas, as e.g. by RUDBERG (1964) in Swedish Lapland. The distances covered by these migrating lobes are, however, obviously minimal: a buried humus layer, 50 cm upward from the front of the lobe, had already a radiocarbon age of 1150 ± 60 y.b.p. (Hv 14089).

To sum it up: we found that the steeper slopes of the research area do not evidence strong geomorphic activity. Even unvegetated slopes are feigning geomorphic activities. They should not be mistaken with active "triangular slopes" ("Dreieckshänge"), as assumed by PRIESNITZ (1981: 150). Denudation or erosion on the surface worth mentioning occurs only in active cuts of the streams.

4 *Pediments*

The steeper slopes are usually situated above a sharp knickpoint, which separates them form gentle, pediment-like forms. In our opinion, the pediments have typically developed over soft shale. The slope angle of these forms decreases with growing distance from the hogback, and increases again towards their lower end. Here, younger pediments have developed, which have cut into the higher and older surfaces. They are thus responsible for the steeper grades. Local steepening in the middle sections of the older pediments can be attributed to more resistant bedrock. The lower boundaries of the pediments are again hogbacks with resistant strata. They are dissected by canyon-like valleys, which indicate the strong erosive impulse from the White Fox River, which functions as local base level. The younger pediments sometimes extend back to the gaps, which divide the main hogback. The streams flowing in these cuts are occasionally aggrading coarse gravel and boulders, where they leave the main hogback and enter the area of the pediments. Never-

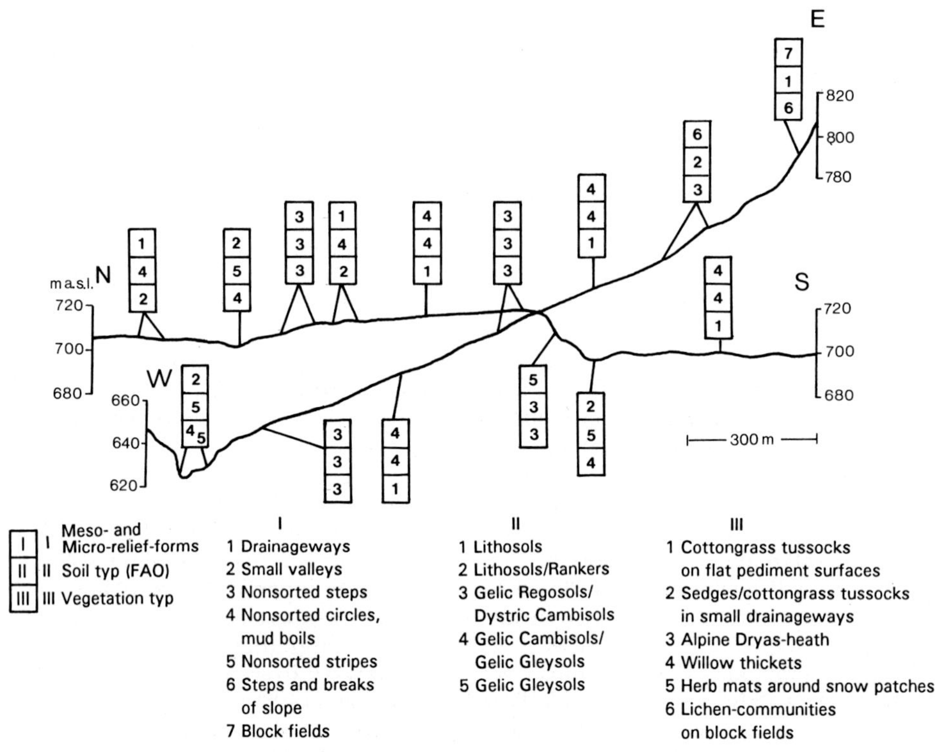

Fig. 4. Section across pediments of different age.

theless, the pediments are without doubt denudational forms, as pointed out by PRIESNITZ (ib. 153). The covering boulders and gravel are only 1 to 2 m tick, and are even totally missing in some parts of the pediments.

The larger part of the pediments is now covered by dense grass-tundra, above all by *Eriophorum vaginatum*. Further components are *Cyperaceae* and species of *Sphagnum*. Shrubs, such as *Salix polaris* and *Betula nana* are more rarely found. The vegetation cover usually amounts to 100%. A surficial humus layer and a root zone of 10 to 15 cm thickness are completely covering the mineral soil. They provide such a good isolation that the permafrost reaches to the surface even in late summer. The soil thaws only very seldom, as thick ice lenses are extending into the uppermost parts of the mineral soil (cf. Fig. 4 in SEMMEL 1987). Nevertheless, at one time cryoturbations must have developed, which destroyed the humus horizons and covered them with mineral soil. Table 1 presents radiocarbon dates (^{14}C) from humus-containing cryoturbations.

In some places we found two or three humus layers situated on top of each other. The chronology of the dated samples in these cases is "normal", i.e. the oldest lies at the bottom, the youngest lies on top. It is generally striking that the greatest ages are only slightly higher than 3000 years. We deduce that the processes of cryoturba-

Table 1 ^{14}C-Dates from humus-containing cryoturbations.

Sample	^{14}C ages (a.b.p.)	
Hv 14085	1,205 ± 60	Excavation 1
Hv 14087	3,090 ± 55	
Hv 14093	780 ± 50	
Hv 14095	2,320 ± 60	Excavation 6
Hv 14090	3,785 ± 130	
Hv 14096	540 ± 65	
Hv 14098	2,095 ± 60	Excavation 10
Hv 14099	3,390 ± 55	

Cf. Fig. 1 for the location of the samples.

tion were less intensive before this date, because otherwise one should also find older buried humus layers. Pollen analyses from all fossil layers show a vegetation, which was practically similar to the present one.

The assumption that the cryoturbation processes were less intensive 4000 years ago can be substantiated by results, which seem to indicate lower temperatures during the younger Holocene in Alaska and north-west Canada (e.g. RITCHIE & HARE 1971, SORENSEN et al. 1971, KAY 1979, BRUBAKER et al. 1984, HAMILTON et al. 1986). In our research area, we did not find proof for a distinct progression of the nowadays nearby tree-line into the present tundra during the middle Holocene. But also the assumption of definitely colder phases in the younger Holcene (SEMMEL 1987: 105) cannot be confirmed. All our radiocarbon dates and palaeobotanical results do not testify distinct climatic changes within the last 4000 years.

One might object that the present dense vegetation indicates the absence of cryoturbation processes, and therefore a milder climate than before. This is contradicted by places, where now active mudpits and related forms can be found. Such locations were already described by PRIESNITZ (ib. 153). According to him, the most mature forms are typical for the peripheral sections of the pediments, which he interpreted as inactive forms. Our mapping demonstrates a prevalence of "mature" patterned ground on those parts of the pediments, which are more exposed to the wind (Fig. 5). The youngest parts of the pediments are thus characterized by the most mature patterned ground. This is valid, however, only for the oldest pediment of our research area. Because of their lower and more sheltered position, the younger pediments show only sporadic patterned ground.

In general, many factors have to be taken into account, when the patterned ground is interpreted with regard to geomorphic implications. The impression of excellent sorting can be feigned, for instance, when a stone pavement is underlain by silty to clayey substratum, which is particularly susceptible to frost action. By frequent partial freezing it may break through the stone pavement and then create the image of a "mature" sorted circle. Where the stones are missing, usually only "immature" mudpits will develop. A statement about geomorphic stability or activity must not be deduced from these occurrences. Quite similar examples can be found on the pediments close to the main hogback. The local prevalence of sorted circles

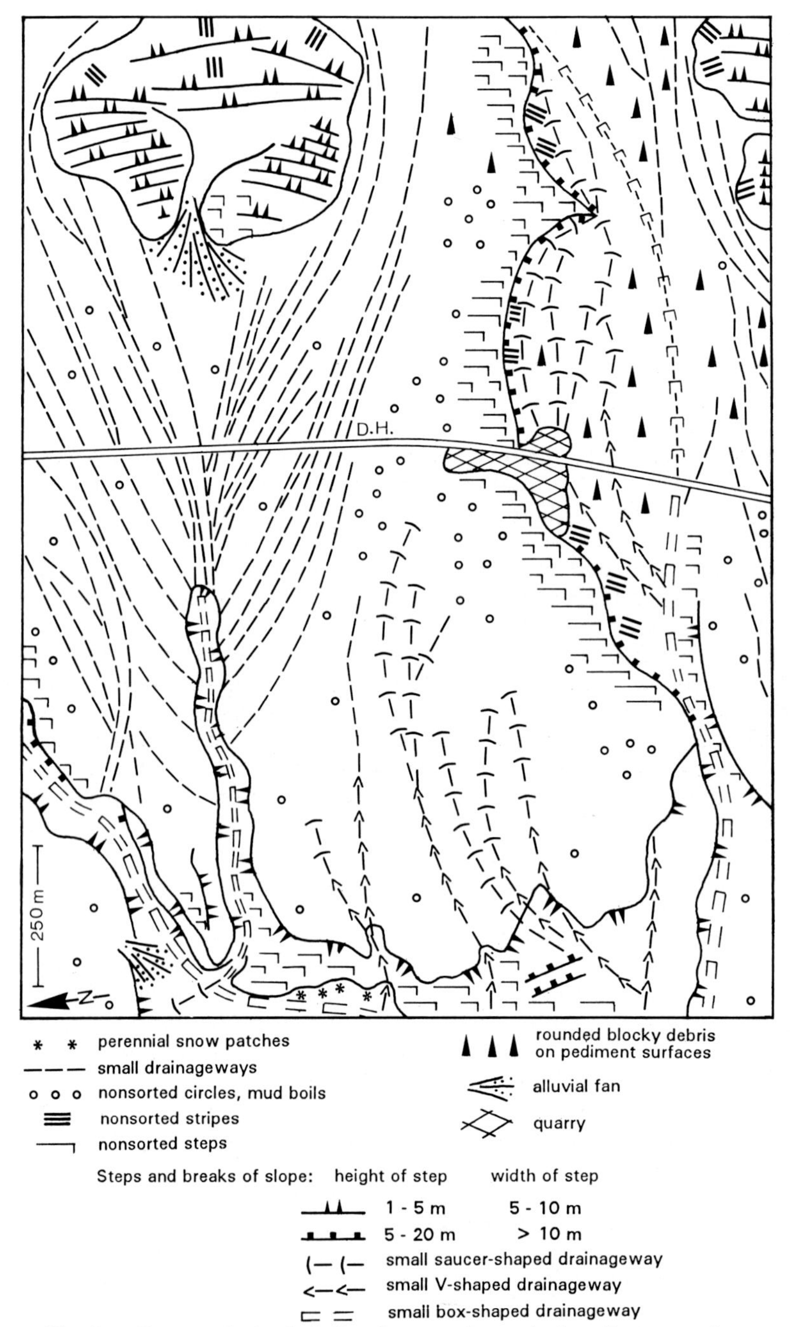

Fig. 5. Geomorphological sketch map of a typical pediment section.

is matched, for instance, in other places without stone pavement by the dominance of densely vegetated solifluction terraces. Immediately below the terrace front, humus samples provided radiocarbon ages of 715±75 a.b.p. (Hv 14105) and 1685±125 a.b.p. (Hv 14104) respectively (excavation 23, Fig. 1). These figures do not testify any essential mass transport, at least none, which could cause distinct changes of the terrain. Again it is confirmed that patterned ground does not indicate intensive denudation (SEMMEL 1987: 98).

We already pointed out that patterned ground is found only sporadically on the pediments. But even in places with dense vegetation, the surface is not absolutely level, but frequently sculptured by numerous shallow rills. They are clearly visible on aerial photographs (cf. Fig. 5; Fig. 11 in PRIESNITZ 1981). Some of the rills were documented by levelling, as presented in Fig. 4. These rills are however, not active. They are densely filled with vegetation and roots. Even here, several fossil humus horizons on top of each other were found, which again indicate a longer period of geomorphic stability. In our opinion, there can be no doubt that on pediments stable conditions have prevailed for a longer period of time. But when were these forms subject to denudation?

In two places we found evidence, which can be interpreted as an indicator of climatic changes. We found former ice wedges, which are now filled with mineral soil. The largest of these wedges extended 3.50 m into the underlying shale. Its maximum width was 50 cm. The fill of the former ice wedge consists of the same alluvial debris, which covers the respective pediment. It is composed of well-stratified fragments of shale and pebbles of quartzite and sandstone. This alluvial debris has, apart from the area of the former ice wedge, a thickness 0.5 to 1 m. It is overlain by 0.5 m of solifluction debris, which did not sink into the former ice wedge. This combination of alluvial pediment debris with ice wedges suggests a development of the pediments under periglacial conditions.

It is surprising that the area at present does not possess ice wedges. The permafrost shows only strong lenticular accumulations of soil ice. A systematic search for recent ice wedges and excavations in all likely locations brought negative results. Also HARRIS et al. (1983: 93) only reported great quantities of ground ice in the permafrost, and the difficulties arising for the construction of the Dempster Highway in this area. Ice wedges were not mentioned.

One reason for their absence might be the lack of extremely low temperatures, which are necessary for the theoretically required shrinking at low temperatures. On the other hand, ice wedges were described from an area with discontinuous permafrost and warmer winter temperatures near Dawson (FRENCH et al. 1983: 54). But these are perhaps fossil ice wedges like those from the Tuktoyaktuk area, which are of Pre-Holocene age (MACKAY et al. 1983: 170). Their thawing may have happened in the course of the Holocene climatic optimum between 5500 and 8500 years b.p., which most likely also affected the Richardson Mountains (RITCHIE & HARE 1971: 339). Pollen analyses of our samples did not, however, confirm the existence of a climatic optimum. On the other hand, none of the samples dated back that far.

In general, one must not forget with these rather speculative considerations the great number of factors, which could influence the existence of permafrost and ice wedges. Less snow cover in winter, for instance, can accelerate the development of the ice wedges. A thickening of the vegetation can assist their preservation. A thinning of the vegetation because of severer frost often brings along greater depths of the active layer with partial or complete thawing of the ice wedges. The hypothesis that ice wedges would testify a colder climate is nevertheless supported by the pollen content, which was found in the fill of the former ice wedges. Contrary to all other samples with a non-arboreal pollen content of 20%, this content rose to 40% in the former ice wedges.

The combination of alluvial pediment debris with syngenetic ice wedges suggests that the pedimentation occurred under a colder climate than at present. This climate allowed the development of ice wedges. We take it for certain that the pediments at that time belonged to the active debris zone, and that they were not covered by tundra vegetation as nowadays. Under these conditions, and contrary to the present situation, denudational processes were active.

We have pointed out already that the surfaces of the individual pediments show differences in elevation. There exists, in other words, a pediment sequence with steps of different ages. The vertical interval at the boundary between the pediment surfaces can be as high as 10 m. The occurrence of pediments from different periods led us to map these forms, and to attempt a correlation with different cold periods. This was not successful. The height differences of the pediment surfaces are most likely decisively determined by the size of the catchment area in the main hogback, which belongs to the respective pediment. The sizes of these catchment areas may change in the course of time by stream capturing. Consequently, the incision of a new pediment is not necessarily related to climatic changes. The same applies to the incision of new pediments from the downstream side by headward erosion. If, for instance, the local base level cuts into more resistant beds, this will undoubtedly influence the denudation and erosion of the upstream area.

In this context one must remember the possibility that streams may assist the development of new pediments through lateral erosion, and possibly also under the present climatic conditions. The small streams flowing from the hogback have usually cut only 1 to 2 m into the younger pediments. Only in their lower reaches, the incision becomes greater. The narrow streams with a maximum width of 2 m transport coarse gravel. Their courses, however, remain to a great extent constantly in the same position. This is testified by more than 100 years old shrubs, which are frequently growing in the sheltered cuts. The local persistence for more than 1000 years could be proven by radiocarbon dates. On both sides of one stream humus layers were intercalated into the alluvial debris, which provided ^{14}C ages of 1280 ± 55 a.b.p. (Hv 14088) and 4085 ± 65 a.b.p. (Hv 14086) respectively (cf. Fig. 1, excavations 4 and 5). Apart from the small lateral erosion, the equally small linear erosion could be dated. The youngest sample with an age of approximately 1300 years lay only 0.7 m above the present stream.

One cannot deny, on the other hand, that short sections of the streams are responsible for lateral erosion, e.g. near undercut slopes. In these places, snow patches can accumulate, which can intensify the slope retreat. A small shift of the stream bed, however, is sufficient to stabilize the slope again. This is achieved by solifluction, which flattens the slope and thus interrupts the superposition for the accumulation of snow patches. In these places, the snow patches are not inducing independent denudational processes, which could lead to cryoplanation and to the development of larger surfaces (SCHUNKE 1974, REGER & PÉWÉ 1976). The streams are therefore at present not responsible for pedimentation. This differs clearly from observations, which were made on the gentle slopes of the debris zone in West-Spitsbergen (Bibus et al. 1976: 34ff.).

A distinctly recent aggradation in the level of the lowest pediment was found only in one place. It was caused, however, by an exceptional situation. An outcrop of resistant rock forced the stream to change its course considerably. Sheltered by this resistant outcrop, a 16 m wide terrace with a gravel accumulation of 1 m thickness contained several fossil humus layers (excavation 14, Fig. 1). The lowest layer had a ^{14}C age of 2045 ± (60) a.b.p. (Hv 14092). The next, which was situated 20 cm higher, was dated at 1360 ± 60 (Hv 14101), and a third, situated 15 cm below the recent Ah – horizon, provided an age of 1080 ± 65 a.b.p. (Hv 14100). The basis of the aggradation, which was almost identical with the oldest humus layer, was situated only 0.5 m above the present stream bed. These dates, as well as the entire geomorphic situation, are excluding the possibility that in the more recent past conditions prevailed, which favoured a distinct pedimentation.

Also the other areas visited by us in the Yukon region did not show clear evidence for active pedimentation. This is supported by Schunke (1987: 163f.), who published similar results from the region of the Brooks Range in northern Alaska. The flat denudational surfaces of the Yukon region, e.g. the cryoplanation terraces described by French et al. from the Klondike Plateau near the Alaskan border (Sixty Mile Highway), must be regarded as inactive. This opinion was also expressed by the authors cited above, because of the incrustation by lichens. The size of these surfaces even suggests Tertiary planation processes. Especially the larger of these surfaces can still be characterized by the words of French et al. (1983: 63): "The existence of the terraces is unquestionable; the processes responsible for their formation and age are debatable."

References

Bibus, E., G. Nagel & A. Semmel (1976): Periglaziale Reliefformung im zentralen Spitzbergen. – Catena **3**: 29–44; Gießen.

Brubaker, L. B., H. L. Garfinkel & M. E. Edwards (1984): A Late Wisconsin and Holocene Vegetation History from the Central Brooks Range: Implications for Alaskan Palaeoecology. – Quat. Res. **20**: 194–214; New York.

Büdel, J. (1948): Die klimamorphologischen Zonen der Polarländer. – Erdkde., II: 22–53; Bonn.

– (1959): Periodische und episodische Solifluktion im Rahmen der klimatischen Solifluktionstypen. – Erdkde., **XIII**: 297–314; Bonn.

Dedkov, A. (1965): Das Problem der Oberflächenverebnungen. – Peterm. geogr. Mitt. **109**: 258–264; Gotha.

French, H. M., S. A. Harris & R. O. Van Everdingen (1983): The Klondike and Dawson. – In: French, H. M. & J. A. Heginbottom (eds.): Northern Yukon Territory and Mackenzie Delta, Canada. – Guidebook 3, IV. Int. Conf. Permafrost, Fairbanks: 35–63; Fairbanks.

Hamilton, T. D., K. M. Red & R. M. Thorson (1986): Glaciation in Alaska – The Geologic Record. – 49 p.; Anchorage.

Hanson, L. W. (1965): Size distribution of the White River ash, Yukon Territory. – Dept. Geol. Edmonton, Alberta: 38 p.; Edmonton.

Harris, S. A. et al. (1983): The Dempster Highway – Eagle Plain to Inuvik. – In: French, H. M. & J. A. Heginbottom (eds.): Northern Yukon Territory and Mackenzie Delta, Canada. – Guidebook 3, IV. Int. Conf. Permafrost, Fairbanks: 87–111; Fairbanks.

Heinrich, J. (1990): Boden- und Vegetationsgeographische Beobachtungen zur rezenten Morphodynamik in der Tundrenzone der Richardson Mountains, Yukon Territory, NW Kanada. – Geoökodyn. **XI**: 116–142; Bensheim.

Karrasch, H. (1970): Das Phänomen der klimabedingten Reliefasymmetrie in Mitteleuropa. – Gött. geogr. Abh. **56**: 229 p.; Göttingen.

Kay, P. A. (1979): Multivariate Statistical Estimates of Holocene Vegetation and Climate Change, Forest-Tundra Transition Zone, N. W. T., Canada. – Quat. Res. **11**: 125–140; New York.

Mackay, J. R., H. M. French & J. A. Heginbottom (1983): Tuktoyaktuk. – In: French, H. M. & J. A. Heginbottom (eds.): Northern Yukon Territory and Mackenzie Delta, Canada. – Guidebook 3, IV. Int. Conf. Permafrost, Fairbanks: 147–177; Fairbanks.

Penck, A. (1912): Über Polygonboden in Spitzbergen. – Z. Ges. Erdkde. Berlin, 1912: 241–249; Berlin.

Péwé, T. L. (1975): Quaternary geology of Alaska. – U. S. Geol. Surv. Prof. Paper **835**; 145 p.; Denver.

Priesnitz, K. (1981): Fußflächen und Täler in der Arktis NW-Kanadas und Alaskas. – Polarforsch. **51**: 145–159; Münster i.W..

– (1988): Cryoplanation. – In: Clark (ed.): Advances in Periglacial Geomorphology: 49–67; London.

Reger, R. D. & T. L. Péwé (1976): Cryoplanation Terraces: Indicators of a Permafrost Environment. – Quat. Res. **6**: 99–109; New York.

Ritchie, J. L. & F. K. Hare (1971): Late Quaternary Vegetation and Climate near the Arctic Tree Line of Northwestern North America. – Quat. Res. **1**: 331–342; New York.

Rudberg, S. (1964): Slow mass-movement processes and slope development in the Norra Storfjällaren, Southern Swedish Lappland. – Z. Geomorph. N. F., Suppl.-Bd. **5**: 192–203; Berlin.

Schunke, E. (1974): Formungsvorgänge an Schneeflecken im isländischen Hochland. – Abh. Akad. wiss. Gött., math.-phys. Kl., 3. F., **29**: 274–289; Göttingen

– (1987): Studien zur periglazialen Reliefformung der zentralen Brooks Range und des Arctic Slope, Nord-Alaska. – Polarforsch. **57**: 149–171; Bremerhaven.

Schwarzbach, M. (1963): Zur Verbreitung der Strukturböden und Wüsten in Island. – Eiszeitalter u. Gegenwart **14**: 85–95; Öhringen.

Semmel, A. (1964): Junge Schuttdecken in hessischen Mittelgebirgen. – Notizbl. Hess. L.-Amt Bodenforsch. **92**: 275–285; Wiesbaden.

– (1969): Verwitterungs- und Abtragungserscheinungen in rezenten Periglazialgebieten (Lappland and Spitzbergen). – Würzb. geogr. Arb. **26**: 82 S.; Würzburg.

– (1985): Periglazialmorphologie. – Ertr. Forsch. **231**: 116 S.; Darmstadt.

– (1987): Periglaziale Formung im nordwestlichen Kanada. – Gött. geogr. Abh. **84**: 91–107; Göttingen.

Sorenson, C. J. et al. (1971): Palaeosols and the Forest Border in Keewatin, N. W. T. – Quat. Res. **1**: 468–473; New York.

Späth, H. (1986): Die Bedeutung der "Eisrinde" für die periglaziale Denudation. – Z. Geomorph. N. F. **61**: 3–23; Berlin/Stuttgart

Tarnocai, C. & K. W. G. Valentine (1989): Relict Soil Properties of the arctic and subarctic regions of Canada. – Catena Suppl. **6**: 9–39; Cremlingen.

Troll, C. (1944): Strukturböden, Solifluktion und Frostklimate der Erde. – Geol. Rdsch. **34**: 545–694; Stuttgart.

WASHBURN, A. L. (1979): Geocryology. A survey of periglacial processes and environments. – 320 p.; London.

WESTGATE, J. A., TH. D. HAMILTON & M. P. GORTON (1983): Old Crow-Tephra: A New Late Pleistocene Stratigraphy Marker across North-Central Alaska and Western Yukon Territory. – Quat. Res. **19**: 38–54, New York.

Address of the authors: Institut für Physische Geographie, Universität Frankfurt/Main, Senckenberganlage 36, D-6000 Frankfurt/Main 11.

Z. Geomorph. N.F.	Suppl.-Bd. 92	71–84	Berlin · Stuttgart	Mai 1993

Permafrost, Gelifluction and Fluvial Sediment Transfer in the Alpine/Subnival Ecotone, Central Alps, Austria: Present, Past and Future

by

Heinz Veit, Bayreuth, and Thomas Höfner, Bamberg

with 7 figures

Summary. Current processes of gelifluction and fluvial sediment transfer have been monitored by our research groups for several years in the periglacial altitudinal belt of the southern Hohe Tauern mountain range, central Alps, Austria. These studies furnish information on boundary conditions and forcing factors of the morphodynamic system under observation, and allow the interpretation of its spatio-temporal patterns and oscillations. As one of the main results, a close connection between cryogenic processes and the destruction of vegetation cover, the intensity of gelifluction and also the amount of sediment transfer in the headwater channels can be demonstrated.

Through a combination of these results with data on Holocene slope, soil and channel dynamics, projections into the past are made possible and allow, by comparison with the vegetational and glacial history of the area, palaeoclimatic and palaeoecologic reconstructions on a Holocene scale.

These palaeoenvironmental reconstructions, in turn, form the starting point of a model which estimates sediment output from the periglacial altitudinal belt as a function of climatic change. This opens up possibilities for the simulation of morphodynamic change, for instance, as a result of global warming.

Zusammenfassung. Messungen zum aktuellen solifluidalen und fluvialen Materialtransport, die von uns seit einigen Jahren in der periglazialen Höhenstufe der südlichen Hohen Tauern in Österreich durchgeführt werden, geben Hinweise auf die steuernden Faktoren und erlauben eine Interpretation der räumlichen und jährlichen Schwankungen der Morphodynamik. Dabei zeigt sich ein enger Zusammenhang von frostdynamischen Vorgängen mit Vegetationszerstörung, Intensität der Solifluktion und Materialfracht in den Gerinnen, der letztlich klimatisch bedingt ist.

Die so gewonnenen Ergebnisse können in Verbindung mit geomorphologischen Untersuchungen zur holozänen Hang-, Boden- und Gerinneentwicklung auf die Vergangenheit übertragen werden und erlauben im Vergleich mit der Vegetations- und Gletschergeschichte eine paläoklimatische und paläoökologische Charakterisierung des Holozäns.

Auf dieser Basis kann ein Modell erstellt werden, das den Sedimentaustrag aus der periglazialen Höhenstufe als Funktion von Klimaänderungen beschreibt. Damit eröffnet sich die Möglichkeit von Simulationen und Prognosen morphodynamischer und geoökologischer Veränderungen z. B. als Folge einer wärmer werdenden Atmosphäre.

0044-2798/93/0092-0071 $ 3.50

1 *Introduction*

The alpine/subnival ecotone can be classified as a potentially unstable ecosystem that reacts very sensitively to climatic change and human impact. This ecotone is closely interconnected with the densely populated valley floors of the subalpine and montane altitudinal belts by mountain torrents. Thus, its influence is far-reaching and not solely restricted to the periglacial zone (VEIT 1989). Investigations concerning these phenomena have been carried out for many years by our research group at selected study sites throughout the central Alps (Fig. 1). STINGL (1969, 1971) started fieldwork by detailed mapping of periglacial surface features. In 1984 an instrumented plot for the monitoring of gelifluction rates and various soil and climatic parameters was established (STINGL & VEIT 1988), and has been supplemented since 1989 by a second instrumented plot. Complementary studies concerning the mechanisms and rates of fluvial sediment transfer are in progress since 1988 (HÖFNER 1989, 1992, HÖFNER & GARLEFF 1992). The distribution and ecology of alpine permafrost in this area have been described by RENNERT (1991).

The monitoring of present geomorphodynamic processes is also prerequisite for an evaluation of changes in the geologic past. Conversely, such studies of landscape genesis allow the validation of models describing currently observable or just postulated relationships within the morphodynamic system. Results concerning these issues have been published by VEIT (1988 a, b, 1989) for the postglacial development of the periglacial altitudinal belt.

In addition, budgeting of current and reconstruction of past periglacial process rates, both within a temporal and a spatial framework, opens up possibilities for extrapolation and modelling with regard to future environmental change.

2 *Study area*

The study area is situated in the eastern central Alps, south of the "Großglockner" ("Hohe Tauern", Austria, Fig. 1). The monitoring plots have been set up in the Glatzbach catchment (2450–2900 m a.s.l.). The relatively small catchment area of 1.3 km^2 is a convenient size for such process work. The catchment comprises parts of the middle and upper alpine belt, characterized by an almost complete cover of alpine tundra (*Curvuletum, Salicetum*), as well as parts of the subnival altitudinal belt above 2600–2750 m a.s.l., which is almost devoid of vegetational cover.

The monitoring plots I and II for gelifluction processes are situated near the upper boundary zone of alpine tundra vegetation (2650–2700 m a.s.l., Fig. 1 and 2). Gelifluction movements are monitored by annual surveys of surface targets during summer. Both wooden dowels and slips of aluminium foil are being used, which not only allow the registration of surficial movements but also of movement trends within the active layer of the soil/regolith column. However, as this can only be done by excavating the target sites down to a depth of 1 m, these deeper movement rates can only be surveyed once at the end of the monitoring period. Up to now, this has only been done on monitoring plot I in 1989. Air and soil temperatures are continuously recorded by a data logging device which receives its signals from different levels above and below ground in discrete time intervals. For the most part, the

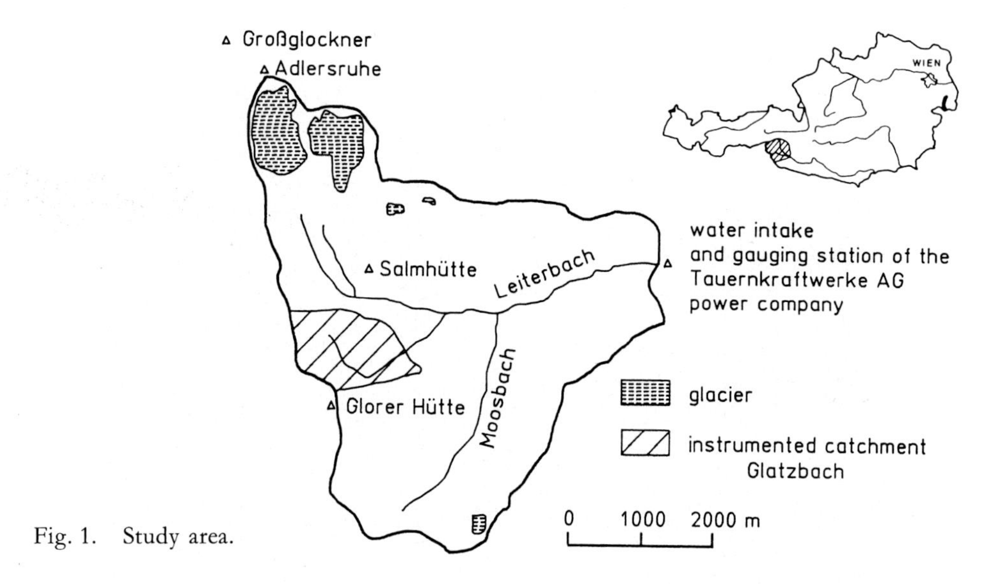

Fig. 1. Study area.

technical setup of monitoring plots I and II is similar to that used by GAMPER (1981, 1983, 1987) in his investigations in Switzerland.

Two instrumented subcatchments have been established, corresponding to the geoecological differentiation of the catchment with its vegetation cover of 63% above the main gauging station III. Subcatchment IIIa is situated almost entirely in the alpine tundra belt (2450–2650 m a.s.l.) and has a vegetation cover of 95%, whereas subcatchment IIIb has been set up in the predominantly vegetation-free subnival frost-shattering zone (2650–2900 m a.s.l., vegetation cover 14%). Broad-crested V-notch weirs, mechanical discharge recorders, automatic pumping samplers and wire-enforced dry-stone walls, which act as bedload traps, are employed in monitoring the fluvial sediment transfer (Fig. 3). In addition, the delimitation of source areas for fluvially transported sediments is attempted by sediment traps set up in debris gullies and up the slopes. This is supplemented by discontinuous on-site measurements and detailed mapping both in the field and from aerial photographs.

The geology of the area is characterized by being part of the "Matreier Schuppenzone" at the southern margin of the "Tauernfenster". Geomorphologically weak and easily weathering calcareous mica schists and phyllites with lenses of dolomite and quartz form the predominant rocks. For the most part, this rock base is covered by morainic material or gelifluction sheets of varying thickness.

3 *Present morphodynamic processes*

3.1 *Monitoring plots I and II*

The recorded gelifluction rates show a strong spatial variability. This is controlled by various factors like vegetation cover, soil type, relief and snow cover, so that soil

Fig. 2. Monitoring plot I for the study of gelifluctional morphodynamics.

Fig. 3. Gauging stations IIIa (above) and IIIb (below).

frost and oil water regime become important variables. Altogether, both the mean (0.2–0.8 m/a) as well as the maximum rates (more than 0.8 m/a) are much higher than rates previously measured with comparable methods (GAMPER 1981, 1983). This is mainly due to the extremely frostsensitive and fine-grained material supplied by the weathering of the phyllites and mica-schists. Gelifluction lobes densely overgrown by vegetation are largely inactive.

Apart from a strong spatial variability, great differences in the rates of movements from year to year can also be noticed (STINGL & VEIT 1988). This depends on changes of the annual temperature and precipitation regime. High movement rates, e.g., during the years of 1988/89, are mostly triggered by cold winters and a delay in snowfall. This enhances frost penetration into the ground, leading to intensive gelifluction processes during spring snowmelt. During mild winters with early snow cover and, because of that, shallow frost penetration, such as in 1990/91, only minimal movements can be detected. On average, the depth of these movements reaches 40–50 cm. Surficial movement rates increase with increasing thickness of the material in motion. This is caused by frost penetration to a greater depth combined with lack of vegetation and thin snow cover, which results in a large number of gelifluction days.

An intensification of the morphodynamics of largely inactive and overgrown gelifluction lobes can mainly be ascribed to cryogenic processes that are responsible for a destruction of the vegetation cover. Thus, gelifluction lobes are important sediment stores for fluvial erosion, as cryogenic processes cause the exposure, loosening and mobilization of materials. However, the lobes themselves hardly ever reach the channels. Years distinguished by high gelifluction rates can also be expected to show considerable surface wash and sediment input into the channels.

At lower altitudes, gelifluctional movements under dense alpine tundra approach zero. This stability is also indicated by alpine podsol soils that become ubiquitous with decreasing elevation, whereas they are lacking on the active or partially active lobes.

3.2 *Fluvial dynamics*

Runoff and sediment load data show a markedly nival regime both for the entire catchment (gauging station III) and for the subcatchments in the alpine (gauging station III a) and the subnival belt (gauging station III b). Precipitation events during the summer months are comparatively unimportant in terms of magnitude and frequency, and account only for about 10–25% of the entire runoff volume.

Predominantly bedload and suspended load are closely connected to spring snowmelt. During this period, approximately 95% of the annual sediment discharge is achieved. Conversely, the amount of exported solutes is relatively high even during the summer because of much higher baseflow concentration levels, but it still doesn't reach the snowmelt values.

Denudation rates for the two subcatchments III a and III b differ by more than an order of magnitude, and are significantly correlated with the percentage of vegetation cover. Estimated values amount to 88 metric tons/km^2/a for the entire catchment, 8.2 t/km²/a for the subcatchment in the alpine tundra zone and 251 t/km²/a for the subcatchment in the subnival zone. According to currently available data sets, changes in vegetation cover have much greater effects on sediment output in the long and medium range than annual variability of winter precipitation or the amount of water stored in the snowpack at the beginning of spring snowmelt. This rule can, at times of course, be reversed by events of extraordinarily high magnitude and low frequency.

High rates of denudation in the subnival zone can mainly be ascribed to the following factors:

- predomination of vegetation-free surfaces susceptible to erosion,
- predisposition for sheetwash processes during spring snowmelt because of late thawing of seasonal and/or perennial soil ice,
- comminution, displacement and destabilization of material by cryogenic and gelifluction processes.

Field observations of sediment transport on the slopes in the subnival zone confirm the importance of wash processes on the surface of, within and below the snowpack. Conversely, meltwater in the zone of dense vegetation cover is hardly able to detach and transport substantial amounts of material. This proves the importance of vegetation cover with respect to the intensity of fluvial erosion.

4 *Holocene morphodynamic oscillations*

4.1 *Phases of gelifluction and soil formation*

In the "Hohe Tauern" area a number of gelifluction phases can be reconstructed, which were interrupted by phases of soil formation (Fig. 4). In agreement with findings from Switzerland (GAMPER 1981) and the "Dolomiten" of southern Tyrol (STEINMANN 1978), the early Holocene down to 5000 B.P. was a time of widespread slope stability in the periglacial altitudinal belt. This resulted in the formation of relatively well developed soils, like podsols on rocks poor in, or devoid of, carbonate and rendzinabrownearths on dolomites and limestones. During the early Holocene a dense vegetation cover of alpine tundra thus reached far higher altitudes than today, these conditions prevailing for a considerable time span. Fig. 5a shows the present vegetation cover and Fig. 5b the reconstructed vegetation cover for the Hypsithermal based on field mapping and infrared aerial photographs. From the close of the late Atlantic to the present, periglacial morphodynamics oscillated between phases of activity and stability. Two gelifluction phases stand out because of the intensity and spatial extension of alpine permafrost, which reached about 250 m further down than under present conditions (3350–2800 B.P. and around 1250 B.P.). Apart from gelifluction processes, intensive wash erosion which is connected with the destruction of alpine tundra vegetation by cryogenic processes can also be traced. Thus, the spatial extent of the largely vegetation-free subnival zone, which corresponds to the area of effective morphodynamics and sediment input into the headwater channels, was considerably expanded (Fig. 5c).

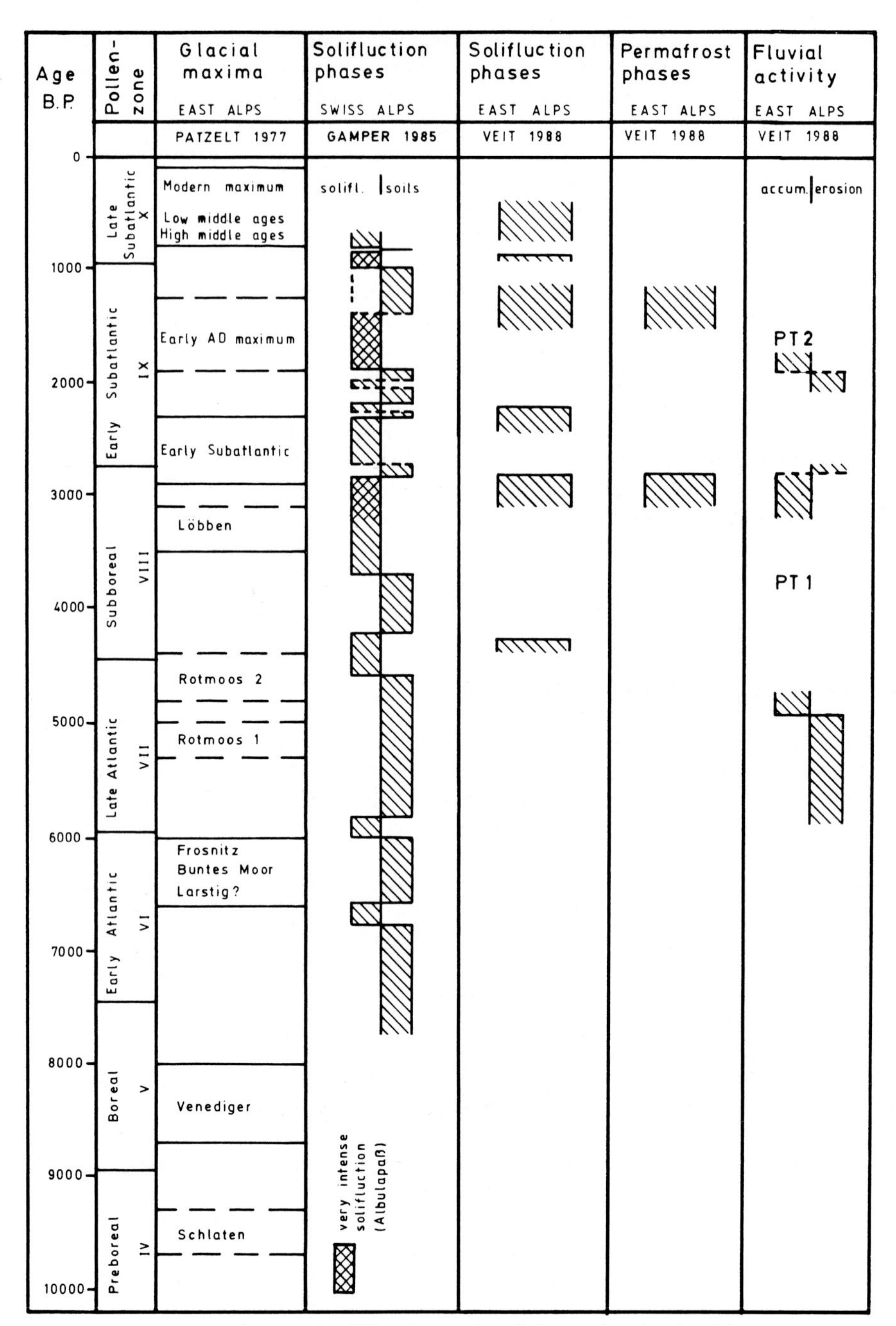

Fig. 4. Phases of solifluction and soil formation in the Alps.

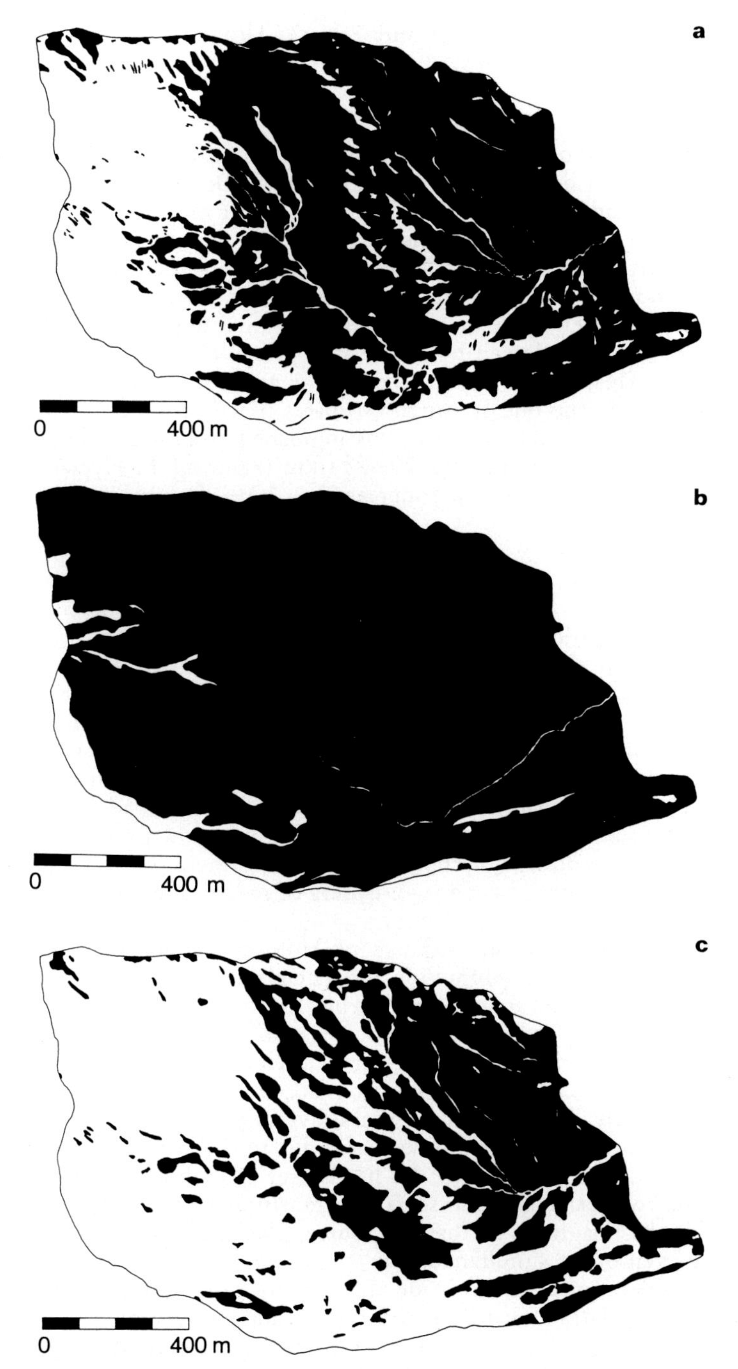

Fig. 5. Vegetation cover in the Glatzbach catchment. 5a Present state (63%) – 5b Hypsithermal (95%) – 5c Permafrost phase (38%).

4.2 *Fluvial dynamics*

The periglacial altidudinal belt forms an important headwater area for many higher order channels further down in the mountain valleys. Because of this, changes in its morphodynamics and sediment supply have important repercussions on the fluvial dynamics of valley floors as low as the subalpine and the montane belt. During the Holocene, two marked phases of valley infilling in various such valleys of the eastern Tyrol correspond well with periods of intensified slope erosion in the periglacial zone (Fig. 4). The older phase of aggradation set in at about 5000 B.P. and lasted until about 2800 B.P., followed by an incision phase in the early Subatlantic until about 1800 B.P. This, in turn, was succeeded by another aggradational phase. The interdependence is borne out by rough estimates based on process data, according to which the Glatzbach catchment was dinstinguished by a mean sediment discharge (bedload and suspended sediment) of 20–25 $t/km^2/a$ during the Hypsithermal. During the permafrost periods this rate increased to 100–200 $t/km^2/a$.

5 *Climatic interpretation of morphodynamic change and sediment transfer modelling*

Our current set of process data show that both slope and fluvial dynamics are essentially controlled by the degree of vegetation cover. Apart from the immediate impact of temperature change on plant metabolism, soil frost regime exerts a special influence that may result in rapid changes in the composition of plant communities and the destruction of vegetation cover. Cold winters with late snowfall are favourable for intensive periglacial soil creep, whereas mild winters with early snowfall only give rise to weak movements. Accordingly, sediment load in the headwater channels varies greatly. Leaving events of extraordinary low frequency apart, however, the changes in sediment discharge induced by annual precipitation variability are on average about one, or even two, orders of magnitude lower than variations triggered by changes of vegetation cover.

Palaeoenvironmental reconstructions of Holocene conditions furnish similar results. Phases of intensified gelifluction and glacial advances, which are well known from the "Venediger" mountain range nearby (PATZELT 1977, PATZELT & BORTENSCHLAGER 1973, Fig. 4), did not always occur simultaneously. Indeed, during cold and dry climatic phases increased periglacial activity and high sediment input into headwater channels are likely, but this may not necessarily imply marked glacial advances. Conversely, cool and moist years favour positive glacial mass balances, while the importance of gelifluction decreases due to the insulating influence of the snowpack. Thus, sediment load of the headwater channels likewise decreases. In between these two extremes, all sorts of transitional combinations of climatic parameters can be imagined, so that, of course, glacial advances and gelifluctional phases can as well occur simultaneously.

These considerations can be made the starting point for the calibration of a partial model that relates climatic change to percentage of vegetation cover (Fig. 6). Cardinal points are the minimum and maximum extension of vegetation cover during the Holocene, while the present condition occupies an intermediate position. Transitions from one environmental and morphodynamic state to another, however,

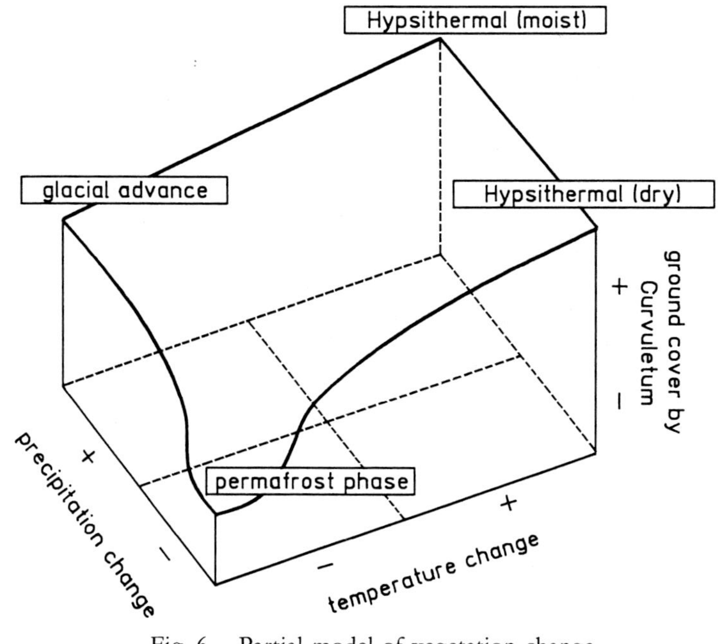

Fig. 6. Partial model of vegetation change.

are not to be expected to behave as linear relationships. Rather, because of different reaction and relaxation times of individual vegetation units (*Curvuletum, Salicetum*, pioneer communities), the possibility of long-range hysteretic loops must be taken into account. For these reasons, temporal resolution is only in the order of decades.

This partial model, which, for simplification, is presented as a linear regression model in the left half of Fig. 7, is an integral part of a more comprehensive conceptual model (Fig. 7), linking palaeoenvironmental reconstructions with modern process data. The model thus allows the estimation of sediment discharge from the periglacial altitudinal belt as a function of climatic change. These results can then be made the starting point for computer simulations concerning predictions of morphodynamic or geoecologic change triggered by global warming.

Palaeodata form an important constituent in the setting up of such a conceptual model. In particular, data sets from palaeoenvironmental reconstructions developed from soil- and vegetation mapping, or from sedimentological and palynological records, must be correlated with palaeoclimatic data estimated from the extent of past glacial advances or depressions of the lower boundary of discontinuous permafrost (Patzelt & Bortenschlager 1973, Patzelt 1977, Haeberli 1982, Kerschner 1985, Gamper 1987, Veit 1988a, 1989, Buchenauer 1990). True enough, this introduces an unspecifiable amount of uncertainty into the model so that this part must be classified as semi-quantitative.

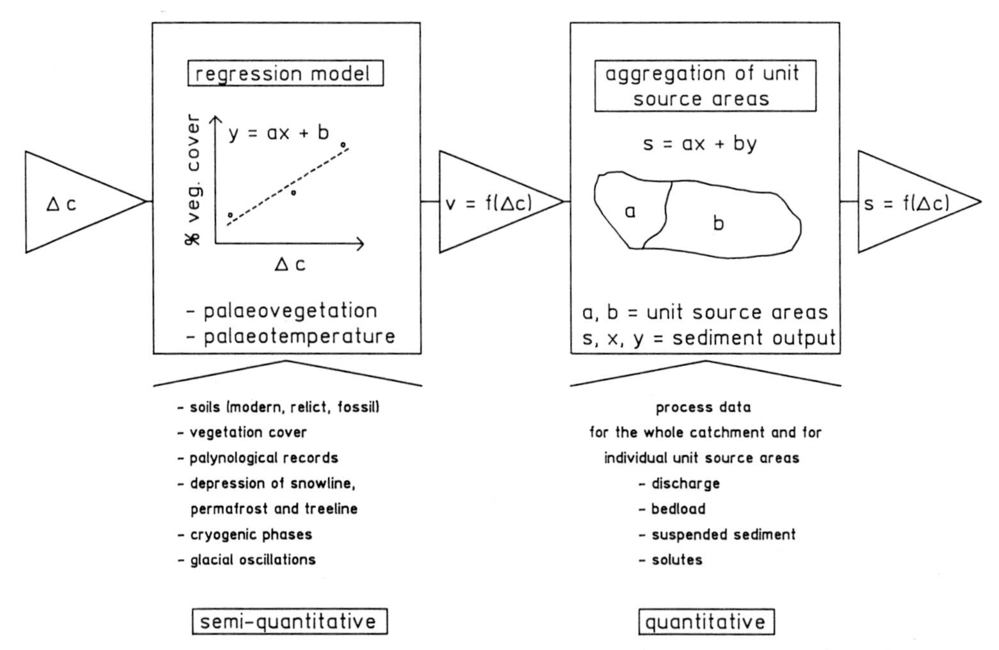

Fig. 7. Conceptual model for the estimation of sediment discharge triggered by climatic change.

The input variable to the model delta c represents the deviation of the long-term (10^1-10^2 years) climatic mean from the present value. Within this context of climatic change, mean annual temperature deviation can be estimated relatively well, so that this parameter is employed as a surrogate indicator for the much more complex mechanisms of climatic change. The output variable of this part of the model then is the degree of vegetation cover as a function of delta c. In a second step, the estimated vegetation cover forms the input variable for the quantitative aggregation model, which estimates sediment discharge from the areally weighted aggregation of unit source area process data, i.e., sediment discharge data from the alpine tundra and the morphodynamically active subnival belt. Thus, sediment discharge is ultimately a function of climatic change represented by delta c, while changes of precipitation volume can be incorporated indirectly via appropriate adjustments in the means and standard deviations of the process data.

In this context, long-term extrapolation of process data must be treated with provisional caution because of the so-called "Hurst-effect". The Hurst-effect is a form of long-term peristence in climatic, hydrological and geophysical data sets (Kirkby 1987). It manifests itself by the fact that, for instance, dry and wet years don't occur in isolation but go in runs. This means that short monitoring periods of 3–4 years' duration may not be able to estimate correctly the mean and standard deviation of the target distribution on a timescale of decades. The long-term mean, in turn, is itself subject to a persistence effect on a timescale of 10^2-10^3 years. Here the persistence is controlled by long-term climatic changes, which can be reconstructed by palaeopedologic and palaeobotanic methods. According to KIRKBY (1987), a combination of data about climatic change reconstructed in such a manner with modern process data could markedly increase the reliability of long-term extrapolations, as such data sets are obtained by independent methods. In the model presented above, these thoughts have been incorporated by combining information about the climatically controlled variable of vegetation cover with process data of fluvial sediment transfer. In addition, in the Glatzbach catchment such a procedure is favoured by the fact that differences in sediment discharge between the alpine and subnival belt on average are substantially larger (usually by an order of magnitude) than interannual deviations within the individual altitudinal belts themselves. Thus, a possible long-term persistence in process data sets, which may mask the true means and standard deviations, will considerably loose weight in the extrapolation process.

Acknowledgements

We express our gratitude to the "Deutsche Forschungsgemeinschaft" and the University of Bayreuth for financial support, to Prof. Dr. H. STINGL and Prof. Dr. K. GARLEFF for initiating and supervising the project, to Dr. M. GAMPER, Dr. K.-H. EMMERICH, Dr. B. JOHN, Dr. H. LIEBRICHT, Dr. F. SCHÄBITZ, R. BEER and generations of undergraduate students for assistance during fieldwork.

References

BUCHENAUER, H. W. (1990): Gletscher- und Blockgletschergeschichte der westlichen Schobergruppe (Osttirol). – Marburger Geogr. Schr. **117**: Marburg/Lahn.

GAMPER, M. (1981): Heutige Solifluktionsbeträge von Erdströmen und klimamorphologische Interpretation fossiler Böden. – Erg. wiss. Unters. Schweiz. Nationalpark, **XV**(79): 355–443.

– (1983): Controls and rates of movement of solifluction lobes in the Eastern Swiss Alps. – Fourth Intern. Conf. Permafrost, Proceedings; Washington.

– (1987a): Postglaziale Schwankungen der geomorphologischen Aktivität in den Alpen. – Geogr. Helvet. **42**(2): 77–80.

– (1987b): Mikroklima und Solifluktion: Resultate von Messungen im schweizerischen Nationalpark in den Jahren 1975–1985. – Gött. Geogr. Abh. **84**: 31–44.

HAEBERLI, W. (1982): Klimarekonstruktionen mit Gletscher – Permafrost – Beziehungen. – Materialien z. Physiogeographie **4**: 9–17; Basel.

HÖFNER, T. (1989): Aspekte fluvialen Sedimenttransfers in der alpinen Periglazialstufe – vorläufige Ergebnisse zu Geröll- und Lösungsfracht im Glatzbach, südliche Hohe Tauern. – Gött. Geogr. Abh. **86**: 95–104; Göttingen

– (1992): Fluvialer Sedimenttransfer in der periglazialen Höhenstufe der Zentralalpen, südliche Hohe Tauern, Osttirol. Bestandsaufnahme und Versuch einer Rekonstruktion der mittel- bis jungholozänen Dynamik. – Bamberger Geogr. Schr. (in print).

HÖFNER, T. & K. GARLEFF (1992): Fluviale Dynamik in der zentralalpinen Periglazialstufe. – Verh. deutschen Geographentages (in print).

Kerschner, H. (1985): Quantitative palaeoclimatic inferences from lateglacial snowline, timberline and rock glacier data, Tyrolean Alps, Austria. – Z. Gletscherkd. Glazialgeo. **21**: 363–369; Innsbruck.
Kirkby, M. J. (1987): The Hurst effect and its implications for extrapolating process rates. – Earth Surface Processes and Landforms **12**: 57–67; Chichester.
Patzelt, G. (1977): Der zeitliche Ablauf und das Ausmaß postglazialer Klimaschwankungen in den Alpen. – In: Frenzel, B. (ed.): Dendrochronologie und postglaziale Klimaschwankungen in Europa: 248–259; Wiesbaden.
Patzelt, G. & S. Bortenschlager (1973): Die postglazialen Gletscher- und Klimaschwankungen in der Venedigergruppe (Hohe Tauern, Ostalpen). – Z. Geomorph. N. F., Suppl. **16**: 25–72; Berlin, Stuttgart.
Rennert, R. (1991): Geoökologische Untersuchungen zur Bodengefrornis an der Untergrenze des alpinen Permafrostes unter Einsatz von Hammerschlagseismik, Geoelektrik und Bodentemperaturmessungen. – unpubl. Dipl.-Arbeit, Lehrstuhl f. Geomorphologie, Universität Bayreuth, 153 p.
Steinmann, S. (1978): Postglaziale Reliefgeschichte und gegenwärtige Vegetationsdifferenzierung in der alpinen Stufe der Südtiroler Dolomiten (Puez- und Sellagruppe). – Landschaftsgenese u. Landschaftsökologie **2**: 91 p.
Stingl, H. (1969): Ein periglazialmorphologisches Nord-Süd-Profil durch die Ostalpen. – Gött. Geogr. Abh. **49**: 115 p.
– (1971): Zur Verteilung von Groß- und Miniaturformen von Strukturböden in den Ostalpen. – Nachr. Akad. Wiss. Göttingen, II. Math. – Physik. Kl. 1971(2): 25–40
Stingl, H. & H. Veit (1988): Fluviale und solifluidale Morphodynamik des Spät- und Postglazials in den südlichen Hohen Tauern im Raum um Kals/Osttirol. – In. Hüser, K. & H. Stingl (eds.): Exkursionsführer Osttirol-Dolomiten: 5–69, 15. Tagung des Deutschen Arbeitskreises für Geomorphologie, Bayreuth.
Veit, H. (1988a): Fluviale und solifluidale Morphodynamik des Spät- und Postglazials in einem zentralalpinen Flußeinzugsgebiet (südliche Hohe Tauern, Osttirol). – Bayreuther Geow. Arb. **13**: 167 p.; Bayreuth.
– (1988b): Postglaziale Schwankungen der periglazialen Morphodynamik in den südlichen Hohen Tauern. – 46. Deutscher Geographentag München, 12.–16. Okt. 1987, Tagungsber. u. wiss. Abh.: 408–413; Stuttgart.
– (1989): Geoökologische Veränderungen in der periglazialen Höhenstufe der südlichen Hohen Tauern und ihre Auswirkungen auf die postglaziale fluviale Talbodenentwicklung – Bayr. Geow. Arb. **14**: 59–66; Bayreuth.

Addresses of the authors: Dr. H. Veit, Lehrstuhl für Geomorphologie, Universität Bayreuth, Universitätsstraße 30, D-8580 Bayreuth. Th. Höfner, Lehrstuhl II für Geographie, Universität Bamberg, Am Kranen 1, D-8600 Bamberg.

Z. Geomorph. N.F.	Suppl.-Bd. 92	85–96	Berlin · Stuttgart	Mai 1993

Pleistocene Glaciations in Eastern and Central Tibet – Preliminary Results of Chinese-German Joint Expeditions

by

J. Hövermann, F. Lehmkuhl, and K.-H. Pörtge, Göttingen

with 7 figures

Summary. As a contribution to the controversial discussion regarding kind and extent of the Pleistocene glaciation on the Tibetan Plateau additional observations and results were obtained during the expedition in 1989. The hypothesis of an extensive plateau glaciation could not be verified. By the example of the eastern margin of the Tibetan Plateau it will be shown that the maximum extent of the last glaciation is limited to glaciation of isolated mountain groups ("Gebirgsgruppen-Vergletscherung") isolated or smaller plateau glaciations.

Zusammenfassung. Zur kontroversen Diskussion über Art und Umfang der pleistozänen Vergletscherungen im Bereich des tibetanischen Plateaus konnten 1989 zusätzliche Beobachtungen und Befunde erbracht werden. Dabei ließen sich für die Hypothese einer umfangreichen Plateauvergletscherung keine Belege finden. Am Beispiel des Ostrandes des tibetanischen Plateaus soll gezeigt werden, daß sich die Ausdehnung der letztglazialen maximalen Vergletscherung auf isolierte Gebirgsgruppen- bzw. kleinere Plateauvergletscherungen beschränkte.

1 *Introduction*

The glaciation of the Tibetan Plateau during the Pleistocene is discussed controversially at the moment. While Kuhle (1982, 1985, 1987, 1988, 1991) is postulating an inland ice sheet up to a thickness of 2.5 km and an extent of 2–2.4 mill. km² (this means nearly a glaciation of the whole Tibetan plateau, see Fig. 1) and derives from these far-reaching results for the global climate (Kuhle 1985, 1987 b), the Chinese investigations yield a more less glaciation (Shi Yafeng et al. 1992, Derbyshire et al. 1991; see Fig. 1): a maximum ice sheet cap of about 297,000 km² or 25% of the region (Derbyshire et al. 1991, from: Li et al., 1983, in Chinese). In Fig. 1 the different interpretations regarding the extent of the pleistocene glaciation were projected on the expedition area (Kuhle – Shi Yafeng et al.).

[1] Co-operation with the Institute of Mountain Disasters & Environment, CAS, Chengdu, the Geographical Institute of the Goettingen University and the Botanical Institute of the Hohenheim University.

0044-2798/93/0092-0085 $ 3.00

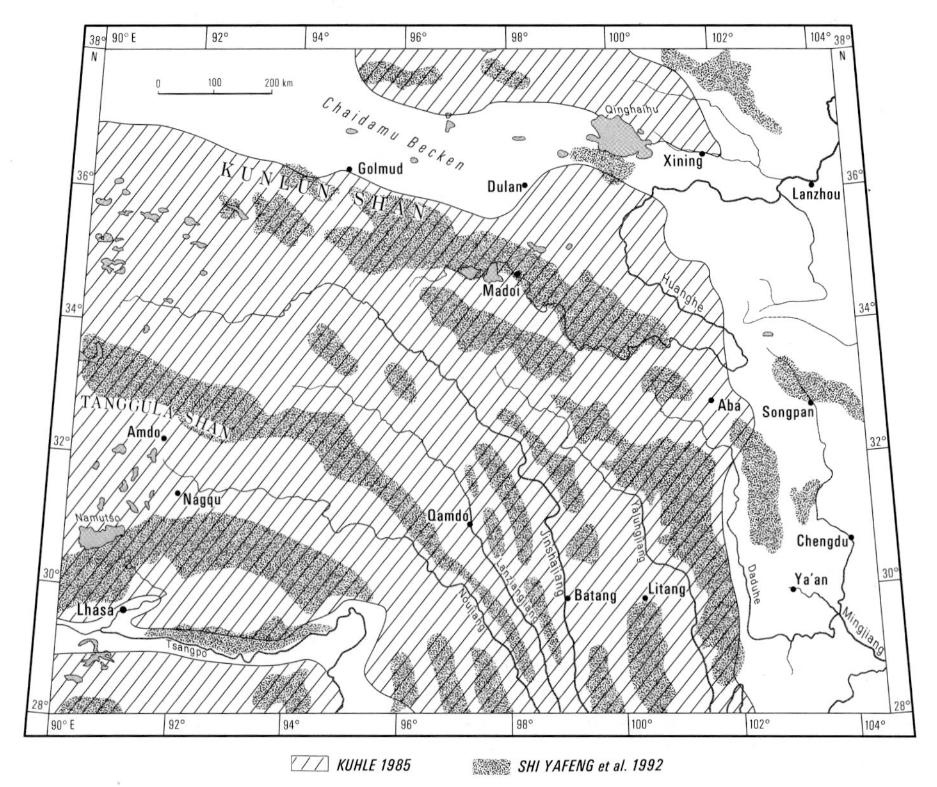

Fig. 1. The extent of the pleistocene glaciation(s) of the Tibetan Plateau (for the expedition area of 1989) by KUHLE (1985) and SHI YAFENG et al. (1992).

During the joint Chinese-German expedition in 1989[1] also investigations were made regarding the prehistorical glacial forms and their extent. Proceeding from this expedition, whose results could only give a survey because of the area's dimension, further expeditions were made to special areas of Eastern Tibet[2]. The most important results will be published in: Göttinger Geographische Abhandlungen, Vol. **95**. The following contribution will show the first detailed results of the Pleistocene glaciation by the example of the eastern margin of the Tibetan Plateau (between 99° and 104°E, between 30° and 34°N; see Fig. 3).

The presentations of the glaciation by different Chinese authors (ZHENG BENXING & JIAO KEQIN 1991, SHI YAFENG et al. 1992; Quaternary Glacier & Environment Center 1991) differ from each other, but all of them show in the area of the Tibetan Plateau only mountain and plateau glaciations of less extent. The map of the

2 In the year 1991 an expedition led to the area of the Bayan Har Shan, where a mountain complex was investigated in detail (Nianbaoyeze Shan: 101–101°20'E, 33–33°30'N). In 1992 a further expedition was made to Eastern Tibet (northeasterly plateau, Tanggula Shan, Chola Shan, about 32°N, 99°E).

Quaternary glacial distribution on the Qinghai-Xizang (Tibet) Plateau shows in addition a small ice sheet (morainic platform) of an older glaciation in northeast Tibet (source of the Huang He).

A uniform division of the ice-marginal grounds'/endmoraines' sequences in the various mountain areas, which are already investigated in detail, is still missing at the moment. The single ice-marginal grounds are named by local terms and often the classification seems not to be ensured. For example the Tanggula Shan (ZHENG BENXING & JIAO KEQIN 1991): Moraines which were directly in front of the recent glaciers were coordinated to the Wisconsin (Würmian) Glacial and therefore the partly clear dams being situated before them were coordinated to more older glaciations. We believe that these moraines are late glacial ones. The last ice age extended, because of the freshness of the forms, up to the margin of the Tanggula Shan and might correspond to the ice-marginal ground "Tanggula glaciation" named by ZHENG BENXING & JIAO KEQIN (1991).

2 *Detailed results from the eastern margin of the Tibetan plateau*

Between 31°N and 34°N and also between 100°E and 104°E cirques can be found in nearly every mountain range with an altitude of more than 4,000 m[3]. The cirque floors are distributed between 3,800 m and 4,500 m, but two clear levels can be found at an altitude of 4,000 m and at 4,300 m. There the northward and eastward exposed cirques are situated deeper than the southward and westward exposed ones. Cirques and cirque stairways are not only visible in the terrain itself, but also on the satellite images. Even under a closed snow cover the round blue cirque lakes can be recognized and on maps clear-cut defined, and you can map their their altitude. Most of the cirques are "armchair" cirques with a marked cirque threshold.

A regular changing of the cirque floors' altitudes cannot be recognized, despite of the difference of latitude of three degrees in the northsoutherly direction. In opposition to this the cirque floors' altitudes rise clearly from east to west by about 200 m inbetween these three longitude degree[4]. This increase corresponds with the fast decrease of the precipitation from east to west, which accordingly was more important during the ice age than now.

The marginal ranges in the upper catchment area of the Minjiang river (Fig. 2 [1]) are distinguished by clear former cirque levels at an altitude of 4,000 m–4,100 m and 4,200 m–4,300 m[5]. The last glacial (Würmian) snow line can be stated for this area at an altitude of 4,000 m (3,800 m ?). This corresponds with the already by v. WISSMANN (1959: Fig. 23) stated altitudes of the glacial isochiones as well as with new Chinese investigations in this region (see TANG BANGXING et al. 1992).

---

[3] The results from the Minjiang river's catchment area to the Huanghe catchment area (see Fig. 2), which are shown here, are completed by first results of a further Chinese-German joint expedition in 1991.

[4] Cirques can be proved here at altitudes between 3,800 m and 4,500 m, mostly at two clear altitudinal levels between 3,800 m and 4,000 m and between 4,200 m/4,300 m and 4,400 m.

[5] Here the results of the field work were completed by an interpretation of the atlas of the Sichuan Province (1981) as well as of some topographical maps on the scale 1:100,000.

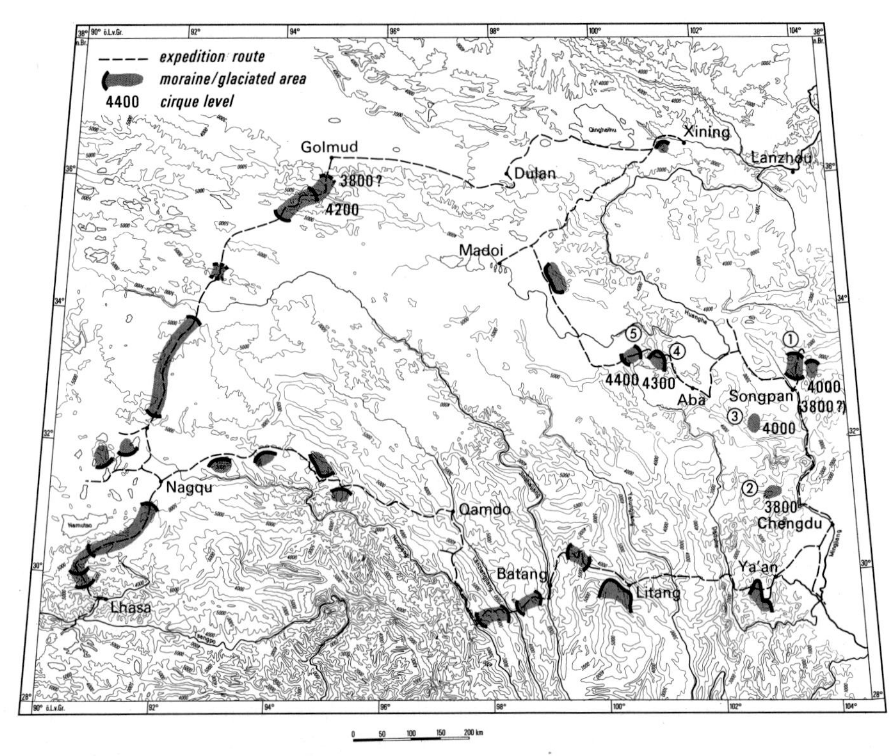

Fig. 2. Own investigations and definite indicators for Pleistocene glaciations (glacial erosion and accumulation features such as englacial till, end moraines – expeditions of 1989 and 1991). The numbers show the regions described in the text. – Draft: F. Lehmkuhl.

Marked ice-marginal grounds, which are signed by glacial sequences, can be found in the longitudinal valley (103°40′E, 33°N), whose southern part is flown through by the upper Minjiang, south of the watershed at 33°N. In this area three ice-marginal grounds must be distinguished. The lowest one is characterized by a large push moraine at an altitude of about 3,200 m (Gami temple: 103°41′E, 32°55′N, see Fig. 3). The basal part of this moraine includes a rose-coloured and red interglacial residual soil as well as red clay. The middle one is a terminal moraine with a following glaciofluviatile gravel field, in which an additional erosional terrace is cut in above the present-day river bed. Here the terraces and the river bed are converging down into the valley. The highest ice-marginal ground is characterized by a moraine, which contains both angular boulders and rounded gravels. This ice-marginal ground developed as a lateral moraine is situated at the valley exit of two valleys, which run into the latitudinal valley from the east. Immediate near by, but not meeting the above-mentioned moraine a glaciofluviatile gravel field with rounded and classified gravels is beginning. Here an erosional terrace is also cut in. The terraces and the river bed are also converging down into the valley. The height

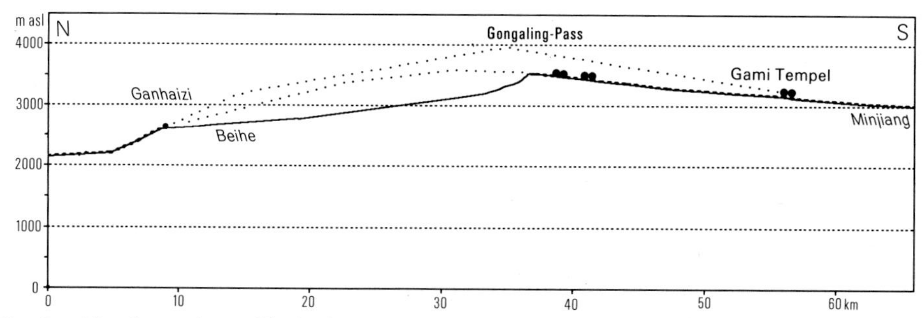

Fig. 3. North-south profile Beihe – Minjiang river over the Gongaling pass with different ice-marginal grounds and following glacial-fluvial terraces. – Draft: F. LEHMKUHL.

of the moraines is up to 40 m. The fault throw of the respectively uppermost terrace is at about 20 m.

Between the lateral moraines of the uppermost ice-marginal ground (at about 3,500 m) a two-termed terrace is continuing in the direction of the mountain area. In the main valley of the Minjiang itself these terrace gravels, in the uppermost part uncut, reach the pass altitude of 3,500 m just north of 33°N. They end at the northerly following moraine covered basin, which is located by about 100 m and where the northern drainage starts. In the small valleys, which run into the latitudinal valley from the west, there was sometimes a glacier advance in the upstream direction. This is especially shown by a 50 m high whaleback worked out from layered glaciofluviatile gravels, whose steep and with moraines covered side is oriented eastward, while its western side slopes away streamlined.

North of the pass (catchment area of the Beihe/Jialingjiang) basal moraine material can be found here and there and at an altitude of 2,600 m–2,650 m a moraine dam is surrounding a nearly complete silt up glacial piedmont lake (see Fig. 3: Ganhaizi, 103°44′E, 33°14′N). Down into the valley a 3-km-long transitional cone is following the final moraine dam with an incline of about 10%. This cone runs out into a glaciofluviatile gravel plain, which characterizes the further valley area.

It seems to be astonishing that south of the pass the glaciers already retreated to the mountain areas at a time when north of the pass, and furthermore at an altitude of about 100 m deeper than on the southern side, the ice still reached the pass. This fact is proved by the gravel accretion which is exposed to the south. An explanation can be found, when one takes into consideration that the altitude of the cirque floors as well as the altitude of the ice-marginal grounds is differing in the north and south exposition. The coherent ice field in the longitudinal valley i.e. ended in the area of the southern slope at an altitude of 3,200 m and in the area of the northern slope at an altitude of 2,650 m. This can be only explained by a stronger ablation of the ice stream into the southern direction than those into the northern direction, because in contrary to that the catchment areas of the valley glaciers south of the pass are higher and more extensive than those north of the pass. But at the same time the ice stream into the southern direction was not as huge as the one into the northern direction. This means that even with the same ablation result the southern glacier would have to disappear earlier than the northern glacier. The remaining ice mass would regenerate during a following small glacial advance, even if it already had become dead ice. These reflections also explain the clearly differing three ice-marginal grounds in the southern part of the longitudinal valley. In the northern part only one ice-marginal ground can be found.

During a second joint expedition in 1991 further ice-marginal grounds were found and mapped in the catchment area of the River Zugunao and the River Heishui, both subsidiaries of the Minjiang. In the catchment area of the River Zagunao (Fig. 2 [2]) the moraine terraces near Lixian (103°11'E, 31°26'N) were mapped. Chinese scientists distinguish here 3–5 stages, which are coordinated with different ice ages. The last stage (deepest moraine 1,950 m–2,060 m = "Zugunao Ice Age") is equilized to the Würm (English summary see TANG BANGXING & SHANG XIANGCHAO 1991). At an altitude of 2,100 m there are dead water depositions of a glacial piedmont lake and moraines, which are coming down to 1,950 m. A U-shaped valley is following above these depositions. When the catchment area is higher than 5,900 m a glacial snow limit of 3,900 m can be supposed.

In 1991 the deepest ice-marginal grounds in the catchment area of the River Heishu were found out to reach an altitude of about 2,840 m (eastern exposition), proceeding from a recent still glacierized granite dome with a maximum altitude of more than 5,200 m (at 101°06'E, 33°17'N; Fig. 2 [3]). Troughs verify an ice stream net for which a snow limit of about 4,050 m can be taken into consideration. By reason of the strong relief intensity the recent morphological processes are very strong, too (shown by many landslides, debris-flows, rockfall and slides; LEHMKUHL & PÖRTGE 1991). Therefore it is possible that deeper moraine depositions were removed or covered. A deeper cirque level at an altitude of 3,900 m–4,000 m points at this possibility.

The glaciation found here is similar to the type of glaciation of isolated mountain groups ("Gebirgsgruppen-Vergletscherung") type. In 1989 this type was found at first in Eastern Tibet at the margin of the Nianbaoyeze (HÖVERMANN & LEHMKUHL 1992). This mountain massif in the Bajan-Har-Shan is situated at the transition to the actual northeastern plateau west of the big Huanghe loop and raises over two peneplains with levels at about 4,200 m and 4,400 m. The centre with a granite core (at 102°45'E, 32°14'N; Fig. 2 [4] reaches a maximum altitude of 5,369 m. Here the altitude of the last glacial snow limit is fallen below by some 100 m and one can find marked forms there.

During the expeditions such glaciations of mountain groups proceeding from higher mountain massifs were also observed in a smaller mountain massif further to the west (Fig. 2 [5]), in the area of the southern Tibetan Plateau southwesterly of Amdo as well as in the mountain areas between Nagqu and Qamdo. They are always proceeding from higher mountain massifs, which rise above the plateau and distinguish themselves by a radial water net. This water net is limited at the margin by annular marginal channels, in which mostly a main ice margin is located (see Fig. 4). The extent of these former glaciations can clearly be seen on satellite images (Atlas . . . 1983, as an example see Fig. 5 mapped in Fig. 6). These glaciations are characterized by glacial erosional types such as U-shaped valleys in the inner mountain massifs, cirques and Karoide in the higher summit areas as well as glacial accumulation forms, such as end moraines and lateral moraines (more seldom than basal

Fig. 4. An ice-marginal ground formed as a "marginal channel". An example from the northern Nianbaoyeze Shan (101°15'E, 33°25'N; at an altitude of 3,800 m). East-west going valley of the Duotzuojian with a clear lateral and basal moraine consisting of coarse granite boulders over cristalline schistes. Photograph by F. LEHMKUHL, Sept.-01-1991.

moraines), mostly of uniform (erratic) rocks, such as granites and basalts. For example in the mountains of the eastern margin (Fig. 2 [3, 4, 5]) one can find terminal and lateral moraines containing of big boulders layered above cristalline schistes. These moraines characterized by large erratica were named "Big-Boulder-Moraine" by us. Its classification into the Late Glacial or into the High Glacial is contested.

In 1989 the ice-marginal ground in the north of the already above-mentioned Nianbaoyeze Shan, in the transition area to the northeastern Tibetan Plateau (33°22'N, 101°01'E-101°16'E) could be investigated at a length of 20 km[6]. The former ice margin is situated at altitudes between 3,920 m and 4,140 m and is traced by the water net, which at first is starting radially at the massif and then takes an annular course as a marginal channel (Fig. 4). This ice-marginal ground is partly reaching the pass (4,140 m). The further drainage of the melting water ensued through some main valleys which flow in northern direction to the Huanghe. A change in the character of the valleys can be observed: On the outside of this marked ice-marginal ground the U-shaped valleys are detached by V-shaped valleys.

[6] This mountain massif was investigated in detail in the framework of a further expedition in 1991, also by reason of its geographical position at the transition of the marginal ranges to the northeasterly plateau, its thick humus and peat layers etc.

Fig. 5. Section of a satellite image (section of Atlas... 1983, Vol. **2**, No. 142-37) with two clearly recognizable massifs (arrows), which were formed by a glaciation of isolated mountain group ("Gebirgsgruppen-Vergletscherung"). – Draft: F. LEHMKUHL.

Inside this ice-marginal ground glacial piedmont lakes can be seen on Chinese maps (1:100,000) as well as on satellite imagery. Fig. 6 was drawn on the basis of the results of the 1989 expedition and the satellite imagery (1:500,000). This mountain massif was investigated in detail during a further expedition in 1991 by reason of its special position at the eastern margin of Tibet with a therefore sufficient precipitation for a supposed continuing peat development at least since the Holocene (shown by first ^{14}C analysis; personal communication with Prof. FRENZEL)

The maximum ice expansion (during the younger Pleistocene), mapped in Fig. 6 from the satellite image, could be confirmed essentially. Deviations are mainly found in the southeast, where the ice expansion was smaller.

This main ice-marginal ground recognized already in 1989 has a supraregional significance as it could be found as a "Big-Boulder-Moraine" in other mountain massifs, too.

As an anticipation to the results of the expedition in 1991 the morainic sequence, until the recent glacier, of a typical valley is introduced (Fig. 7). It concerns the valleys of the Ximen(tzuo) and the Jiukuhe at the northern slope following up the highest culmination (5,369 m) in the area of the Nianbaoyeze. Before the recent

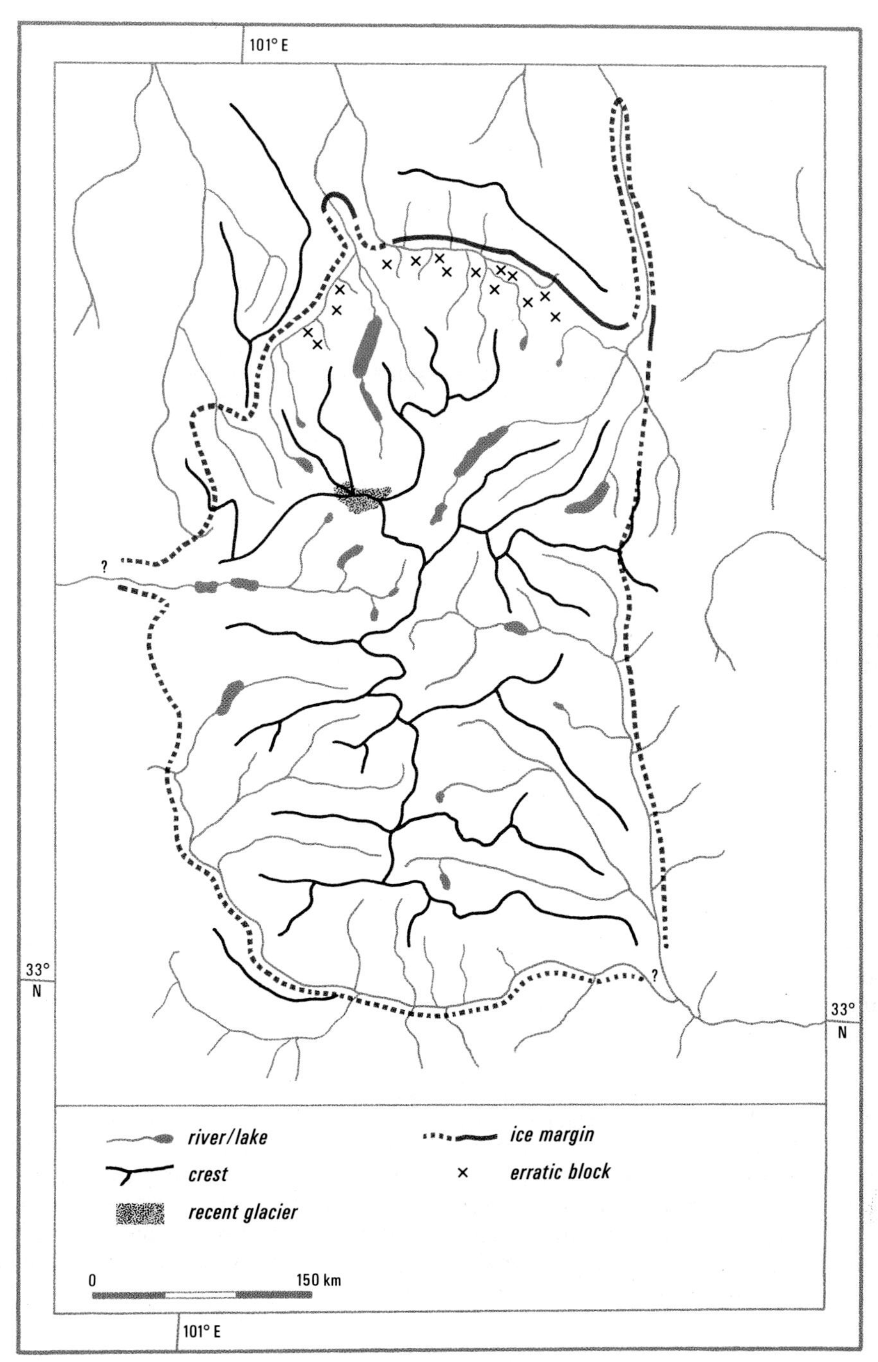

Fig. 6. Glaciation of isolated mountain group of Nianbaoyeze Shan according to the analysis of the satellite image 1:500,000. – Draft: F. LEHMKUHL.

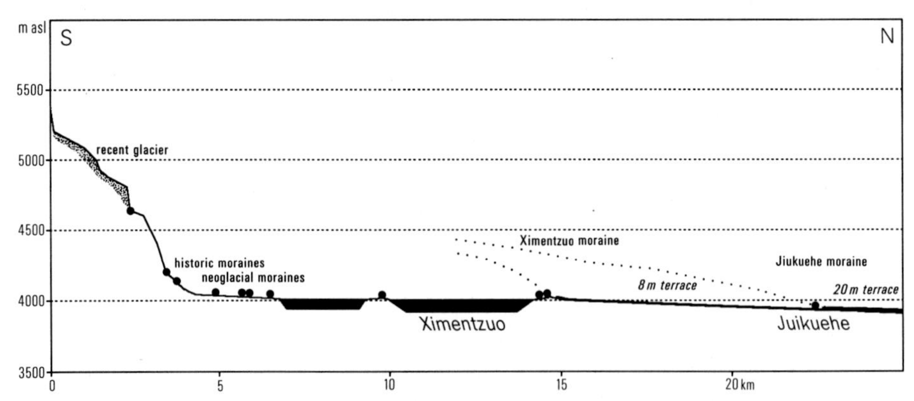

Fig. 7. Valley longitudinal profile (northern slope Nianbaoyeze Shan) with the different ice-marginal grounds. – Draft: F. LEHMKUHL.

glacier end, at an altitude of 4,600 m, is a probably historical complex with nearly uncovered final and lateral moraines between 4,210 m and 4,220 m. Before this complex there is a presumable neoglacial or younger late glacial complex at an altitude between 4,060 m and 4,130 m as well as a clear marginal ground at the northern margin of the largest glacial piedmont lake (Ximentzou) at an altitude of 4,040 m. This marginal ground, one of the both main ice-marginal grounds, surrounds nearly all of the outer big lakes in the central mountain areas and is mostly consisting of two characteristic moraine dams.

This ice-marginal ground is followed up by a small terrace of a fault throw of 6–8 m. It has a thin layer (20–50 cm) of sandy loess, which contains in the upper 15–20 cm a humus layer. Outside of this marginal ground erratic granite boulders (also above cristalline schistes) could be mapped up to an altitude of about 3,920 m (Jiukuehe moraine in Fig. 7). Subsequently the valley character is changing: the plain and wide trough (included into the peneplain with altitudes of about 4,000 m–4,100 m) is detached by a V-shaped valley. Here and there a higher terrace can be observed. A temporal classification of this erratica and gravel area into the last or penultimate glaciation would be possible.

Summarized an increase of the prehistorical snow limit from east to west (everywhere supposed as the last glacial maximum) could be recognized in the east of the expedition area: From 3,800 m–4,000 m in the marginal ranges of the Red Basin (Mingshan) [1] and [2] over 4,200 m north of Aba (proved by the cirque levels) to 4,300 m [4] and 4,400 m [5] further to the west (see Fig. 2). This increase, which can also be observed regarding the recent altitudinal limits (timber line, snow limit, periglacial low limit), is in principle already mapped by v. WISSMANN 1959 (Fig. 23, p. 223); but with higher values of 200 m–400 m.

3 *Prospect*

Regarding the controversial discussion on kind and extent of the pleistocene glaciation in the area of the Tibetan Plateau additional observations and results were found in 1989.

There it was shown that proofs for a total glaciation in the expedition area are not existent. Further detailed investigations on kind, extent as well as on the temporal classification of the pleistocene glaciations are still necessary.

In the eastern part of the Tibetan Plateau single complexes of glaciations could be investigated in detail. These investigations were continued during two further Chinese-German expeditions. They show glaciations of isolated mountain groups ("Gebirgsgruppen-Vergletscherungen") or also smaller glaciations of the plateau, which are supposed to be the maxima of the last glaciation and with this fact most of the results of our Chinese colleagues are confirmed.

The existence of larger plateau glaciations, also supposed by SHI YAFENG et al. (1992) for the area of the source of the Huang He west of the Anyêmaqên massif for an older glaciation, can be proved by means of further detailed studies only. These investigations should exhaust all possibilities of the dating.

4 *Acknowledgements*

We thank the DFG (Deutsche Forschungsgemeinschaft, Bonn), the GTZ (Gesellschaft für Technische Zusammenarbeit, Eschborn) and the MPG (Max-Planck-Gesellschaft) for their financial support of the expeditions.

References

Atlas of False colour Landsat images of China (1983): Compiled by: Institute of Geography, Academia Sinica, Beijing. (Scale: 1:500,000).

DERBYSHIRE, E., SHI YAFENG, LI JIJUN, ZHENG BENXING, LI SHIJIE & WANG JINGTAI (1991): Quaternary Glaciation of Tibet. The geological evidence. – Quat. Sci. Rev. **10**: 485–510.

HÖVERMANN, J. & F. LEHMKUHL (1993): Die vorzeitlichen Vergletscherungen in Ost- und Zentraltibet – Ergebnisse einer chinesisch-deutschen Expedition 1989. – Göttinger Geogr. Abh. **95** (im Druck).

HÖVERMANN, J. & WANG WENYING (Hrsg., 1986): Reports of the Qinghai-Xizang (Tibet) Plateau. – 510 p.; Science Press, Beijing.

KUHLE, M. (1982): Was spricht für eine pleistozäne Inlandvereisung Hochtibets. – Sitzungsber. u. Mitt. Braunschw. Wiss. Ges., Sonderh **6**: 68–77.

– (1985): Ein subtropisches Inlandeis als Eiszeitauslöser. Südtibet und Mt. Everestexpedition 1984. – Georgia Augusta, Nachrichten aus der Universität Göttingen, Mai 1985: 1–17.

– (1987 a): The Problem of a Pleistocene Inland Glaciation of the Northeastern Qinghai-Xizang Plateau. – In: J. HÖVERMANN & WANG WENYING (Eds.): Reports of the Qinghai-Xizang (Tibet) Plateau: 250–315; Peking.

– (1987 b): Subtropical mountain and highland glaciation as Ice Age triggers and the waning of the glacial periods in the Pleistocene. – Geo Journal **13**: 1–29.

– (1988a): Geomorphological Findings on the Built-up of Pleistocene Glaciation in Southern Tibet and on the Problem of Inland Ice. – Geo Journal **17**(4): 457–512.
– (1988b): The pleistocene glaciation of Tibet and the onset of ice ages – an autocycle hypothesis. – Geo Journal **17**(4): 581–511.
– (1991): Observations Supporting the Pleistocene Inland Glaciation of High Asia. – Geo Journal **25**(2/3): 133–231.
Lehmkuhl, F. & K.-H. Pörtge (1991): Hochwasser, Muren und Rutschungen in den Randbereichen des tibetanischen Plateaus. – Z. Geomorph. N.F., Suppl.-Bd. **89**: 143–155.
Li, B., P. Wang, Q. Zhang, Y. Yang, Z. Yin & K. Jil (1983): Quaternary Geology of Xizang. – Science Press, Beijing. [chin]
Mahaney, W. C. & N. W. Rutter (1992): Relative Ages of Moraines of the Dalijia Shan, Northwestern China. – Catena **19**, 179–191.
Quaternary Glacier & Environment Research Center, Lanzhou University (1991): Quaternary glacial distributation map of Qinghai-Xizang (Tibet) Plateau 1:3,000,000. Scientific advisor: Shi Yafeng; Chief editors: Li Binyuan, Li Jijun.
Shi Yafeng (1992): Glaciers and glacial geomorphologie in China. – Z. Geomorph. N.F., Suppl.-Bd. **86**: 51–63.
Shi Yafeng, Zheng Benxing & Li Shijie (1992): Last glaciation and maximum glaciation in the Qinghai-Xizang (Tibet) Plateau: A controversy to M. Kuhle's ice sheet hypothesis. – Z. Geomorph. N.F., Suppl.-Bd. **84**: 19–35.
Tang Bangxing, Liu Shijian & Liu Suqing (1990): Recent Disaster and Prevention of Debris Flow in the east Area of Qingzang Plateau. – Vortrag auf einem Tibet-Symposium am 19.7.1990 in Göttingen, Manuskript 9 p (Gött. Geogr. Abh. **95**; im Druck).
Tang Bangxing & Shang Xiangchao (1991): Geological Hazards on the Eastern Border of the Qinghai-Xizhang (Tibetan) Plateau. – Excursion Guidebook XIII. INQUA 1991, XIII International Congress, 28 p.; Beijing.
Wissmann, H. v. (1959): Die heutige Vergletscherung und Schneegrenze in Hochasien mit Hinweisen auf die Vergletscherung der letzten Eiszeit. – Akad. Wiss. Lit., Abh. Math.-Nat. wiss. Kl.: **14**: 1103–1407; Mainz
Zheng Benxing & Jiao Keqin (1991): Quaternary Glaciations and Periglaciations in the Qinghai-Xizhang (Tibetan) Plateau. – Excursion Guidebook XI. INQUA 1991, XIII International Congress, 54 p.; Beijing.

Address of the authors: Geographisches Institut der Universität Göttingen, Goldschmidtstraße 5, D-3400 Göttingen.

Z. Geomorph. N.F.	Suppl.-Bd. 92	97–111	Berlin · Stuttgart	Mai 1993

Permafrost in Glaciofluvial Sediments of the Late Pleniglacial of the Last Glaciation – and Some Conclusions to Draw *

by

Karl Albert Habbe, Erlangen

with 8 figures

Summary. The paper demonstrates, that – and how – permafrost of the underground penetrated the glaciofluvial sediments of the late pleniglacial of the last glaciation, and that it affected the fluvial processes in an erosion-hampering and accumulation-promoting manner. This explains the high longitudinal gradients of the transition cones in front of the terminal moraines of the alpine foreland glaciers and the accumulation bodies of the late pleniglacial at all. On the other hand ceasing permafrost influence at the end of the late pleniglacial explains the downward erosion starting at the same time. Thus the alternation of accumulation and downward erosion in spite of unchanging longitudinal gradients of the perialpine valleys, that formed the terrace flights, on which the subdivision of the Pleistocene of the alpine foreland is based since more than 100 years, finds a satisfying explanation as well as the fluvial redeposition processes during late glacial and early postglacial times.

Résumé. L'article démontre que – et comment – le permafrost du sous-sol pénétrait les sediments glaciofluviaires du pléniglacial tardif de l'ultime période glaciaire, et qu'il affectait les processus fluviaux en entravant l'érosion et en provoquant l'accumulation. Ceci explique somme toute les hauts gradients longitudinaux des cônes de transition situés devant les moraines terminales des glaciers des terrains pré-alpins et les corps d'accumulation du pléniglacial tardif en général. À l'opposé l'arrêt de l'influence du permafrost à la fin du pléniglacial tardif explique le début simultané de l'érosion en profondeur. De cette façon l'alternance de l'accumulation et de l'érosion en profondeur en dépit des gradients longitudinaux invariables dans les vallées péri-alpines qui a créé les gradins en terrasse, sur lesques la subdivision du pléistocène des Alpes et des terrains pré-alpins est basée depuis plus de cent ans, trouve une explication satisfaisante tout comme les processus de redéposition fluviaux pendant les périodes du glacial tardif et du début de l'holocène.

Zusammenfassung. Es wird gezeigt, daß – und wie – der Permafrost des Untergrunds die glazifluvialen Sediments des Hochglazials der letzten Eiszeit erfaßte und daß er auf die fluvialen Prozesse erosionshemmend und akkumulationsfördernd wirkte. Das erklärt die hohen Längsgefälle der Übergangskegel vor den hochglazialen Endmoränen der alpinen Vorlandgletscher und die hochglazialen Akkumulationskörper überhaupt, der Wegfall des

---

* English version of a paper read at the annual meeting of the Hugo-Obermaier-Gesellschaft at Trier on 18th April, 1990, and first published in German in QUARTÄR **41/42**, 1991: 7–18.

0044-2798/93/0092-0097 $ 3.75

Permafrosteinflusses am Ende des Hochglazials umgekehrt die gleichzeitig einsetzende Tiefenerosion. Damit findet auch der pleistozäntypische Wechsel von Akkumulation und Erosion bei gleichbleibenden Längsgefällen, der die Terrassentreppen schuf, auf die sich seit über 100 Jahren die Grundgliederung des Pleistozäns im Alpenvorland gründet, ebenso eine befriedigende Erklärung wie die fluvialen Umlagerungsvorgänge des Spät- und frühen Postglazials.

The photograph Fig. 1 was taken in a sand pit west of Vojens (Sönder-Jylland). The pit is situated near the proximal end of a sandur, which had been aggradated by meltwaters of the maximum stage of the Weichsel glaciation, that came up the Haderslev tunnel valley (the situation of the site within the North German sandur belt is shown by Woldstedt (1961): 138). The face of the exposure runs approximately west-east, the aggradation aimed to the southwest, i.e. obliquely leftward from the face of the exposure. The sediment series is penetrated by a frost-crack pseudomorph, more precisely: the pseudomorph of a syngenetic frost-crack, which grew up with the sediment series, thus being permanently rejuvenated. The frost-crack proves not only, that the aggradation took place under permafrost conditions, but also, that during the pleniglacial of the last glaciation the glaciofluvial sediments were incorporated into the permafrost body immediately after deposition, the permafrost table thus growing up simultaneously with the aggradation and the whole process lasting until the end of the deposition. This means, that during the late pleniglacial of the last glaciation permafrost penetrated not only – something which has been known for a long time (Poser 1947, Kaiser 1960) – the older sediments outside the glaciated areas and their runoffs, but also the sediments aggradated by the alpine glaciers and the North-European inland-ice themselves.

In principle this is not a new piece of knowledge. Gallwitz (1949) already presented corresponding observations and conclusions and Ehlers & Grube (1983: 252) concerning the meltwater deposits of Northern Germany tersely stated: "The sands were deposited under permafrost conditions". But the implications and consequences of this statement are not discussed in the pertinent literature (Jopling & McDonald 1975, Reineck & Singh 1980, Starkel & Thornes 1981, Hinze et al. 1989). I myself have read a paper on the chronological classification of the phenomenon at the meeting of the Hugo-Obermaier-Gesellschaft six years ago on the basis of observations in the area of the Iller glacier of the Würm glaciation (Habbe 1985a). It could be shown then, that permafrost conditions were a characteristic feature of the late pleniglacial of the last glaciation and that permafrost influence rapidly dwindled away in the late glacial – in terms of the terminology used in the Rhine glacier area: the period after the second advance to the Inner Young Terminal Moraines – (and this criterion can be used to distinguish pleniglacial from late glacial landforms and sediments).

Thus the phenomenon is in itself indubitable. But one may ask: is it traceable – and if so: how is it traceable – in exposures in which – and that is the rule – the former existence of permafrost is not proven by frost-cracks? And one may further ask: Do the exposures allow further statements on the effect of permafrost on the sedimentation process?

To find an answer one might enter into the particulars of Fig. 1. The exposure confronts with at least two problems, viz.: 1. How can be explained the occurrence

Fig. 1. Eastern section of the north face of the Vojens sand pit with intersecting frost-crack pseudomorph (K.A.H. 21.8.1980).

of frost-cracks – which require temperatures far below 0° – and meltwater sediments – which indicate temperatures above 0° – in the same position? and 2. How can the sedimentation pattern be explained? The photograph does obviously not show a concordant sediment series. Rather the different strata show

a) highly differing grain sizes – which indicates differing carrying power of the meltwaters in spite of equal stream gradients, i.e. heavily changing meltwater discharge –, and
b) different bedding angles – partly surface parallel, partly oblique –, which signalizes differing flow conditions, – and they have been deposited partly in channels, partly on a plain surface, but the latter case prevails (Fig. 2).

The bed boundaries clearly cut the older sediment structures, i.e. they indicate erosional disconformities. The sediment sequence therefore obviously represents an alternation of accumulation and removal repeated many times in spite of an unchanged stream gradient (in the present case approximately 3‰).

A third problem is, that sand is the easiest removable sediment of all, and there is no simple explanation, why – if removal was actually possible – there was not much more removal of the sands.

To find an explanation, the conditions under which the sediment series came into existence have to be reconstructed. This cannot be done without hesitation by falling back on observations from recent permafrost areas in European or North American arctic regions (Büdel 1969, Church 1972, McCann et al. 1972, Bibus et al. 1976,

Fig. 2. Central section of the north face of the Vojens sand pit. Alternation of oblique and cross-bedding with surface parallel bedding (K.A.H. 21.8.1980).

SCOTT 1978, BARSCH & KING 1981, SCHUNKE 1981, 1989), even if they had been made on glaciofluvial (and not merely periglacial-fluvial) systems. The counterarguments are: 1. the necessary long-term observations are lacking: 2. the arctic glaciers of today and their meltwater runoff are of other (and much lesser) dimensions than the nordic inland-ice and the alpine foreland glaciers in Pleistocene Middle Europe; 3. the arctic regions were subject to isostatic earth-crust movements (with serious influence on the stream gradients), and 4. the high latitudes are subject to climatic (especially radiation) conditions other than Middle Europe during the Pleistocene cold periods. Thus one has to try to get the necessary informations from the sediment series itself.

One fact is certain, that the sediments have been deposited under a glacial discharge régime (PARDÉ 1933, 1947). Every glacier runoff shows a characteristic hydrograph (Fig. 3): three quarters to nine tenths of the year's discharge come down during the summer months June to September, correspondingly only one quarter to one tenth of the discharge is registered in the other eight months of the year. The main reason for the concentration of the discharge on merely a few months is snow melting during the warm season. Differing from strictly periglacial areas, where snow meltwater often comes down during a mere two or three weeks (MCCANN et al. 1972, SEMMEL 1985) snow melting lasts – because of the upward motion of the snow-line (GROSS et al. 1976) on the glacier surface – for the whole warm season[1]. Much additional meltwater comes from the thawing ice masses in mid-summer. It

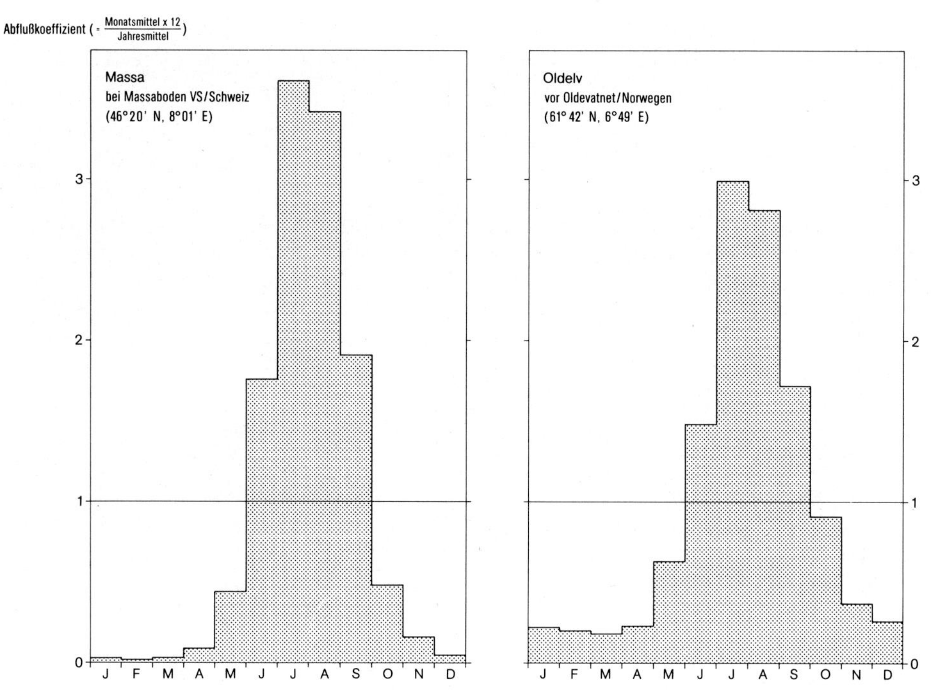

Fig. 3. Hydrographs of two glacier runoffs after PARDÉ (1933, 1947). The river Massa is the runoff of the Großer Aletsch-Gletscher/Switzerland, the river Oldelv one of the runoffs of the Jostedalsbre/Norway. For better comparison the discharge is represented by the discharge coefficients. Coefficient 1 corresponds with the arithmetic mean of the monthly discharges.

has surely played an even more important role under pleniglacial conditions in Middle Europe than under present-day conditions in arctic regions (FLÜGEL 1981). In any case a glacial discharge régime means, that the flood plain before the ice-front will be flooded by meltwaters from June to September – especially during the peak discharge in early summer –, whereas it lies dry in large parts during the other months. Similar conditions must be assumed for the runoff of the alpine foreland glaciers and the inland-ice of the late pleniglacial, perhaps – because of the generally lower temperatures – with a still heavier concentration on the warmest months.

This means for the explanation of a pleniglacial glaciofluvial sediment series: it is the result of extremely intensive alternations in discharge, which led to two countermoving tendencies of landforming: viz. a growing erosion tendency during

---

[1] The process is – depending on wheater conditions – subject to intensity fluctuations: it may be concentrated or extended over a longer period, in some years earlier, in other years later. Single precipitation events may heavily influence the discharge (CHURCH 1972, FLÜGEL 1981, SCHUNKE 1989).

Fig. 4. Mid-summer aspect of two glaciofluvially controlled braided river systems on Plaun Segnas near Flims GR/Switzerland in approximately 2100 m a.s.l. The runoff goes to the right. The system in the middle shows the typical ground plan with elongated gravel bars and interjacent channels, the system coming from the opposite valley flank in the back-ground contours the typical cross section with the small-spaced alternation of gravel bars and channel deposits (K.A.H. 6.8.1988).

increasing discharge in early summer and – in contrast to this – a growing accumulation tendency during decreasing discharge in mid- and late summer.

Additionally this process was influenced by permafrost effects. On the one hand permafrost prevented the meltwaters from seeping into the underground: in permafrost areas there is no ground-water. The total discharge therefore had to run off on the surface. This is why on sandur and foreland gravel surfaces, which are well drained at present and therefore lie dry all the year round, fluvial erosion and accumulation process could take place at all.

On the other hand permafrost also influenced the periodic alternation of erosion and accumulation. In detail the processes throughout the year can be subdivided into four phases:

1. After the short but intensive accumulation period in late midsummer, which resulted – as in recent flood plains, – in a broadly extended system of elongated gravel and sand bars and interjacent slightly deepened channels gradually also falling dry, i.e. a braided-river system (LEOPOLD et al. 1964, MIALL 1977, RUST 1978; Fig. 4), the dry-lying areas were completely penetrated by frost during the long period of low discharge between autumn and late spring, i.e. they became part of the per-

mafrost body, in other words: the seasonal permafrost-table grew up to the relief surface. In the late phase of frost-penetration frost-cracks could develop.

2. Snow melting in early summer resulted in a rapidly growing waterlevel in the glacier forefield, but – because of the still high-lying permafrost-table – there was relatively little transport of material.

3. While the water-level increased during the warm season, the meltwater masses began to melt down the permafrost-table. Thus the erosion preventing effect of the ground-ice was restricted to deeper parts of the underground, the now boundless sands and gravels could be taken away by the running water, material transport began on the whole flooded plain. The process was not – as in subpolar and polar regions of today (KOUTANIEMI 1984, BARSCH 1981) – restricted to mere transport of material, but resulted – apparently regularly every year – in downward erosion. This erosion obviously (see Fig. 1 and 2) was capable of removing the sand and gravel bars that had been left by the accumulation period of the previous year almost completely, thus extinguishing the typical sedimentation pattern of a braided river – the alternation of bars and channels on short distances with its characteristic cross-bedding (MIALL 1977) – and to establish the surface parallel structures that characterize the exposures of pleniglacial fluvial sediments. On the other hand the permafrost-table has also hampered downward erosion in the warm season; erosion only worked easily in thoroughly thawed and therefore quickly removable material, whereas it was obviously confined to narrow limits in the still frozen underground. Otherwise it would be impossible to explain why, as a balance of the periodic alternation of downward erosion and accumulation an accumulation body came into existence. Besides that only the erosion hampering effect of the permafrost makes understandable, why quickly running meltwaters transporting coarse grains – and therefore in principle of high erosion capacity – did not remove the underlying sandy strata.

4. Once the peak of summer flooding had passed the accumulation of – according to the position towards the streamline – deposits of very different grain size and also different thickness began to take place anew. Again the sand and gravel bars typical of braided rivers came into existence with interjacent, gradually dry-falling channels, thus forming an aggradation body of non-frozen material above the seasonal permafrost-table, which became a member of the permafrost body not before the following cold season.

This periodic alternation of accumulation and downward erosion, frost penetration and thawing took place every year. The position of the thawing boundary (and therewith the erosion boundary) differed – according to weather conditions – from year to year. As a result of this frequently repeated process those mid-summer permafrost-tables appear in the exposures as bedding boundaries or rather disconformities, that were reached no more in later times, i.e. primarily – as is usual in Pleistocene deposits – the youngest members of the deposition series. They can be interpreted as a sequence of years with particularly limited thawing depth, in other words: most unfavourable climatic conditions.

In summing up the following can be said: 1. The – as a rule remarkably surface-parallel – bed boundaries of pleniglacial glaciofluvial deposits represent former mid-summer permafrost-tables, or – conversely and as an answer to the first of the questions put at the beginning – permafrost is traceable in pleniglacial glaciofluvial deposits by the sedimentation pattern: the surface parallel bedding characteristic of these deposits – and contrasting in a remarkable manner with the surface aspect of the flood plains before recent glaciers – is a direct consequence of the permafrost in the underground during the time of deposition[2], and 2. – in answer to the second question – the permafrost of the Pleniglacial hampered downward erosion and therefore promoted accumalation, and this under relatively high stream gradients, too[3].

The erosion-hampering and accumulation-promoting effect of the pleniglacial permafrost caused a whole series of consequences. I want to discuss three of them briefly:

1. The high longitudinal gradient of the transition cones before the terminal moraines of the ice-age glaciers of the alpine foreland (Fig. 5) has usually been explained as a consequence of the coarser grain sizes of near-glacier deposits (DU PASQUIER 1891, PENCK & BRÜCKNER 1901–09, SCHAEFER 1950, TROLL 1977). Actually these deposits as a whole are not particularly coarse-grained (Fig. 6). On the other hand the river Iller as a subsequent river of a former meltwater course with a stream gradient of 1.5‰ in the terminal moraine zone is capable of remarkable downward erosion in flood times (HABBE 1985 b). Under these premises the longitudinal gradients of the transition cones of the last glaciation – they generally exceed 10‰ and may reach 17‰ (HABBE & RÖGNER 1989), thus being ten times steeper than the gradient of a river that erodes under present-day conditions – would be inexplicable, if the erosion-hampering and accumulation-promoting effect of pleniglacial permafrost had not been given. It not only explains the particularly conspicuous phenomenon of the transition cones, and not only, why accumulation in the downstream following glaciofluvial drainage channels took place – which had already been emphasized by I. SCHAEFER (1950)[4] – only under late pleniglacial conditions at all. About that this permafrost effect seems to have been the reason, why in the late pleniglacial fluvial environment – not only in glaciofluvial, but also in periglacial-fluvial systems – locally unproportionally large detritus masses could be accumulated (a particularly conspicuous example are the oversized alluvial cones of the late pleniglacial at the mouth of tributary valleys into the main valleys) and therefore a remarkably unbalanced longitudinal profile – especially in the upper parts

[2] This means, that – at least in typical cases – pleniglacial (permafrost-affected) and late glacial or postglacial (not permafrost-affected) fluvial deposits can be distinguished by their sedimentation pattern. K. HEINE (1982) has described an illustrative model case comparing the Older (pleniglacial) and the Younger Lower Terrace (dated into the Younger Dryas period) of the Middle Rhine near the mouth of the river Ahr: The Older Lower Terrace reveals surface parallel bedding, the Younger Lower Terrace oblique and cross-bedding, i.e. the sedimentation pattern of recent braided-rivers (MIALL 1977, RUST 1978).
Admittedly HEINE explained the difference of the sedimentation pattern in a different manner: he interprets it as a consequence of different flowing velocities during the deposition process. But the rushing discharge assumed for the deposition of the pleniglacial sediments
a) seems inconceivable under the stream gradient given, and would have – if nevertheless given –
b) transported (and not accumulated) the easily removable gravels and sands of the Older Lower Terrace, thus leading to downward erosion in the end.

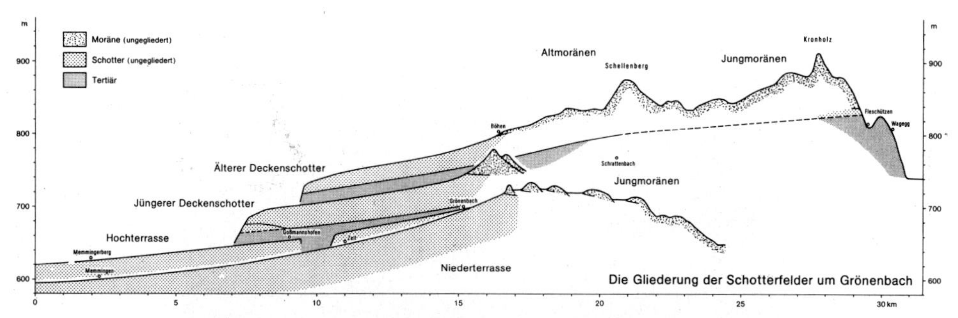

Fig. 5. Terminal moraines and outwash plains on both flanks of the Memmingen dry valley (Allgäu/Germany), the type region of A. PENCK for the quadripartition of the alpine ice-age. Each of the four gravel bodies shows the enlarged longitudinal gradient of the transition cones before the terminal moraines (from HABBE 1986b).

of the middle course of present-day rivers – be developed. This discontinuity could not be eliminated until late glacial and postglacial times. The detritus masses removed during this process have substantially contributed to the deposition of late glacial and postglacial accumulation bodies in the lower parts of the middle course of present-day rivers (see below p. 108).

2. The typical late pleniglacial accumulation was replaced by downward erosion in glaciofluvial as well as in periglacial-fluvial environments in late glacial times. This downward erosion removed – at least along larger streams – a large part of the late pleniglacial sediments down to their base and locally cut deeply into the prequaternary underground (STARKEL & THORNES 1981). This can be proven most impressively in the Danube valley between Ulm and Vienna (FINK 1977, BUCH 1987 et seq., BUCH & HEINE 1988). M. BUCH (1987) has delineated the chronological course of this late glacial erosion and the following accumulation phase (Fig. 7; see also KOZARSKI & ROTNICKI 1977). BUCH's main interest admittely was focused on the

---

[3] This statement seems to be contradictory to BÜDELS (1969, 1977) ideas of the influence of his "Eisrinden-Effekt" ("ice-crust effect"). According to BÜDEL permafrost below glaciofluvial streams should lead not to accumulation but to downward erosion. BÜDEL founded his "subpolar zone of excessive valley formation" on this concept. But except for the fact, that the "Eisrinden-Effekt" in itself is still disputed (BIBUS et al. 1976, BARSCH & KING 1981, STÄBLEIN 1983, WEISE 1983, SEMMEL 1985, SPÄTH 1986), the glaciofluvial meltwater streams of pleniglacial Middle Europe with their tremendous discharge and high load of coarse material seem to be a special case, which cannot be generalized without hesitation.

[4] SCHAEFER himself explained the exclusively late pleniglacial accumulation as a consequence of a gradually increasing unpropitious relation between meltwater discharge and solid load "in a late phase of the glacial development" (1950: 73). On the other hand he made clear, that it was "an uniform, nearly non-interrupted, homogenous large-scale process affecting the whole width of the valley floor, which took its course relatively rapidly" (1950: 80). This second statement – substantiated by numerous observations also by SCHAEFER himself – is incompatible with gradually diminishing meltwater discharge and increasing solid load. Therefore SCHAEFER's explanation of the late pleniglacial accumulation was already questionable in 1950. The problem has been open since then.

Fig. 6. Southwest face of the Eichelsteig gravel pit north of Biberach/Riß (Upper Swabia/Germany) within the transition cone of the Rhine glacier of the forelast glaciation. Alternation of coarser and finer grained material, markedly coarse material is an exception. Parallel bedding also prevails in this case (K.A.H. 5.4.1966).

postglacial accumulation processes. He described the preceding erosion phase as an "initial disturbing impulse", which might be "caused by the large cycles of climatic development during the Quaternary" (1988 b; 132). The character of this disturbing impulse is not discussed. But BUCH's diagram (Fig. 7) clearly shows, that the erosion tendency begins to have consequences at the turn from late pleniglacial to late glacial, i.e. at the same time, in which – as has been shown in the Iller glacier area (HABBE 1985 a et seq.) – the permafrost begins to dwindle away. It is natural to suspect a connection between the two processes. In quite another location – viz. in the periglacial environment of the Nuremberg Reichswald (HABBE 1980, 1981) – I made the plausible suggestion ten years ago, that the beginning of the downward erosion of present-day watercourses coincides with the dwindling away of permafrost (equally: BROSE & PRÄGER 1983). Admittedly positive proof of this is difficult to furnish.

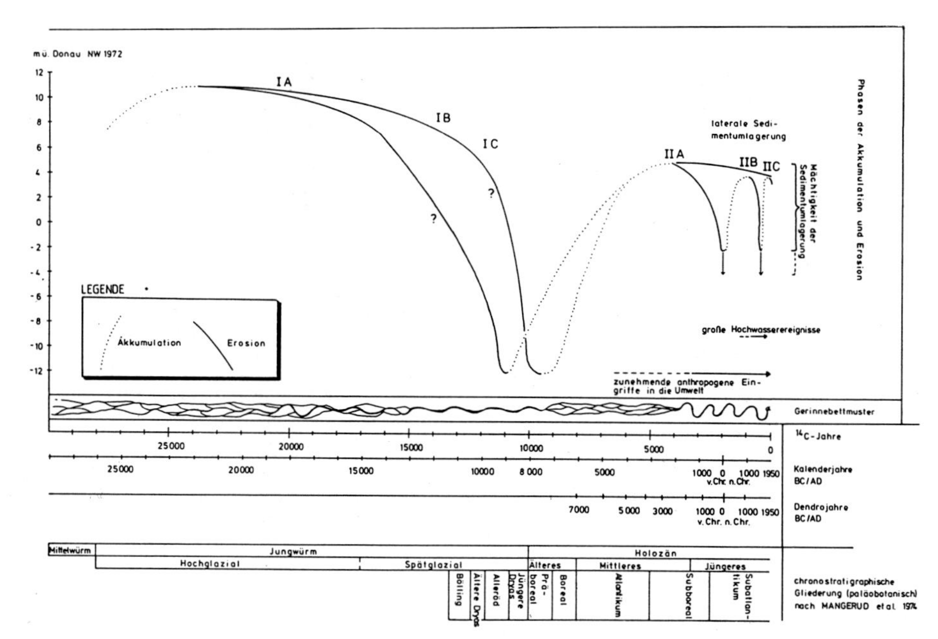

Fig. 7. Downward erosion and accumulation of the river Danube in the Regensburg area since the late pleniglacial of the last glaciation after BUCH 1988 a (detail).

But the existence of permafrost during the high pleniglacials and its absence during all other (and not merely during the distinctly warm) periods of the Quaternary may plausibly explain, why during the Pleistocene accumulation on the one hand and vigorous downward erosion on the other hand took place in the glaciofluvial drainage channels (and in the periglacial-fluvial ones as well) under quite equal stream gradient conditions, – in other words: how the terrace sequences came into existence, by means of which the subdivision of the Pleistocene of the alpine foreland into cold and warm periods has been justified since more than 100 years (PENCK 1882).

3. A detailed investigation of the postglacial fluvial deposits containing fossil tree trunks (FRENZEL 1977, BECKER 1982) shows that (Fig. 8)

a) they have not been deposited continuously, but in batches during only a few separate sedimentation events at quite different times, and
b) the intensity of the process ceased abruptly during the upper Holocene – in the area of the Iller alluvial cone near Ulm during the late Subboreal –, which is manifested by a change from coarse to fine grain sizes in the younger deposits.

The sediment sequences as a whole are usually interpreted as a consequence of either climatic variations or the influence of the land use by prehistoric men (SCHIRMER 1983, STARKEL 1985). But it can be shown, that the discernable sedimentation events are contemporary neither in different sections of the same river system (FINK 1977) nor in different river systems (BUCH 1988 b), which means, that the two

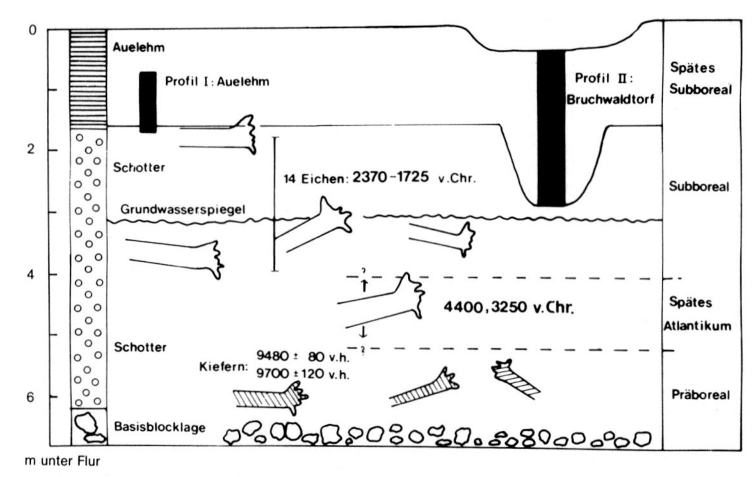

Fig. 8. Postglacial accumulation sequence of the alluvial cone of the river Iller near Ulm (from BECKER 1982).

explanations are valid regionally at most. As a general explanation they are unsatisfactory. Therefore it should be remembered, that

a) every deposition is not only a matter of the releasing mechanism, but also of the material available, and
b) a potential of easily removable loose sediments had been accumulated in the upper sections of the middle course of present-day rivers during the late pleniglacial, i.e. a non-equilibrium in relation to present-day rivers' longitudinal profiles (see above, p. 105), which had to be balanced under late glacial and postglacial conditions.

The sedimentation in batches characteristic of the Older Holocene therefore can be interpreted as an expression of single phases of that equalization process, the conspicuous change from coarse to fine grained deposits in the Late Holocene as a signal for the termination of the process. Admittedly this is a hypothesis, which must be subject to further examination. But if this hypothesis is correct – and until now there is no evidence against it –, it would mean, that the after-effects of the accumulation of the late pleniglacial under permafrost conditions influenced the land-forming process far into the Holocene period.

References

BARSCH, D. (1981): Terrassen, Flußarbeit und das Modell der exzessiven Talbildungszone im Expeditionsgebiet Oobloyah Bay, N-Ellesmere Island, N.W.T., Kanada. – In: D. BARSCH & L. KING (Hrsg.): Ergebnisse der Heidelberg-Ellesmere Island-Expedition. – Heidelberger Geogr. Arb. **69**: 163–201.

BARSCH, D. & L. KING (1981): Ergebnisse der Heidelberg-Ellesmere Island-Expedition. – Heidelberger Geogr. Arb. **69**.

BECKER, B. (1982): Dendrochronologie und Paläoökologie subfossiler Baumstämme aus Flußablagerungen – Ein Beitrag zur nacheiszeitlichen Auenentwicklung im südlichen Mitteleuropa. – Mitt. Komm. Quartärforsch. Österreich. Akad. Wiss. **5**.

BIBUS, E., G. NAGEL & A. SEMMEL (1976): Periglaziale Landformung im zentralen Spitzbergen. – Catena **3**: 29–44.

BROSE, F. & F. PRÄGER (1983): Regionale Zusammenhänge und Differenzierungen der holozänen Flußgenese im nordmitteleuropäischen Vergletscherungsgebiet. – In: H. KLIEWE et al. (Hrsg.): Das Jungquartär und seine Nutzung im Küsten- und Binnentiefland der DDR und der VR Polen. – Petermanns Geogr. Mitt. Erg.-H. **262**: 164–175.

BUCH, M. (1987): Spätpleistozäne und holozäne Geomorphodynamik im Donautal östlich von Regensburg – ein Sonderfall unter den mitteleuropäischen Flußsystemen? – Z. Geomorph. N.F., Suppl-Bd. **66**: 95–111.

– (1988a): Spätpleistozäne und holozäne Geomorphodynamik im Donautal zwischen Regensburg und Straubing. – Regensburger Geogr. Schriften **21**.

– (1988b): Zur Frage einer kausalen Verknüpfung fluvialer Prozesse und Klimaschwankungen im Spätpleistozän und Holozän – Versuch einer geomorphodynamischen Deutung von Befunden von Donau und Main. – Z. Geomorph. N.F. Suppl.-Bd. **70**: 131–162.

BUCH, M. & K. HEINE (1988): Klima- oder Prozeß-Geomorphologie – Gibt das jungquartäre fluviale Geschehen der Donau eine Antwort? – Geogr. Rundsch. **40**/5: 16–26.

BÜDEL, J. (1969): Der Eisrinden-Effekt als Motor der Tiefenerosion in der exzessiven Talbildungszone. – Würzburger Geogr. Arb. **25**.

– (1977): Klima-Geomorphologie. – Berlin/Stuttgart.

CHURCH, M. (1972): Baffin Island sandurs – a study of arctic fluvial processes. – Geol. Surv. Canada Bull. **216**.

EHLERS, J. & F. GRUBE (1983): Meltwater deposits in north-west Germany. – In: J. EHLERS (Ed.): Glacial deposits in north-west Europe. – Rotterdam: 249–256.

FINK, J. (1977): Jüngste Schotterakkumulationen im österreichischen Donauabschnitt. – In: B. FRENZEL (Hrsg.): Dendrochronologie und postglaziale Klimaschwankungen in Europa. – Erdwiss. Forsch. **13**: 190–211.

FLÜGEL, W. A. (1981): Hydrologische Studien zum Wasserhaushalt hocharktischer Einzugsgebiete im Bereich der Oobloyah-Tals, N-Ellesmere Island, N. W. T., Kanada. – In: D. BARSCH & L. KING (Hrsg.): Ergebnisse der Heidelberg-Ellesmere Island-Expedition. – Heidelberger Geogr. Arb. **69**: 311–382.

FRENZEL, B. (Hrsg.) (1977): Dendrochronologie und postglaziale Klimaschwankungen in Europa. – Erdwiss. Forsch. **13**.

GALLWITZ, H. (1949): Eiskeile und glaziale Sedimentation. – Geologica **2**.

GROSS, G., H. KERSCHNER & G. PATZELT (1976): Methodische Untersuchungen über die Schneegrenze in alpinen Gletschergebieten. – Z. Gletscherkunde u. Glazialgeologie **12**: 223–251.

HABBE, K. A. (1980): Die äolischen Sandablagerungen vor dem Albtrauf in Franken und ihre Bedeutung für die Rekonstruktion der jungpleistozänen Klimaentwicklung. – Tagungsber. u. wiss. Abh. **42**. Dt. Geographentag Göttingen 1979: 276–278.

– (1981): Über zwei ^{14}C-Daten aus fränkischen Dünensanden. – Geol. Blätter f. Nordost-Bayern **31**: 208–221.

– (1985a): Das Späthochglazial der Würm-Eiszeit im Illergletscher-Gebiet – Ergebnisse einer geomorphologischen Kartierung. – Quartär **35**/**36**: 55–68.

– (1985b): Erläuterungen zur Geomorphologischen Karte 1:25000 der Bundesrepublik Deutschland – GMK 25 Blatt 18/8127 Grönenbach. – Berlin.

– (1986a): Zur geomorphologischen Kartierung von Blatt Grönenbach (I) – Probleme, Beobachtungen, Schlußfolgerungen. Erlanger Geograph. Arbeiten 47 = Mitteil. Fränk. Geogr. Ges. 31/32, 1984/85: 365–479.

– (1986b): Bemerkungen zum Altpleistozän des Illergletscher-Gebietes. – Eiszeitalter u. Gegenwart **36**: 121–134.
HABBE, K. A. & K. RÖGNER (1989): Bavarian Alpine Foreland between Rivers Iller and Lech. – In: O. SEUFFERT (Ed.): Manual of field trips in and around Germany. – Geoöko-Forum **1**: 181–222.
HEINE, K. (1982): Das Mündungsgebiet der Ahr im Spät-Würm und Holozän. – Erdkunde **36**: 1–11.
HINZE, C., H. JERZ, B. MENKE & H. STAUDE (1989): Geogenetische Definitionen quartärer Lockergesteine für die Geologische Karte 1:25000 (GK 25). – Geol. Jahrb. **A 112**.
JOPLING, A. V. & B. C. MCDONALD (Ed.) (1975): Glaciofluvial and glaciolacustrine sedimentation. – Soc. Econ. Palaeontol. Mineralog. Spec. Public. **23**.
KAISER, K. (1960): Klimazeugen des periglazialen Dauerfrostbodens in Mittel- und Westeuropa. – Eiszeitalter u. Gegenwart **11**: 121–141.
KOUTANIEMI, L. (1984): The role of ground frost, snow cover, ice break-up and flooding in the fluvial processes of the Oulanka river, NE Finland. – Fennia **162**: 127–161.
KOZARSKI, S. & K. ROTNICKI (1977): Valley floor and changes of river channel patterns in the North Polish Plain during the late Würm and Holocene. – Quaestiones Geograph. **4**: 51–93.
LEOPOLD, L. B., M. G. WOLMAN & J. P. MILLER (1964) Fluvial processes in Geomorphology. – San Francisco/London.
MCCANN, S. B., P. J. HOWARTH & J. G. COGLEY (1972): Fluvial processes in a periglacial environment – Queen Elizabeth Island, N. W. T., Canada. – Inst. Brit. Geograph. Transactions **55**: 69–92.
MIALL, A. D. (1977): A Review of the Braided-River Depositional Environment. – Earth-Science Rev. **13**: 1–62.
PARDÉ, M. (1933): Les cours d'eau glaciaires – Débit, régime, charge en matières solides et dissoutes. – Comptes Rendus Congr. Intern. Géographie Paris 1931, t. II, 1: 359–370.
– (1947): Fleuves et Rivières. – Paris.
DU PASQUIER, L. (1891): Über die fluvioglazialen Ablagerungen der Nordschweiz. – Beitr. Geolog. Karte der Schweiz **31**: 1–128.
PENCK, A. (1882): Die Vergletscherung der deutschen Alpen – Ihre Ursachen, periodische Wiederkehr und ihr Einfluß auf die Bodengestaltung. – Leipzig.
PENCK, A. & E. BRÜCKNER (1901/09): Die Alpen im Eiszeitalter. – 3 Bde., Leipzig.
POSER, H. (1947): Dauerfrostboden und Temperaturverhältnisse während der Würm-Eiszeit im nicht vereisten Mittel- und Westeuropa. – Die Naturwissenschaften **34**: 10–18.
REINECK, H. E. & I. B. SINGH (1980): Depositional Sedimentary Environments – with Reference to Terrigenous Clastics. – 2nd ed., Berlin/Heidelberg/New York
RUST, B. R. (1978): The interpretation of ancient alluvial successions in the light of modern investigations. – In: R. DAVIDSON-ARNOTT & W. NICKLING (Ed.): Research in fluvial systems. – Proceedings of the 5th Guelph Symposium on Geomorphology, 1977, Norwich: 67–105.
SCHAEFER, I. (1950): Die diluviale Erosion und Akkumulation – Erkenntnisse aus Untersuchungen über die Talbildung im Alpenvorland. – Forsch. dt. Landeskunde 49.
SCHIRMER, W. (1983): Die Talentwicklung an Main und Regnitz seit dem Hochwürm. – Geol. Jahrb. **A 71**: 11–43.
SCHUNKE, E. (1981): Abfluß und Sedimenttransport im periglazialen Milieu Zentral-Islands als Faktoren der Talformung. – Die Erde **112**: 197–215.
– (1989): Schneeschmelzabfluß, Aufeis und fluviale Morphodynamik in periglazialen Flußgebieten NW-Kanadas. – Erdkunde **43**: 268–280.
SCOTT, K. M. (1978): Effects of permafrost on stream channel behavior in arctic Alaska. – U.S. Geolog. Survey Profess. Pap. 1068.
SEMMEL, A. (1985): Periglazialmorphologie. – Erträge der Forschung 231, Darmstadt.

SPÄTH, H. (1986): Die Bedeutung der „Eisrinde“ für die periglaziale Denudation. – Z. Geomorph. N.F., Suppl.-Bd. **61**: 3–23.

STÄBLEIN, G. (1983): Polarer Permafrost – Klimatische Bedingungen und geomorphodynamische Auswirkungen. – Geoökodynamik **4**: 227–248.

STARKEL, L. (1985): Lateglacial and postglacial history of river valleys in Europe as a reflection of climatic changes. – Z. Gletscherkunde u. Glazialgeologie **21**: 159–164.

STARKEL, L. & J. B. THORNES (Ed.) (1981): Palaeohydrology of river basins. – Techn. Bullet. Brit. Geomorphol. Research Group **28**.

TROLL, C. (1977): Die „fluvioglaziale Serie“ der nördlichen Alpenflüsse und die holozänen Aufschotterungen. – In: B. FRENZEL (Hrsg.): Dendrochronologie und postglaziale Klimaschwankungen in Europa. – Erdwiss. Forsch. **13**: 181–189.

WEISE, O. R. (1983): Das Periglazial – Geomorphologie und Klima in gletscherfreien kalten Regionen. – Berlin/Stuttgart.

WOLDSTEDT, P. (1961): Das Eiszeitalter – Grundlinien einer Geologie des Quartärs. Erster Band: Die Allgemeinen Erscheinungen des Eiszeitalters. – 3. Aufl., Stuttgart.

Address of the author: Prof. Dr. KARL ALBERT HABBE, Institut für Geographie der Universität Erlangen-Nürnberg, Kochstraße 4, D-8520 Erlangen.

Z. Geomorph. N.F.	Suppl.-Bd. 92	113–125	Berlin · Stuttgart	Mai 1993

Periglacial overlaying strata in the Bavarian Forest Methods for their stratigraphic division and questions about their ecological significance

by

JÖRG VÖLKEL, Regensburg

with 2 figures and 3 tables

Summary. Character and stratigraphic division of the overlaying strata of the Bavarian Forest are described. The differentiation of stratum boundaries carried out in the field is verified by means of a wide spectrum of laboratory analyses. Different methodological approaches are discussed. A special problem offers the identification of loess admixtures of local origin, which are mineralogical similar to other local substrata. The knowledge of the constitution of overlaying strata and the origin of material is of great importance to the pedogenetic interpretation and the ecological assessment of soil sites in subdued mountain areas. This is demonstrated by means of a special soil type, the humic Lockerbraunerde (loose brown earth), which formation is restricted to periglacial overlaying strata of certain altitudes.

Zusammenfassung. Ausbildung und Gliederung der Deckschichten im Bayerischen Wald werden beschrieben. Die im Felde ausgehaltenen Schichtgrenzen werden mittels einer breiten Laboranalytik zu belegen versucht. Unterschiedliche methodische Ansätze werden diskutiert. Ein besonderes Problem ist in den oberen Deckschichtengliedern der Nachweis von Lößbeimengungen, die lokalen Ursprungs sind und sich mineralogisch kaum von den anderen Substraten unterscheiden. Die Kenntnis vom Aufbau der Deckschichten und deren Materialherkunft ist von großer Bedeutung für die pedogenetische Interpretation und die ökologische Beurteilung der Standorteigenschaften von Böden in einem Mittelgebirgsraum. Dies wird am Beispiel eines besonderen Bodentypus vorgestellt, der humosen Lockerbraunerde, welcher nur in bestimmten Höhenlagen auf periglazialen Deckschichten entwickelt ist.

1 *Introduction*

Almost all the soils of the Bavarian Forest have formed on quaternary overlaying strata of mostly periglacial origin. Their stratigraphic division, origin of material, and the percentage of aeolian admixtures etc. are basic knowledge for special ecological studies. In the following, analytical methods are discussed which seem apt to document layer and substratum boundaries additional to the field diagnostic differentiation.

These studies are fundamental to special research projects on the ecology of soils and landscapes carried out by the author and his study group in the Bavarian Forest.

0044-2798/93/0092-0113 $ 3.25

Models on how to fix caesium in soils of forest ecosystems which have been highly contaminated by the Chernobyl fallout are being developed (Völkel 1992). Furthermore we investigate the question of atmogenic input of anthropogenic pollutants into forest ecosystems. Points of emphasis are soil acidification and heavy metal load. For the latter, the differentiation between geogenic and anthropogenic factors is of crucial importance.

2 *The stratigraphic division of overlaying strata in the Bavarian Forest*

2.1 *Criteria for diagnosis and differentiation in the field*

The division of overlaying strata in subdued mountain areas has remained a point of issue till today. Since basic research studies were carried out by Schilling & Wiefel (1962) and Semmel (1964, 1968) for the German highlands, extensive studies have been put forward repeatedly. An essential aspect of all these studies is the comparability of pedogenetic field criteria established in different study areas and carried out by different soil scientists. Currently, the study group "soil systematics" of the German Pedological Association (DBG) is taking final steps in generalizing the division of overlaying strata with regard to terminology and criteria. Those applied terms and criteria for defining layers have been recently published in Schilling & Spies (1991). The author follows the recommendations of the DBG study group for the stratigraphical division of overlaying strata closely for the work in his study area.

To some extent, constitution and distribution of overlaying strata vary for the different subunits of landscape of the East Bavarian Crystalline. Thereby change the criteria of stratigraphical division, showing that even within a homogenous major landscape the fixation of general criteria poses problems.

The Basislagen are composed of markedly compacted soil flow material or debris largely free of aeolian admixtures. The stony material is aligned. In the socalled Hinterer Bayerischer Wald and in the Hoher Bogen area the stones in the base layers are regularly topped with silt or loam caps, whereas the bottoms are bare. In the socalled Vorderer Bayerischer Wald and in the Naabgebirge this characteristic feature is often missing. The developing of loam caps seems to be bound to in-situ downwash of fine material and thus is attributed to higher percolating rates as a consequence of higher precipitation rates in high altitude areas. The Basislagen (base layers) may show a multi-layered substructure and are largely levelling the relief because of their thickness and their wide-spread occurrence. Though compacted, the Basislagen (base layers) in slope positions rarely operate as water impermeable deposits, which is a consequence of their sandy-loamy character and their high content of skeleton material. When composed of debris, the base layers are important paths for interflow. In the Hinterer Bayerischer Wald and at the Hoher Bogen massif, the Basislagen (base layers) in high altitude positions are regularly free of fine material as an effect of in-situ downwash by drainage. In-situ downwash is a current process, as field observations during snowbreak in spring and after periods of abounding precipitation during summer have shown.

The Mittellagen are of wide-spread occurrence too and are bound to levels lower than 1100 m a.s.l. This phenomenon is caused by relief, as extended slopes without zones of extreme steepening are widely dominating the relief of the Bavarian Forest. With regard to the composition of fine material, there exists a distinct difference between Mittellagen (medium layers) and Basislagen (base layers). Characteristic for the Mittellagen (medium layers) is a considerably high content of aeolian material; the skeleton material is not aligned. Compared to the hanging Hauptlagen (main layers) (see below), the Mittellagen (medium layers) are usually closer grained, yet without exerting a water-damming effect, and show a higher content of stones. Another diagnostic feature is the characteristic change of colour, occurring even when similar pedodynamic processes have worked on both layers (forming Bv-horizons). The Mittellagen (medium layers) are of special importance to soil formation and soil ecology, as they are considerably enlarging the physiological soil depth. Although the material of the Mittellagen (medium layers) is clearly related to that of the Hauptlagen (main layers), important analytical differences do exist, which are verifying the differentiation into Haupt- und Mittellagen (main and medium layers) and are supporting the validity of diagnostic field criteria. Differences are, for example, the marked change of mineral composition within the clay fraction to more illitic spectra, which is independent of the pedogenetic transformation of clay minerals and the differentiation of the Mittellage (medium layer) into soil horizons (II Bv, II CvBv and II BvCv) but is regularly bound to the alternation of layers. A great number of respective analyses already completed is statistically validating this assertion.

The Hauptlagen are especially rich of fine material. Characteristic is the low content of stones, although they are often rich of boulders. High percentages (>50%) of silt and clay declare the fine material to be largely of aeolian origin. The same is true for the Hauptlagen (main layers) of medium and high altitudes in the Bavarian Forest. Unlike the loess along the Danube marginal fault in direct neighbourhood of the vast loess deposits of the Danube valley and the eastern Tertiary hills of Lower Bavaria (described in STRUNK 1989), where the loess undoubtedly has to be classified as foreign material by its mineral composition, the loess of medium and high altitudes in the Bavarian Forest is of local origin (VÖLKEL 1991).

The transition from Hauptlage to Mittellage (main to medium layer) in the soil profile is marked by a fluid boundary. Often the zone of transition may be classified as a layer of its own, being of intermediate compactness when compared to Hauptlage and Mittellage (main and medium layer), while colour plus type and content of skeleton material shows its close relation to the Mittellage (medium layer). The transition from Basislage to Mittellage (base layer to medium layer) is always a clear boundary within the soil profile. Besides the above described characteristic features, it is especially by constitution of fine material that makes a doubtless distinction possible. Moreover, the alternation of layers always coincides with a horizon boundary. The Basislagen (base layers) usually have formed into Cv-horizons and thus present marked ecological boundaries. In cases of a missing Mittellage (medium layer), the differentiation of main and base layer is just as clear as it is for Mittellage and Basislage (medium and base layer).

Crystalline disintegration occurs up to high altitudes as well, usually buried under overlaying strata. The disintegration may be of different character. The texture ranges between loam and sand, depending on the type of parent material. Distribution, properties and origin shall be matter of another publication.

2.2 *Analytical validation of field characteristics and possibilities of assessing types and percentages of contributing substrata*

Numerous methods for detecting layer boundaries have been discussed in the literature. Questions on how to establish aeolian admixtures in the main and medium layers and on the differentiation between both layers are of special interest within this study.

As grain size parameters are being established in the course of the soil ecological site evaluation, they present a first data base for establishing layer boundaries. In literature, grain size parameters have been repeatedly the key for differentiation; Stahr (1979), for example, confirms strata boundaries in the Black Forest which were established in the field and specifies additional ones by applying the so-called quantitative stratification evidence. Frühauf (1990) succeeds in classifying loess into foreign and local fractions in connection with rising altitude of site with the aid of silt-quotients. But for the Bavarian Forest, methods applying grain size parameters have not yielded acceptable results. This is not at least a concequence of high silt and clay percentages produced by the in-situ weathering in form of Pre-Quaternary disintegration. Simple cluster analyses yet may underline the main stratification boundaries.

With the help of heavy minerals, qualitative and quantitative parameters may be found for identifying alternations of layer and material. This may basically be done by establishing special indicator minerals. In other regions of Western and Central Europe the so-called Laacher spectrum of volcanic heavy minerals presents a key for differentiating between Hauptlagen and Mittellagen (main and medium layers). As generally known, the Laacher spectrum has not been found in loess profiles of Eastern Bavaria so far. By fall 1991, the author had already gathered a broad database of heavymineral analyses of overlaying strata profiles (n = 79), which will be completed soon. As expected, the Laacher spectrum was not established in the samples. Singular mineral findings which correspond to the Laacher spectrum have proved to be of local origin.

Nevertheless, the presence of a foreign component of aeolian material can easily be established by analysing the heavy mineral spectrum. The loess along the Danube marginal fault originates from the Danube valley area and the Tertiary hills. Here, the admixture of loess becomes obvious by a distinct qualitative change within the overlaying strata profile (cf. Fig. 1). Yet, the expected alpine component in form of the minerals disthene and staurolite is missing, plus the heavy mineral spectra of pure loesses of the Regensburg area seem to be very heterogeneous in general. Special studies on these problems are being in progress. Much more difficulties presents the establishment of loess admixtures with the help of heavy mineral spectra when investigating high-altitude sites of the Bavarian Forest, because the loess is of local origin here and usually lacks qualitative features for the distinction within the mineral spectrum. As the methodological scatter within homogeneous substrata is relatively low, a distinction of individual layers may be possible by mere quantitative differences between the mineral spectra of the horizons as a result of the mineral sorting by segregation during aeolian transport.

Just as useful and essential for the stratigraphic division of overlaying strata with their varying material features has proved to be the X-ray diffractometrical analysis

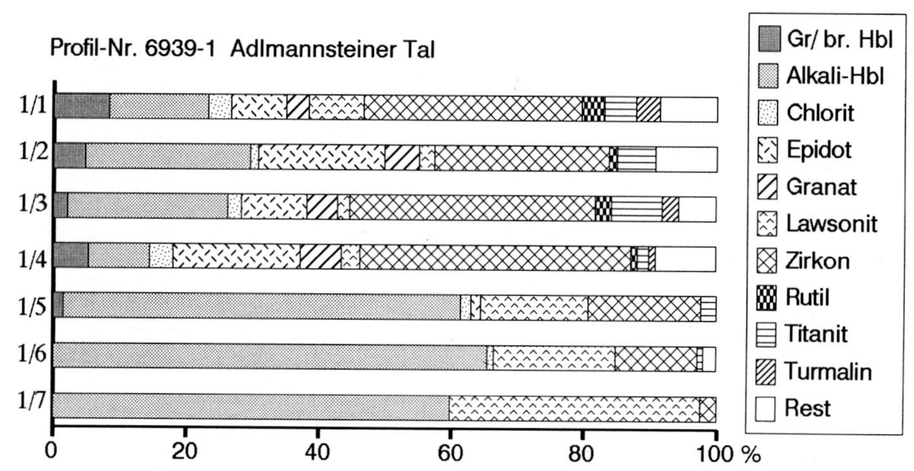

Fig. 1. Heavey mineral analysis on the fraction 200–63 μm. The profile is from the area of the Danube marginal fault near Regensburg. Compared with the loess-free layer (sample 1/5), the admixtures of loess in Hauptlage and Mittellage (main and medium layer) becomes obvious by qualitative (mainly rutile and garnet) and quantitative features. The lying disintegrated crystalline rock is mineralogically similar to the material of the Basislage (base layer).

of fine-material fractions. To begin with, the silicates of the clay fraction (<2 μm) are subject of consideration. It was expected that overlaying strata and Crystalline disintegration show differences in their spectra of clay minerals (cf. Fig. 2). The spectra of overlaying strata material often show a dominance of caolinite. Even between the Basislage (base layer) and the zone of disintegration should be a difference in mineral composition due to the periglacial pedogenesis of the first, except for horizons only superficially affected by solifluidal reworking of disintegrated material in form of the so-called outcrop bending. As a matter of fact a difference in clay mineral spectra was established analytically by the presence of vermiculite and illite and their interstratification within the Basislage (base layer). Mittellage (medium layer) and base layer usually show quantitative differences. An essential qualitative difference is the occurrence of secondary Al-chlorite in the Mittellage (medium layer), which is always a result of pedogenetic processes and is therefore independent from the origin of material and its stratification. Primary chlorites are rare within the soil sample collection. Due to stratification, there is a strict qualitative difference between Hauptlage and Mittellage (main and medium layer). Especially kaolinite is significantly reduced in its occurrence or is missing altogether within the Hauptlage (main layer). The peaks at 0.7 nm stand for halloysite and pedogenetic Al-chlorite. Illite is only of subordinate importance within the spectra or is missing as well. The clay minerals are thus an important aid in confirming the stratigraphic division of overlaying strata of the study area with view to significant changes of material composition, an aspect which, as far as I know, hasn't been described in literature yet.

In case of an alternation of rock within down-slope catenas, X-ray diffractometrical analysis of pulverized fine material (grain size <2 mm) may deliver basic

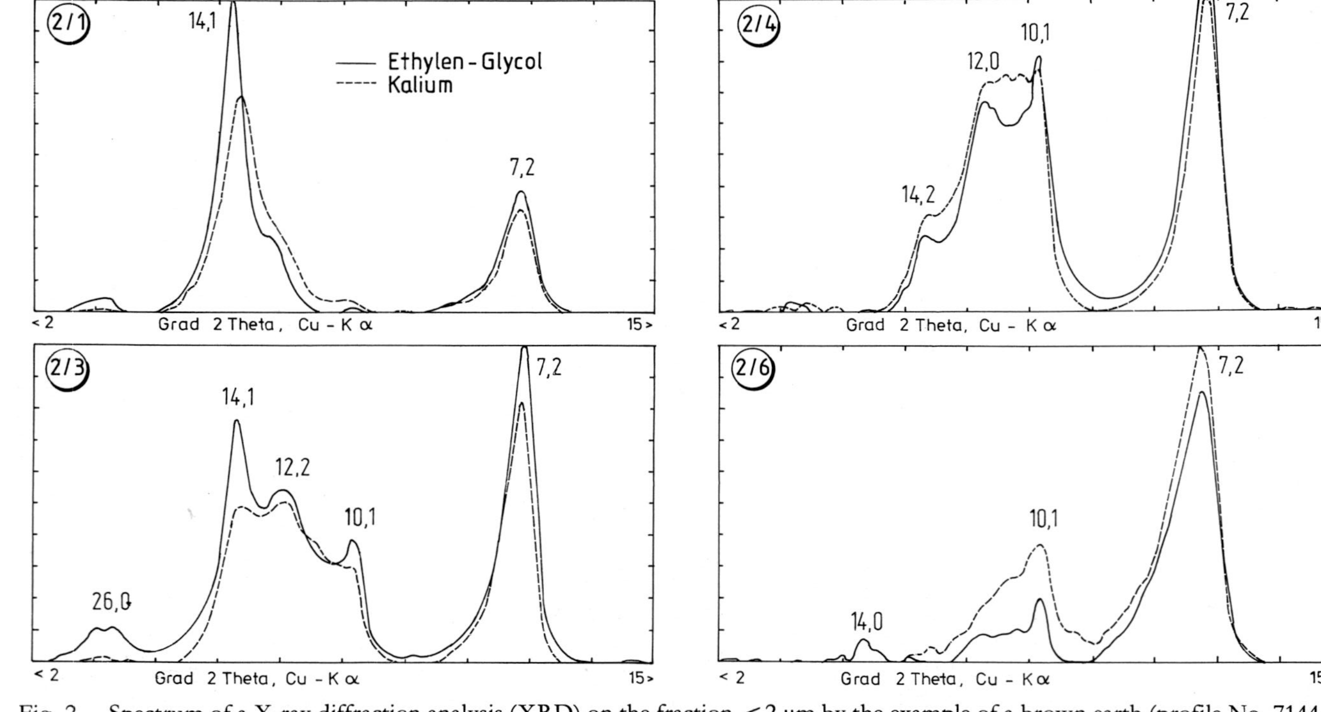

Fig. 2. Spectrum of a X-ray diffraction analysis (XRD) on the fraction <2 µm by the example of a brown earth (profile No. 7144-2, Dattinger Berg). Shown are a) the ethylenglycole preparations (Gly) and b) the potassium preparations (K). The extreme secondary chloritization of the 1.4 nm minerals becomes obvious in the Bv-horizons. The sample 2/1 (Bv) represents the Hauptlage (main layer), the sample 2/3 (IIBv) the Mittellage (medium layer), the sample 2/4 (IIIBvCv) the Basislage (base layer) and the sample 2/6 (IVCv) the disintegrated crystalline rock.

Table 1 X-ray diffraction analysis on textured preparations of the fraction 6–20 µm (medium silt) by example of a Hauptlage (main layer) containing local loess material, followed by a loess-free Basislage (base layer) (IIBvCv), disintegrated crystalline rock (IIICv) superficially reworked by solifluction which is stressed by solifluction, and the lying zone of disintegrated crystalline bedrock itself (IVCv). At the right side of the table the results of X-ray diffraction patterns of the fraction <2 µm (clay) are outlined. Explanation in the text.

Horizon	Fraction 20–6.3 µm (medium silt)					Fraction <2 µm (clay)						
	Quartz	Fsp. Alkali	Fsp. Plagio.	Illite	Kaol.	1.4 nm	Illite	M. L.	Al-Chl.	Verm.	Smectite	Kaol.
AhBv	++++	○	+	++	+	++	+	+++	+	++	–	+++
Bv1	++++	○	+	+	+	++	+	+++	+	++	–	++++
Bv2	++++	○	+	++	++	++	+	+++	+	++	–	++++
IIBvCv	+++	○	+++	++	++++	+	++	+	○	+	–	++++
IIICv	+++	○	++	++	++++	+	++	○	–	○	–	++++
IVCv	+++	++	+++	+	++	–	++	–	–	○	+++	++++

++++, very many; +++, many; ++, moderate; +, few; ○, in traces; –, absent.

Table 2 Fluorescent X-ray spectrographic analysis on fine ground (fractions <2 mm) of a multilayered profile of overlaying strata. The main elements are shown in Table 2. The total sum delineates the ignition loss, which is considerably higher in the upper horizons; the reason is the high content of organic matter in the Bhv-horizons of the Lockerbraunerde (loose brown earth) and X-ray amorphous alumosilicates.

Horizon	7145			
	38/3 Bhv %	38/5 IIBhv %	38/6 IIICvBv %	38/7 IVCv %
SiO_2	63.00	61.75	64.44	64.79
Al_2O_3	13.20	15.57	15.10	15.15
Fe_2O_3	5.81	6.42	6.25	6.56
CaO	0.72	0.76	0.82	0.97
Na_2O	1.06	1.08	1.39	1.57
K_2O	1.71	1.97	2.14	2.39
TiO_2	0.89	0.94	0.86	0.90
MgO	1.18	1.66	1.82	2.15
P_2O_5	0.19	0.20	0.19	0.18
Sum	87.76	90.36	93.01	94.65
Si/AI	**8.11**	**6.74**	**7.26**	**7.27**

information on stratification, origin and, possibly, distances of transport. Yet this rather simple method is only applicable when index minerals of key rocks are significant within the diagram, which is rarely the case for the carried-out analyses. The Bavarian Forest widely lacks such key rocks.

Good results are gained by the beforehand subfractioning of silt into coarse silt (63–20 μm), medium silt (20–6.3 μm), and fine silt (6.3–2 μm). X-ray diffractometrical analysed texture preparations of soil samples with view to their mineral content show different strata remarkably well (cf. Table 1). The medium-grained silt (20–6.3 μm) seems to be the silt subfraction delivering the best results. Remarkably, the sample-to-sample scatter of mineral content within the coarse silt fraction usually is too high to establish alternations of layers or change of material. The simple example presented here shows a Hauptlage (main layer) with a local loess component; it is underlain by a bipartite loess-free base layer covering autochthonous Crystalline disintegration material. Within the medium silt fraction especially quartz, plagioclase and caolinite are tracing the change of material from Hauptlage to Basislage (main to base layer). The disintegrated material contains microcline, which is absent in the overlaying strata. The clay fraction shows likewise the alternation of layers and material by qualitative changes in the mineral spectra.

Another approach to the detection of strata boundaries is the fluorescent X-ray analysis (RFA) of main and trace elements on fine material (<2 mm) or the clay fraction (>2 μm). Within the profile, the depth function of the main elements' mole-ratio, considering especially silicon and aluminium, gives basic information on possible stratification. In case the field distinction of Hauptlage and Mittellage (main and medium layer) wasn't quite clear, the mentioned analytical method may be very

helpful, as the depth function of moleratios quotients is contrary to the one typical for weathering and the respective soil dynamics. An example is given on Table 2. Here, the mole-ratios of Al and Si already signify inhomogeneities in material which can't be explained pedogenetically but by stratification. Only samples 38/6 and 38/7 taken from the horizons III CvBv and IV Cv are similar; the III CvBv is a transitional horizon, where the underlying Basislage (base layer) (IV Cv) was reworked during the deposition of the Mittellage (medium layer) and thus is very similar in composition to the material of the Basislage (base layer).

Of special interest are the trace elements in the fine material and in the clay fraction (<2 µm). Besides their profile-specific and stratigraphic vertical distribution, indicator elements may help to document inhomogeneities in material. The extent of aeolian influence of the overlaying strata may be quantified by an increase or decrease in concentration of geogenic trace elements, like heavy minerals. Additional to the fluorescent X-ray analysis, the trace elements are subjected to a selective acid desintegration and are then analysed by way of atomic adsorption spectrometry. First results have been gained for the work area. To ensure statistically save data, the evaluation of results will be subject of a later publication.

3 *Soil formation and ecology of Lockerbraunerden (loose brown earths) as an example for soils having formed on overlaying strata*

Overlaying strata strongly influence the pedogenetic progress, as a preliminary weathering of solid rock had already taken place. Admixtures of aeolian fine material produce better textural and structural characteristics. Layer thickness and abundance of fine material are important for the ecological advantages of site. Remarkably is the occurrence of the Mittellage (medium layer) up to high altitudes of the Bavarian Forest (see above). The Mittellage (medium layer) is always submitted to Bv-dynamics, so that a continuous cover of brown earths reaches up onto elevated positions of the Bavarian Forest. A comparatively low compaction facilitates rooting, so that Hauptlage and Mittellage (main and medium layer) together present a one meter thick zone of potentially rooting. The silt and clay content in the fine material of the overlaying strata is of great importance to the water conservation qualities of soils in areas of sandy disintegrating rocks like granite or coarse-crystalline orthogneiss. Field measurements carried out during summer 1991 have shown that even during longer-lasting periods of summer dryness the overlaying strata show a sufficient soil-moisture content. On the other hand, the overlaying strata have a good drainage effect, so that hydromorphous dynamics are rare in soils developed on slopes.

In the Bavarian Forest and also in the Upper Palatinate humic Lockerbraunerden (loose brown earths) occur in altitudes of more than 800 to 850 m a.s.l. Field characteristics of the Lockerbraunerde (loose brown earth) are its bright brown colour, its coffee powder-like texture and the well-known greasing effect. Origin of substrata and pedogenesis are raising a special field of problems. The term originates from SCHÖNHALS (1957), who draws a link between the silt-rich substrata of the Lockerbraunerden (loose brown earths) in the Hessian hills and a loess-like sediment of the Younger Dryas. For the Bavarian Forest, BRUNNACKER (1959, 1965) was the first to describe Lockerbraunerden (loose brown earths). BRUNNACKER (1965) names

the special climatic conditions of altitudes between 800 to 1150 m a.s.l. and the connected increase of ferrum release as being the cause of the particular texture characteristics. The discussion on possible causes of the particular texture formation has remained to be of interest up to the present, since the approach of Brunnacker, which he called theoretical himself, has neither been refuted nor confirmed so far.

The loose Bv-horizons with a pore space volume of >60% have exclusively developed on Hauptlagen (main layers) which are poor in stones and particularly rich in silt, so that they are drained very well. Within the work area, the forming of loose Bv-horizons has various causes. An essential factor is the type of substratum. Without the high silt content of the fine material, the loose texture would not have developed. Against Brunnacker (1965), the author assumes that aeolian material decisively contributes to the composition of substratum. The aeolian character of the fine material particularly becomes obvious in boulder-rich variations of the Hauptlage (main layer); fine-grained skeleton material is often missing altogether, a feature by which the Hauptlage (main layer) differs from the underlying layer. The lower boundary of the loose Bv-horizon always coincides with the layer boundary. As it is loess of local origin, the analytical proof of field diagnostic results is difficult. First results show that heavy mineral spectra for example are of a striking homogeneity. The loess probably originates from the wide valleys of the inner Bavarian Forest, for which a low gradient and a corresponding sedimentation of clayey-silty material is typical (e.g. the large intramount depression of the river Regen). Below 800 m a.s.l. loess occurs as well, yet the soil type of the Lockerbraunerde (loose brown earth) is missing here as the necessary pedochemical preconditions are absent (see below). The highest altitudes of the Bavarian Forest are too exposed for the sedimentation and preservation of aeolian material.

Because of its special soil structure, the development of Lockerbraunerden (loose brown earths) can only be explained by pedogenetic processes, i.e. induced by the soil environment. The soil environment is, with view to the discussed soil type, mainly governed climatically, so that substratum and climate are the most important factors of soil formation. Acid parent rock and climate-induced inhibition of the decomposition of organic litter create unfavourable types of humus, causing an extremely acidic soil environment, in which the mobilization of iron, manganese, aluminium and silicon is intensified. Correspondingly, the Bv-horizons of the Lockerbraunerden (loose brown earths) show raised contents of pedogenic oxides (in DCB-extract analysed Fe_2O_3 3.0–3.4%, Al_2O_3 0.9–1.3%) and extremely high contents of organic matter of 6–10%; the latter is concealed by the formation of chelates and highly polymerous humates and thus is not perceived in the field. Besides raised Feo/Fed-quotients of 0.6–0.9, a favourite bonding of Fe and Al with the organic matter (oxides solulable in pyrophosphate) was established by way of sequential extraction. Low molecular fulvic acids solulable in water accumulate in the Bv-horizons. Moreover, large burrowing animals carry organic litter from the surface downward. In the Lockerbraunerden (loose brown earths) foliage litter is found down to the lower boundary of the Bhv-horizons in about 40 cm depth. Pedogenic iron and aluminium oxides and the organic matter are all apt to form extremely stable complexes, especially in combination with each other. High percentages of silt provide additional bonding and are the cause of cementation. About 90% of the adsorpted cations of the cation-exchanging minerals (CECeff) are constituted

Table 3 Soil physical and soil chemical data of two Lockerbraunerden (loose brown earths): a) profile No. 6845-3, NE slope of the Hochberg near Bayerisch Eisenstein (Rear Bavarian Forest), 870 m a.s.l. b) profile No. 7145-38, E slope of the Brotjacklriegel (Front Bavarian Forest), 925 m a.s.l.
Methods: – Al_2O_3 and SiO_2 after HASHIMOTO & JACKSON (1960), changed by FÖLSTER et al. (1979), determined on the clay fraction <2 µm. – Fe_d, Al_d after MEHRA & JACKSON (1960); Fe_0, Al_0 after SCHWERTMANN (1964), determined on the fine ground <2 mm.

Horizon	depth cm	pH $CaCl_2$	% org. matter	C/N ratio	Fe_0 %	Fe_d %	Al_0 %	Al_d %	Al_2O_3 %	SiO_2 %	grain size distribution (weight %)							
											>2 mm	–630	–200	–63	–20	–6.3	–2.0	<2 µm
A(e)h	00– 05	3.16	07.9	06	0.82	1.29	0.30	0.30	n.b	n.b.	00.1	13.3	28.9	16.9	15.5	10.2	04.9	10.3
AhBv	05– 15	3.80	08.8	17	1.50	1.98	0.72	0.77	17.3	14.5	00.3	14.7	22.0	14.4	11.2	11.9	09.4	16.4
Bhv	15– 50	4.07	05.7	20	1.82	3.04	1.27	1.66	25.2	14.3	00.0	12.7	19.6	14.6	10.9	13.4	10.3	18.5
IIBvCv	50–100	4.35	01.1	15	0.42	1.16	0.82	1.03	21.5	14.5	82.2	16.2	18.7	13.4	11.9	14.6	07.7	17.5
IIICv	100–140	4.37	00.7	13	0.25	0.82	0.46	0.61	20.4	18.3	76.6	20.0	27.6	17.0	11.3	08.5	04.0	11.4
Oh	+7	3.20	12.8	0.6	0.76	0.96	0.34	0.39	20.2	16.3	00.0	18.0	13.0	09.1	25.9	16.9	04.0	13.1
Aeh	00– 01	n.b.	n.b.	n.b.	n.b.	n.b.	n.b.	n.b.	n.b.	n.b.	n.b.	n.b.	n.b.	n.b.	n.b.	n.b.	n.b.	n.b.
Bsh	01– 03	3.30	10.8	15	1.59	1.93	0.50	0.49	31.8	18.1	03.9	12.4	14.9	13.6	14.9	16.9	07.0	20.3
Bhv	03– 35	3.92	07.0	21	1.48	1.62	0.61	0.73	60.6	14.2	04.7	13.3	14.3	12.5	16.9	13.6	10.1	19.3
IIBhv	35– 50	4.15	05.1	18	1.12	1.56	0.76	0.85	44.1	03.0	04.4	13.5	14.9	14.6	13.7	14.1	09.0	20.2
	50– 80	4.31	02.3	12	0.68	1.22	1.07	1.06	56.6	06.8	04.6	12.1	15.5	14.8	15.7	14.1	07.6	20.2
IIICvBv	80–110	4.35	02.2	14	0.42	0.74	0.83	0.88	48.9	04.9	07.2	15.5	19.1	18.2	15.9	11.7	05.8	13.8
IVCv	110–150	4.40	00.7	10	0.18	0.77	0.35	0.40	48.8	00.0	13.1	27.9	20.8	18.4	13.3	10.3	03.6	05.7

by Al^{3+}-ions, which presumably originate largely from organic Al-complexes (Bloom et al. 1979). Thereby, the coagulation is considerably promoted. Important reasons for the stability of the loose structure have been mentioned now, yet they are not necessarily explaining the cause. In a soil environment of pH_{CaCl_2} 3–4.5 with distinctively raised concentrations of organic acids, silicon is strongly mobilized and is at disposal as cementation substance for the formation of soil structure. The contents of silicon in the NaOH-extract amount to 15–30%, whereas in brown earths silicon concentrations of only <10%, usually 3–5%, are measured (Table 3). Furthermore, high concentrations of aluminium-hydroxous compounds have been established, occurring as amorphous Al-hydroxides, Al-hydroxous sulfates or Al-hydroxous cations. This compounds support coagulation as well and favour the origin of a fine and stable structure of aggregates. The complete secondary chloritization of 2:1 clays of the mica group of 1.4 nm interlayer distance appears remarkably. Because of the interlayer occupancy with Al-hydroxide cations, the clay fraction of the concerned horizons has almost totally lost its swelling property, which is contributing to the stability of structure. This explains why Lockerbraunerden (loose brown earths) keep their characteristic structure regardless of alternating moisture contents, even through periods of edaphic dryness during the summer.

The Lockerbraunerden (loose brown earths) of the Odenwald and other Hessian subdued mountain areas described by Stöhr (1963) and others owe their structure to a high content of volcanic glass from the Alleröd Laacher See-volcanism (Eifel). The volcanic glasses weather into allophanes and other X-ray amorphous minerals. Volcanic glasses are completely missing in the soil substrata of the Bavarian Forest. Yet it is possible that allophanes, imogolites etc. are being formed in the above described soils. The formation of X-ray amorphous minerals is described for similar soil types in Scotland, for example, which occurrence there was associated with the formation of the special soil structure (Smith & Mitchell 1984, and others). A number of samples from the Bavarian Forest were already subjected to the differential thermal analysis (DTA) and the infrared spectroscopical analysis (IR-spectroscopy) in order to detect X-ray amorphous minerals. There is a tendency towards the presence of X-ray amorphous substances, but a definite proof has not been furnished yet. So far it can't be evaluated to what extent the occurrence of amorphous minerals contributes to the formation and the stability of soil structure.

By the example of the Lockerbraunerde (loose brown earth) it has been demonstrated to what extent the pedogenetic interpretation and ecological evaluation of site depend on the knowledge of the geomorphological situation. Without knowing about the good drainage effect of overlaying strata and the high percentage of admixted aeolian material, the formation of the above described soil type cannot be explained. Generally speaking, the necessary parameters for the ecological evaluation of sites in the Bavarian Forest remain unclear without understanding the origin and constitution of overlaying strata. Accordingly, regarding a major landscape unit, the ecological characterization of site has always to proceed from geomorphogenesis.

References

BLOOM, P. R., M. B. MCBRIDE & R. M. WEAVER (1979): Aluminum organic matter in acid soils: Buffering and solution aluminum activity. – Soil. Sci. Soc. Am. J. **43**: 488–493.
BRUNNACKER, K. (1959): Zur Kenntnis des Spät- und Postglazials in Bayern. – Geologica Bavarica **43**: 74–150.
– (1965): Die Lockerbraunerde im Bayerischen Wald. – Geol. Bl. NO-Bayern **15**: 65–76.
FÖLSTER, H., H. HASE & R. ULRICH (1979): Freisetzung von Aluminium in mitteldeutschen sauren Braunerden aus Löß-Sandstein-Fließerden. – Z. Pflanzenernähr. Bodenkd. **142**: 185–194.
FRÜHAUF, M. (1990): Neue Befunde zur Lithologie, Gliederung und Genese der periglazialen Lockermaterialdecken im Harz: Fremdmaterialnachweis und Decksedimenterfassung. – PGM 4/1990: 249–256.
HASHIMOTO, J. & M. L. JACKSON (1960): Rapid dissolution of Allo-phane and Kaolinite-Halloysite after dehydration. – Clays and Clay Minerals, 7th Conf., 102–113, Pergamon Press, N.Y., Monograph No. 5.
MEHRA, O. P. & M. L. JACKSON (1960): Iron oxide removal from soils and clays by a dithionite-citrate system buffered with sodium bicarbonate. – Clays Clay Min. **7**: 317–327.
SCHILLING, B. & E.-D. SPIES (1991): Die Böden Mittel- und Oberfrankens. – Bayreuther Bodenkdl. Ber. **17**: 68–82.
SCHILLING, W. & H. WIEFEL (1962): Jungpleistozäne Periglazialbildungen und ihre regionale Differenzierung in einigen Teilen Thüringens und des Harzes. – Geologie **11**: 428–460.
SCHÖNHALS, E. (1957): Spätglaziale äolische Ablagerungen in einigen Mittelgebirgen Hessens. – Eiszeitalter u. Gegenwart **8**: 5–17.
SCHWERTMANN, U. (1964): Differenzierung der Eisenoxide des Bodens durch Extraktion mit Ammoniumoxalat-Lösung. – Z. Pflanzenernähr., Düngung, Bodenkd. **105**: 194–202.
SEMMEL, A. (1964): Junge Schuttdecken in hessischen Mittelgebirgen. – Notizbl. hess. L.-Amt Bodenforsch. **92**: 275–285.
– (1968): Studien über den Verlauf jungpleistozäner Formung in Hessen. – Frankfurter Geogr. Hefte **45**: 133 S.
SMITH, B. F. L. & B. D. MITCHELL (1987): Characterization of poorly ordered minerals by selective chemical methods. – In: WILSON, M. J. (Ed.): A handbook of determinative methods in clay mineralogy: 275–294.
STAHR, K. (1979): Die Bedeutung periglazialer Deckschichten für Bodenbildung und Standorteigenschaften im Südschwarzwald. – Freiburger Bodenkundl. Abh. **9**: 273 S.
STÖHR, W. T. (1963): Der Bims (Trachyttuff), seine Verlagerung, Verlehmung und Bodenbildung (Lockerbraunerden) im südwestlichen Rheinischen Schiefergebirge. – Notizbl. hess. L.-Amt Bodenforsch. **91**: 318–337.
STRUNK, H. (1989): Aspects of the Quaternary in the Tertiary Hills of Bavaria. – Catena Suppl. **15**: 289–295.
VÖLKEL, J. (1991): Bodentypen und -genese auf jungpleistozänen Deckschichten im Bayerischen Wald. – Mitt. Dtsch. Bodendl. Ges. **66**/II: 877–880.
– (1992): Radioaktive Kontamination der Böden im Bayerischen Wald im Raum Zwiesel-Bayerisch Eisenstein. – Dreiländertreffen „Radiocäsium in Wald und Wild", 23.–24. 6. 92 St. Oswald, Proc. StMLU: in press.

Address of the author: Dr. rer. nat. JÖRG VÖLKEL, Lst. Physische Geographie, Universität Regensburg Universitätsstraße 31, D-8400 Regensburg.

Z. Geomorph. N. F.	Suppl.-Bd. 92	127–143	Berlin · Stuttgart	Mai 1993

Fission-track Analysis and Geomorphology in the Surroundings of the Drill Site of the German Continental Deep Drilling Project (KTB)/Northeast Bavaria

by

RALF BISCHOFF, Heidelberg, ARNO SEMMEL, Hofheim a. Ts., and GÜNTHER A. WAGNER, Heidelberg

with 8 figures

Summary. In the crystalline basement of northeastern Bavaria, a part of the Variscan fold mountains, apatite fission-track analyses were for the first time applied to determine the influence of tectonics on landform development. This extension of the application of the fission-track method seemed to be promising, as the respective information from fission-track data has distinctly increased due to track lengths measurements (GLEADOW et al. 1986, WAGNER 1988). In addition, a larger number of datings from the western part of the Bohemian Massif, which had been undertaken in order to determine the site of the KTB deep drilling (WAGNER et al. 1989), fitted very well to the geomorphologic results published by LOUIS (1984).

The data from 22 apatite fission-track analyses (tables of ages and dating-parameters cf. BISCHOFF et al. in print and WAGNER et al. 1989) show that the studied area was subject to a strongly differentiated tectonic development in the course of the entire Caenozoic. The Steinwald area, a ridge between Fichtelgebirge and Oberpfälzer Wald, is tectonically independent of the rest of the KTB-surroundings, which can be characterized by a steady uplift/denudation history since the Cretaceous/Tertiary boundary. During the Palaeogene two tectonic blocks can be distinguished which show an decreasing age-gradient from northeast to southwest. This means stronger uplift in the southwest because of younger ages. The dominant fault direction is northwest-southeast. At the turn from Palaeogene to Neogene, again several tectonic blocks can be distinguished which show decreasing ages from southeast to northwest. The principal fault direction has turned by 90° to southwest-northeast. This faulting which was induced by the activities of the Eger graben, has led to highly differentiated rates of uplift (young fast uplift). This is responsible for the development of the fault-scarp topography. The results of this study clearly prove that the apatite fission-track analysis is a promising method also with regard to geomorphologic problems, such as determination of tectonic influence on the development of topography.

Zusammenfassung. Im nordostbayerischen Grundgebirge, einem Teil des Variszischen Faltengebirges, wurde die Apatit-Spaltspuranalyse erstmals zur Lösung von Fragen des tektonischen Einflusses auf die Reliefgenese eingesetzt. Diese Erweiterung des Anwendungsbereiches der Spaltspurmethode erschien insofern als erfolgversprechend, da sich einerseits der Informationsgehalt von Spaltspurdaten durch Heranziehung von Längenmessungen deutlich vergrößert hat (GLEADOW et al. 1986, WAGNER 1988) und andererseits großräumig angelegte Datierungen im Westteil der Böhmischen Masse, die im Rahmen der Standortsuche für die

0044-2798/93/0092-0127 $ 4.25

Tiefbohrung (KTB) durchgeführt worden waren (Wagner et al. 1989), sehr gute Übereinstimmungen mit den geomorphologischen Befunden von Louis (1984) erbrachten.

Anhand der Daten von 22 Apatit-Spaltspuranalysen (Übersichtstabellen mit Altern und Meßparametern s. Bischoff et al. in print und Wagner et al. 1989) läßt sich zeigen, daß das Untersuchungsgebiet im ganzen Känozoikum eine stark differenzierte tektonische Entwicklung erfahren hat. Der Steinwald, ein Höhenzug zwischen Fichtelgebirge und Oberpfälzer Wald, besitzt gegenüber dem restlichen KTB-Umfeld, das eine seit dem Übergang Oberkreide/Tertiär gleichmäßige Hebungs-/Abtragungsgeschichte aufweist, eine tektonische Eigenständigkeit. Im Paläogen lassen sich zwei tektonische Blöcke mit einem Altersgefälle von Nordost nach Südwest ausgliedern, d.h. stärkere Hebungsbeträge in Südwest aufgrund der jüngeren Alter. Die dominante Bruchrichtung ist Nordwest-Südost. Auch im Übergang Paläogen/Neogen können mehrere tektonische Blöcke ausgegliedert werden, wobei nun eine Altersabnahme von Südost nach Nordwest auftritt. Die Hauptbruchrichtung hat sich um 90° auf Südwest-Nordost gedreht. Diese durch die Egergraben-Aktivität ausgelöste und sehr differenzierte Hebungsbeträge aufweisende Bruchtektonik – junge beschleunigte Hebung – ist für die Ausbildung eines Bruchstufenreliefs verantwortlich. Wie die Ergebnisse dieser Untersuchung verdeutlichen, ist die Apatit-Spaltspuranalyse auch im Hinblick auf geomorphologisch relevante Fragestellungen – tektonischer Einfluß auf die Reliefgenese – eine aussagekräftige Methode.

1 *Introduction*

Geomorphologists, which – according to Davis (1912) – are concerned with the explaining description of landforms, are frequently confronted with problems that result from tectonic processes. It is often very difficult, or even impossible, to make precise statements about the extent and the chronology of uplift or downfaulting. Without them, however, the explaining description of the landforms will only be fragmentary, with negative consequences not only for geomorphology, but for other geoscientific disciplines. Geology, geophysics, mineralogy, especially volcanology in this aspect, and geodesy and related sciences are often requiring geomorphologic results as a basis for well-founded statements. Questions about the extent and the temporal sequence of tectonic movements are playing a major role (Semmel 1979, Fuchs et al. 1983 and Semmel 1984 and 1991).

The geoscientific project "German Continental Deep Drilling Project (KTB)" (Emmermann & Rischmüller 1990) also included a geomorphological analysis of the surroundings of the drill site near Windischeschenbach in the Oberpfalz. Its aim was to determine to what degree the landforms are reflecting the more recent tectonic events. The geomorphologic research was part of the study group "Geologic Environment" of the "KTB"-project. First results were published elsewhere (Bischoff et al. 1990, Bischoff 1991).

The special attraction for geomorphologists to participate in this major project was the possibility to apply the fission-track method for the first time on a large scale to geomorphologic research. Wagner (1969) had already achieved plausible datings for the tectonic uplift of the crystalline Odenwald by means of the fission-track method. In the course of preliminary studies for the KTB-program he had realized striking parallels between fission-track data and the geomorphologic results published by Louis (1984) from the surroundings of the KTB-site. It was obvious to examine these parallels by means of a more detailed geomorphologic analysis and a denser sampling. Related to this was the question, whether the characteristic topography of this crystalline area could be explained by erosional scarps, by fault scarps, or as the expression of different bedrock.

The working program, which was carried out essentially by R. Bischoff as a Ph. D. thesis, consisted of a detailed geomorphologic mapping of the area (Fig. 1), of the subsequent sampling in the defined morphologic units, and of the laboratory analyses in the Max-Planck-Institut in Heidelberg. The studies comprised the years 1990 and 1991.

2 *The study area – location and geomorphologic overview*

The study area is situated in the crystalline basement of northeastern Bavaria and lies in the transitional zone between the two larger upland regions Fichtelgebirge and Oberpfälzer Wald. The area stretches from the KTB-site near Windischeschenbach in the southeast to the Steinwald ridge in the northwest (Figs. 1 and 2). The area is crossed by the Fichtelnaab river, which functions as local base level, from northwest to southeast. From a geomorphologic point of view, three characteristic landform units can be distinguished (Fig. 1):

1. A slightly undulating planation surface in ±500 m a.s.l. extends from the Fichtelnaab in northeastern direction far beyond the boundaries of the study area. Numerous remnants of tropical weathering material in form of plastosols and saprolite can still be found covering various types of bedrock. The clay-fraction of these soils possesses high contents of kaolinite. Characteristic are also the decreasing Si-Al ratios from the recent soils to the saprolite zone, which show that fresh material was worked into the upper parts of the solum in the course of the Pleistocene (periglacial cover layers). The prevailing soil types are loamy to clayey – in the eastern granite area also sandy-brown soils of medium development. Their acidity is low to medium, and they are partly pseudogleyed. Depressions and areas with remnants of Tertiary soil development have pseudogley soils. Occasionally, gentle domes with unweathered bedrock are rising above the regolith, comparable to shield inselbergs.

2. The Steinwald ridge rises in the northwest of the study area. Its culmination reaches 946 m a.s.l., and its surface is gently inclined to the northeast. Its southeastern flank shows numerous remnants of planation surfaces, more or less large, above all at elevations near 600 m, 700 m, and 850 m a.s.l. Remnants of intensive Tertiary weathering can be found only as allochthonous material in the lower areas bordering the Steinwald, or in hollows and depressions. Strongly gritty cover layers with brown soils and andosols of low to medium acidity are prevailing. Because of the steep slopes, their thickness is usually low. Only on the larger remnants of the planation surfaces, thicker profiles can be found. The landforms of this unit are dominated by numerous tors and cliffs, which rise from a few metres up to tens of metres above their surroundings.

3. The third unit extends from the Fichtelnaab (valley floor in the study area between 475 and 420 m a.s.l.) to the Hessenreuther Forst (710 m a.s.l.) in the southwest, and to the marginal hills near the Franconian Line (±620 m a.s.l.). The area is also characterized by several planation surfaces in different elevations, it is, however, not as strongly dissected as the southeastern flank of the Steinwald. The

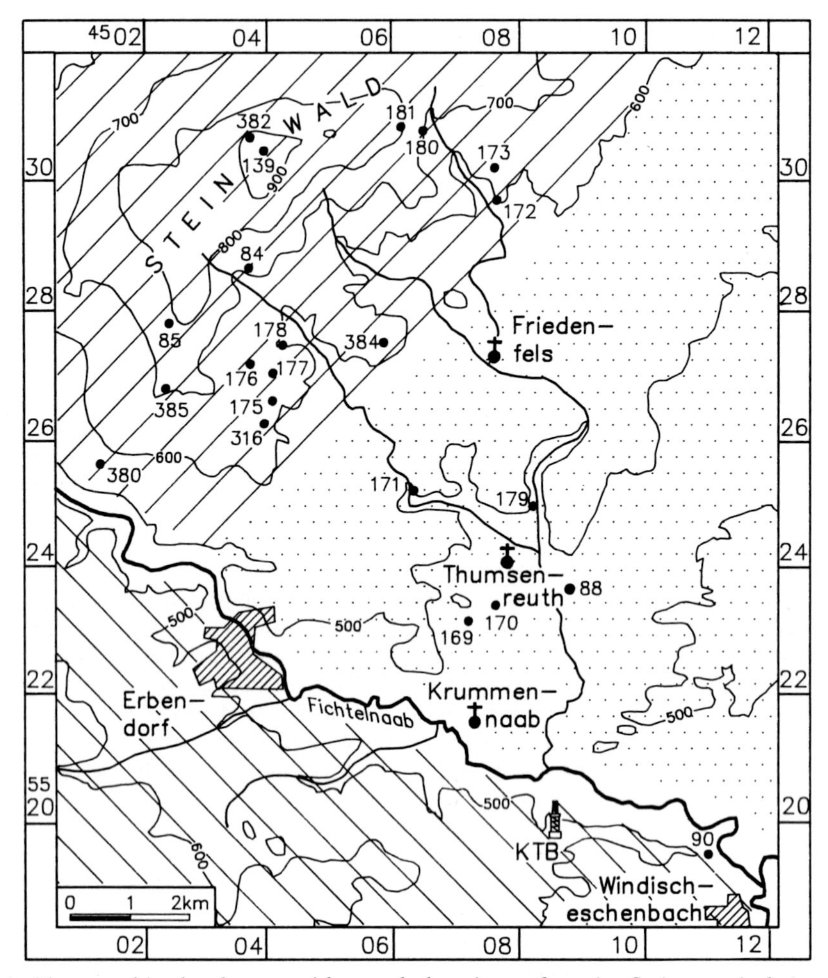

Fig. 1. Topographic sketch map with sample locations of apatite fission-track dating (dots with sample number) and morphographic unit 1 (:::): area northeast of the Fichtelnaab valley (planation surface; ±500 m a.s.l.), 2 (///): Steinwald area (3 even steps between 530 and 946 m a.s.l.) and 3 (\\\): area southwest of the Fichtelnaab valley (slightly stepped between 500 and 640 m a.s.l.).

prevailing soils are again sandy to loamy brown soils of medium development. It could be shown by means of refraction seismics (DEKORP - ISO '89) that the rocks of the units 1 and 3 are deeply weathered. There is, however, no observable orientation along petrologic or tectonic features. Field work in the Steinwald area – unit 2 – showed considerably thinner regolith.

3 *Geological and tectonic setting*

The location of the KTB drill site is situated on the western border of the Bohemian Massif, a horst-like, uplifted part of the Variscan fold mountains (Fig. 2). Three important parts of the crystalline basement are overlapping here: Moldanubian, Bohemian, and Saxothuringian. A few kilometres to the southwest follows the usually prominent scarp, which marks the downfaulting to the foreland of the South-German Mesozoic caprock. The Moldanubian zone – Moldanubian and Bohemian – consists chiefly of high-grade polymetamorphic gneisses. The adjacent Saxothuringian to the north is made up predominantly of lower Palaeozoic metasediments and metavolcanics that were only affected by a monophase metamorphism. The Late-Variscan mobilisation and intrusion of granites (325 to 290 Ma) along the Saxothuringian-Moldanubian borderzone (Erbendorf line) is of great importance, as demonstrated by their large areal extent.

The immediate vicinity of the deep drill (Fig. 2, framed area) is characterized by problematical conditions along the boundaries between Saxothuringian in the north, Moldanubian in the east, and the *Zone of Erbendorf-Vohenstrauß* (ZEV) in the southwest and the south. The ZEV – the geologic site of the deep drill – represents an overthrust block above the autochthonous Moldanubian – whose origin is still discussed: overthrusting from northeast, or upthrusting from southwest. Interlayered volcanic and sedimentary rocks were metamorphosed to garnet-disthene-biotite-gneiss, and to eclogitic and amphibolic rocks. A zone of mylonitic gneiss – products of the overthrusting – terminates the ZEV to the north. It is followed on the one hand by the *Erbendorf Greenschist Zone* with basic to ultrabasic eruptive rocks of low metamorphism – most likely a section of the ocean floor from the western foreland of the Bohemian Massif, which was upthrusted together with the ZEV, and on the other hand, by the *Wetzldorf Sequence* from the Early Palaeozoic, a joint body complex with low-grade metamorphism (phyllitic mica schists, quartzites, metacherts). The northeastern part of the study area, especially the Steinwald ridge, is occupied by fine-grained to porphyric granites.

Late-Variscan to Tertiary tectonics brought along considerable faulting. Of special importance is the *Fichtelnaab-Fault Zone* – a number of parallel faults along the Fichtelnaab valley. Along its northern side, distinctly deeper levels of the bedrock are exposed. Although the *Franconian Line* consists of a complex fault system (*Hirschmann* 1992), the bedrock in the southwest of the studied area is essentially upthrusted with 40° to 50° over the Upper-Cretaceous foreland. Finally, the entire study area possesses smaller and larger occurrences of Late-Miocene basalts (K/Ar age about 22 Ma; two samples kindly dated by Prof. Dr. H. J. Lippolt, Laboratorium für Geochronologie der Universität, Heidelberg), which are directly related to the tectonics of the Eger graben.

4 *Apatite fission-track analysis*

The calcium phosphate mineral apatite – the best known mineral in fission-track dating – is a powerful and sensitive geochrono- and geothermometer in the low

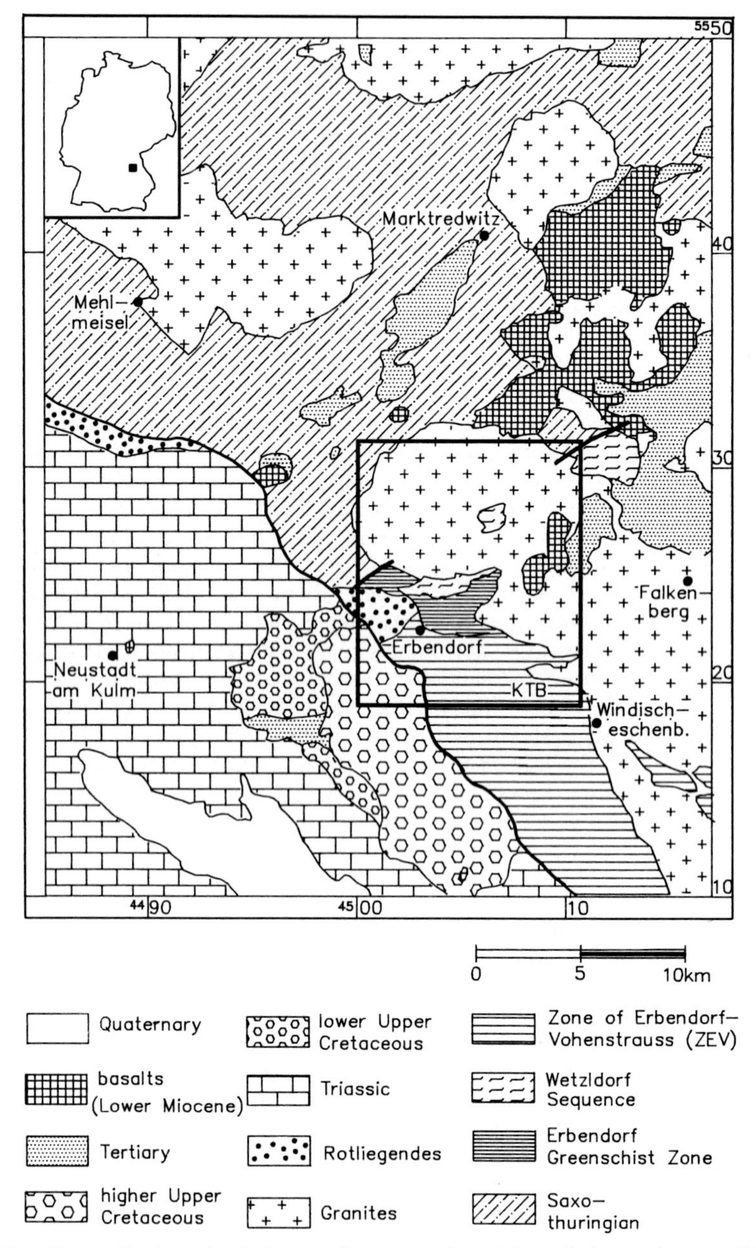

Fig. 2. Generalized geological-tectonic map and position of the study area (framed).

temperature region (<140 °C) which makes it possible to reconstruct the thermal-tectonic history of rocks.

A fission track is the damage zone formed through the spontaneous fission of the nuclide ^{238}U in insulating solids like minerals and glasses. The track-forming mechanism is explaned by the ion explosion spike model (FLEISCHER et al. 1965): the nucleus of ^{238}U breaks up into two lighter nuclei of approximately equal mass and releases a large amount of energy (200 MeV). The lighter fission fragments recoil from each other and ionize the atoms in the mineral lattice along their path. The ions repulse themselves and from a new track which is about 10–20 μm long and only 50 Å wide. By chemical etching these damage zones can be made visible in an optical microscope (PRICE & WALKER 1962).

The number of fission tracks within a mineral is proportional to the uranium concentration in the mineral, and to the time which has elapsed since the closing of the uranium-fission track-system. For the age calculation it is necessary to determine the uranium content and the fission-track density (tracks per square centimetre). The calculation follows the well known formula for radioactive decay with the mother nuclide being ^{238}U and the fission tracks being equivalent to the daughter nuclides. The uranium concentration can also be determined by fission-track counting. According to the grain population technique (WAGNER 1968) the separated apatite grains are divided into two fractions. One fraction is being used for spontaneous tracks counting, the other is heated to anneal all fossil tracks. Later, these grains are irradiated in a nuclear reactor with thermal neutrons. The fission of ^{235}U produces induced fission tracks which physically equal spontaneous tracks. After having counted these tracks, the uranium content is calculated by the known isotopic abundance ratio$^{235}U/^{238}U$.

Already, the first apatite fission-track datings (WAGNER 1968) show a clear relation between the ages and the thermal history of the dated rocks. The number

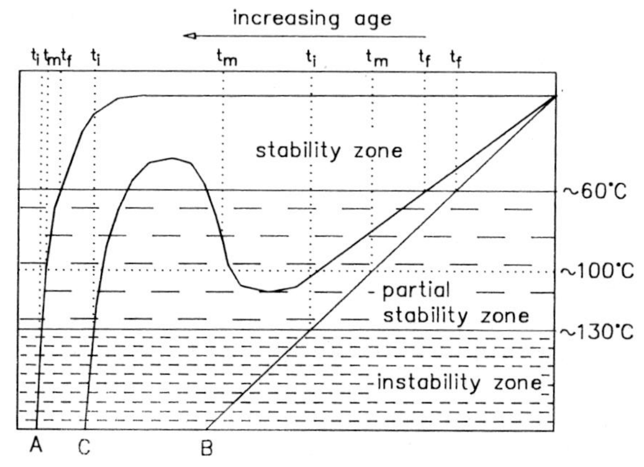

Fig. 3. Concept of the *partial stability zone* (WAGNER 1972) respectively *partial annealing zone* (PAZ); t_m measured fission-track age, t_i moment of cooling below ca. 130 °C, t_f moment of cooling below ca. 60 °C, *A* formation age type, *B* cooling age type, *C* mixed age type (after HEJL & WAGNER 1990).

of the tracks and their lengths are temperature-dependent. Originally, this was considered as a disadvantage of this dating method, however, it leaves a wide range of geological interpretation of fission-track data. WAGNER (1972) developed his concept of the *partial stability zone* respectively the *partial annealing zone* (PAZ) by means of annealing experiments and investigations on drill cores for the different types of age interpretations (Fig. 3). The temperature increasing downward in the diagram, is divided into three zones: A high-temperature zone with complete track erasure ("instability zone"), a medium-temperature zone with partial track fading ("partial stability zone") and a low-temperature zone with full track storage ("stability zone"). For a more detailed interpretation, three different schematic thermal histories are assumed: fast cooling (formation age type; curve *A* in Fig. 3), steady, but slow-cooling (cooling age type; curve *B* in Fig. 3) and a complex history with a thermal overprint event (mixed age type; curve *C* in Fig. 3). Owing to the limited track stability, the fission-track clock does not start immediately at the time of rock formation. Therefore, the fission-track age (t_m; Fig. 3) dates a later point of time, namely the moment at which ca. 50 percent of the track lengths are preserved, i.e. the middle of the *partial stability zone* (ca. 100 °C). This is only be valid for type *A* and *B*, not for type *C* (Fig. 3).

Lengths measurements on fission tracks can be used as a diagnostic tool for deciphering the type of thermal history – fast or slow cooling and complex reheating. The lengths of fission tracks in apatite will be measured either on tracks intersected by and projected on a polished internal face (*projected length of surface tracks*), or on horizontal tracks confined within a crystal and intersected by host tracks, cracks or cleavage planes (*confined track length*). The mean values of confined track lengths of type *A* (rapidly cooling after formation; Fig. 3) lie between 14.0 and 15.6 µm and the standard deviations of the distributions range from 0.8 to 1.2 µm. The mean lengths of type *B* (steady slow cooling) range from 12.0 to 14.0 µm with standard deviations between 1.0 and 2.0 µm. Distributions of type *C* – samples with thermal overprinting – show mean lengths smaller than 13.0 µm and standard deviations greater than 2.0 µm (GLEADOW et al. 1986).

The interpretation of projected track lengths is based on the assumption that the spontaneous track density used for dating is composed of two fractions, one produced within the *partial stability zone* and the other in the *full stability zone* (Fig. 3). It is possible to separate these two fractions by means of track length criteria. Experimental annealing studies and direct observations of the age vs. temperature profile in deep drill holes reveal that spontaneous tracks with projected track lengths ≥ 10 µm are effectively not stable at temperatures > 60 °C. The t_f-age calculated from the c_s/c_i-ratio (with c_s and c_i being the respective fractions of spontaneous and induced fission tracks ≥ 10 µm; $t_f = c_s/c_i * t_m$) is interpreted as the time during which rock temperature cooled to the *full stability zone* of fission tracks in apatite (Fig. 3) which corresponds to ca. 60 °C (WAGNER 1988).

For type *B* (Fig. 3) from the ca. 100 °C (t_m) and ca. 60 °C (t_f) cooling ages the time vs. temperature path can be reconstructed for each sample. Such time-temperature- respectively age-depth-paths *do not reflect the tectonic uplift*, i.e. uplift of a rock column towards sea level, *or the morphological uplift*, i.e. uplift of the surface towards sealevel. *They represent only the movement of the rock column compared to the isotherm of the effective closure temperature.*

Another important aspect of the geological interpretation of fission-track ages is the role of the geothermal gradient. Cooling ages varying significantly from one region to another may principally be explained by different uplift rates in an environment of an equal geothermal gradient (*tectonic model*) or by different geothermal gradients for equal uplift rate (*thermal model*; WAGNER et al. 1989). All gradual transitions between these two extreme cases are also conceivable.

5 *Results and discussion*

As this study aimed at an exemplary demonstration of the possibilities, but also of the limitations of the apatite fission-track analysis for geomorphologic research in old orogenic areas – i.e. in truncated uplands – a very dense sampling was chosen. The entire height interval between 500 m and 946 m a.s.l. was covered. We could not, however, take samples from all remnants of planation surfaces, as we needed, above all, fresh, unweathered rock material for the dating. This was not available on all of these surfaces. The possible problems caused by weathered apatite (GLEADOW & LOVERING 1974) were thus avoided. The 18 dated rock samples consisted of fine-grained to slightly porphyric two-mica granites. We also used four additional dates by WAGNER et al. (1989). From the 18 apatite samples, all of them from the landform units 1 and 2, the following data were determined:

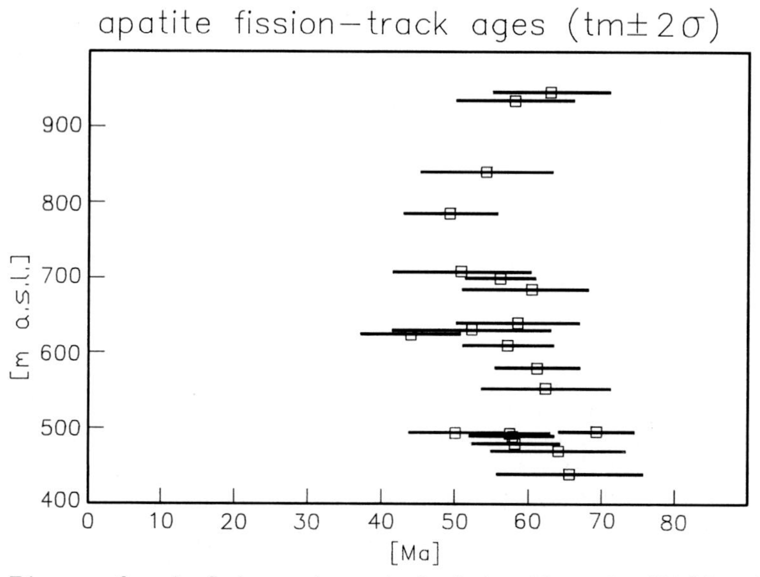

Fig. 4. Diagram of apatite fission-track ages in the Steinwald area (ca. 100 °C cooling ages $\pm 2\ \sigma$). Age/elevation relation reveals a wide distribution of age from 69 to 44 Ma and an age-decrease with increasing elevation, typical for regions with complex tectonic development.

a) the ca. 100 °C cooling age (t_m-age),
b) the age type (formation age type, cooling age type, or mixed age type) by means of confined track lengths measurements, and
c) the ca. 60 °C cooling age (t_f-age) by means of projected lenghts measurements of surface tracks.

(Detailed information of dating conditions and listed ages cf. BISCHOFF et al. in print).

The form of the lengths distribution of the confined tracks and an average length of 14 ± 1.3 µm (average of 7 representative samples) provided the important prerequisite for geologic and geomorphologic interpretations, namely, that the achieved datings are indeed belonging to the type *B* (steady slow cooling; Fig. 3). The great scattering of the ca. 100 °C cooling ages between 69 and 44 Ma is striking. On the other hand, the ages do not increase with topographic height (Fig. 4). Studies in the Alps, which had demonstrated the height-dependence of apatite fission-track ages for the first time (WAGNER & REIMER 1972) had shown, however, that in "en bloc" uplifted areas increasing ages corresponded with greater heights. This is, because the topographically higher rocks pass earlier through the isotherm of the effective closure temperature than the underlying ones. The track accumulation therefore begins earlier. Another striking result is the age gradient between samples from the southwestern Steinwald, which have partly distinctly younger ages, and the northeastern section with larger ages. As there exist significant age differences as well on the individual planation surfaces as between samples from different heights (Fig. 4), the age distribution permits conclusions about tectonics and rates of uplift. Although in large areas of the truncated Variscan uplands of Central Europe the relief is too small, so that the systematic age differences are smaller than the statistical scattering of the fission-track ages, because of random errors in measurement, (HEJL & WAGNER 1990), this is not the case in this area.

According to our findings, the study area was subject to a small-scale, highly differentiated tectonic development already in the course of the Palaeogene. The closer Steinwald area has distinctly younger t_m-ages (44–63 Ma) than the rest of the KTB-surroundings (62–69 Ma); for the wider surroundings cf. WAGNER et al. (1989). This underlines the tectonic independence of the Steinwald area (Fig. 5). The youngest ages appear in the southwestern part of the Steinwald, the oldest ones in the northeastern section; an indication for stronger uplift in the southwest. The northeastern slope of the ridge is most likely caused by this fact. The tectonic model gained by the age distribution correlates well with the geologic-tectonic field results. Important faults are separating the tectonic Steinwald area from the rest of the KTB-surroundings. Furthermore, they subdivide it into two tectonic blocks (Fig. 5). The dominant fault direction is from northwest to southeast, i.e. almost parallel to the Fichtelnaab Fault Zone.

To get some – although rather crude and generalized – information about the rates of uplift, a closer look at the age distribution of the southwestern part of the Steinwald ridge will be helpful. Northwest of Erbendorf, age differences from 62 (Fig. 5, area ///) to about 50 Ma (Fig. 5, area ⁼=) exist over a horizontal distance of only a few kilometers. Assuming a constant geothermal gradient according to the tectonic model (i.e. 30 °C/km, as found in the KTB pilot hole), both sample areas (Fig. 5, area /// and ⁼=) should be separated by about 600 m of rock thickness. As

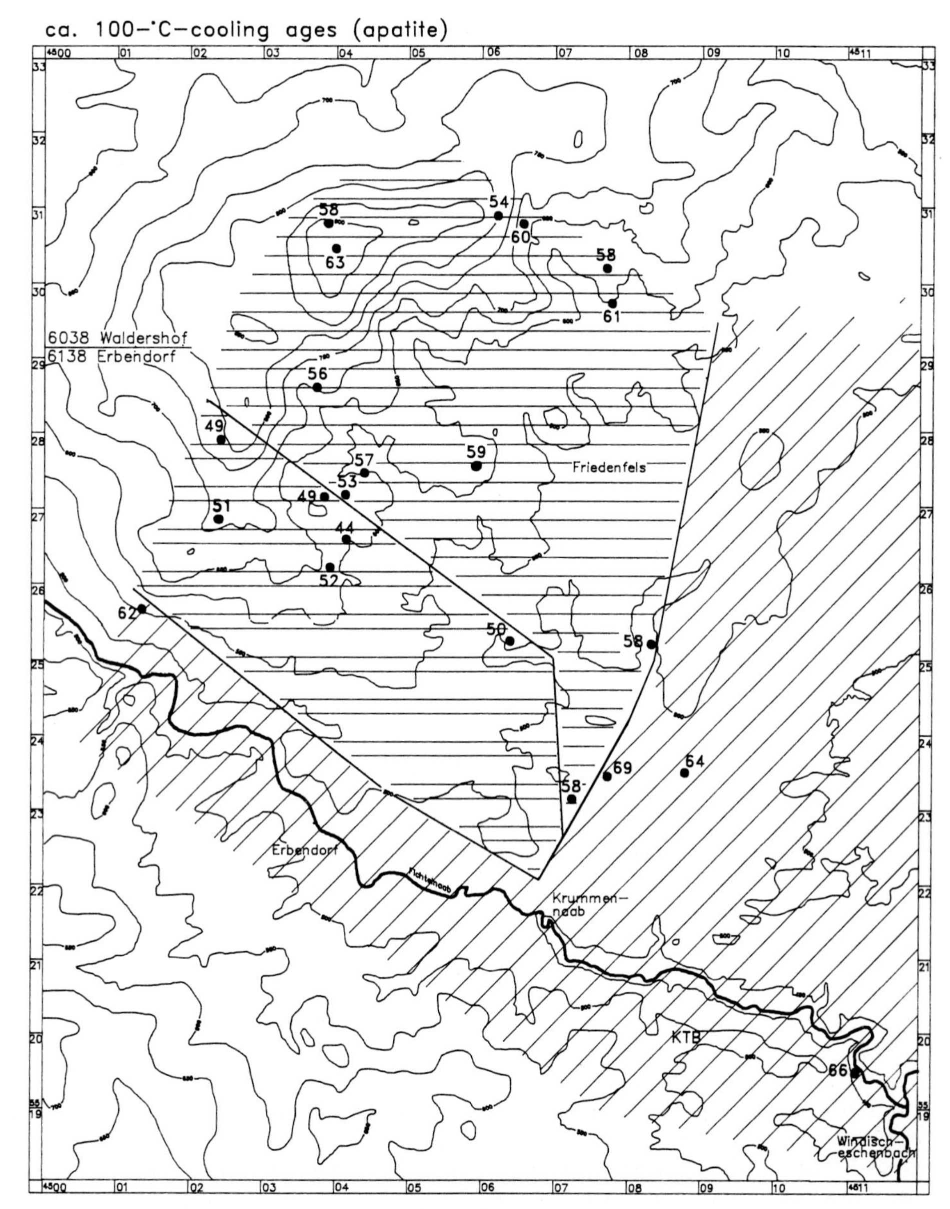

Fig. 5. Regional distribution of the ca. 100 °C cooling ages (dots indicate sample locations with ages in Ma). The Steinwald region shows much younger ages than the other KTB-surrounding (62–69 Ma [///]). Two tectonic blocks can be separated: 63–53 Ma [=] and 52–44 Ma [$^{=}$ =]; the main fault strike is northwest-southeast. The tectonic Steinwald region is limited in the southwest and east by mapped faults (STETTNER 1989).

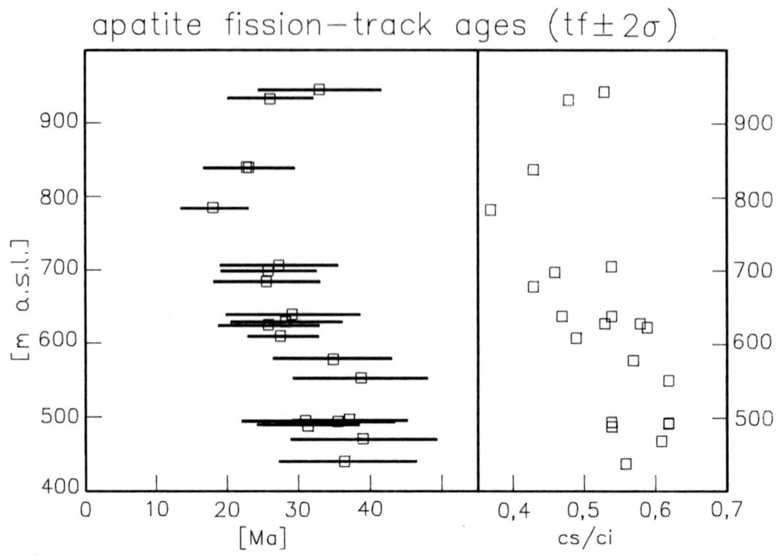

Fig. 6. Diagram of apatite fission-track ages in the Steinwald area (ca. 60 °C cooling ages ±2σ). Age/elevation relation reveals an age-decrease with increasing elevation. c_s/c_i-ratios between 0.5 and 0.7 are typical for constant cooling behaviour; smaller values represent acceleration as a general trend of steady cooling.

both sampling sites are nowadays situated at approximately the same elevation, this difference must have been balanced within the last 50 Ma. This indicates a distinctly greater rate of uplift in the southwestern part of the Steinwald (Fig. 5, area = =).

Also the ca. 60 °C cooling ages (t_f-ages) are indicating the special situation of the Steinwald area (Figs. 6 and 7): the ages between 33 and 18 Ma are distinctly younger than in the rest of the KTB-surroundings (39–36 Ma). The ages are decreasing from southeast to northwest, i.e. with increasing topographic height. The regional age distribution and the mapped pattern of faults is pointing at three tectonic blocks, which are separated by southwest-northeast-striking faults. The shift of the main fault-direction by 90 °C between the 100- and 60 °C-cooling ages can be related to the Eger graben tectonics, which was especially active at the turn from Palaeogene to Neogene, as apart from stratigraphic evidence is indicated by the basaltic volcanism of this period. To this phase belong also the youngest t_f-ages (23 and 18 Ma). The samples from the culminating ridge of the Steinwald (Platte, 946 m and Katzentrögel, 935 m) evidently cooled earlier to ca. 60 °C – as evidenced by higher t_f-ages – than the samples from the flanks of the ridge. These sampling sites are possibly located immediately below the pre-basaltic surface, where the rocks had almost cooled to the surface temperatures 20 to 22 Ma ago. A distinctly higher rate of erosion along the flanks – youngest ages and highest uplift rates – compared to a lower rate in the watershed area must have caused the exposure of the different rock levels, as indicated by their different ages.

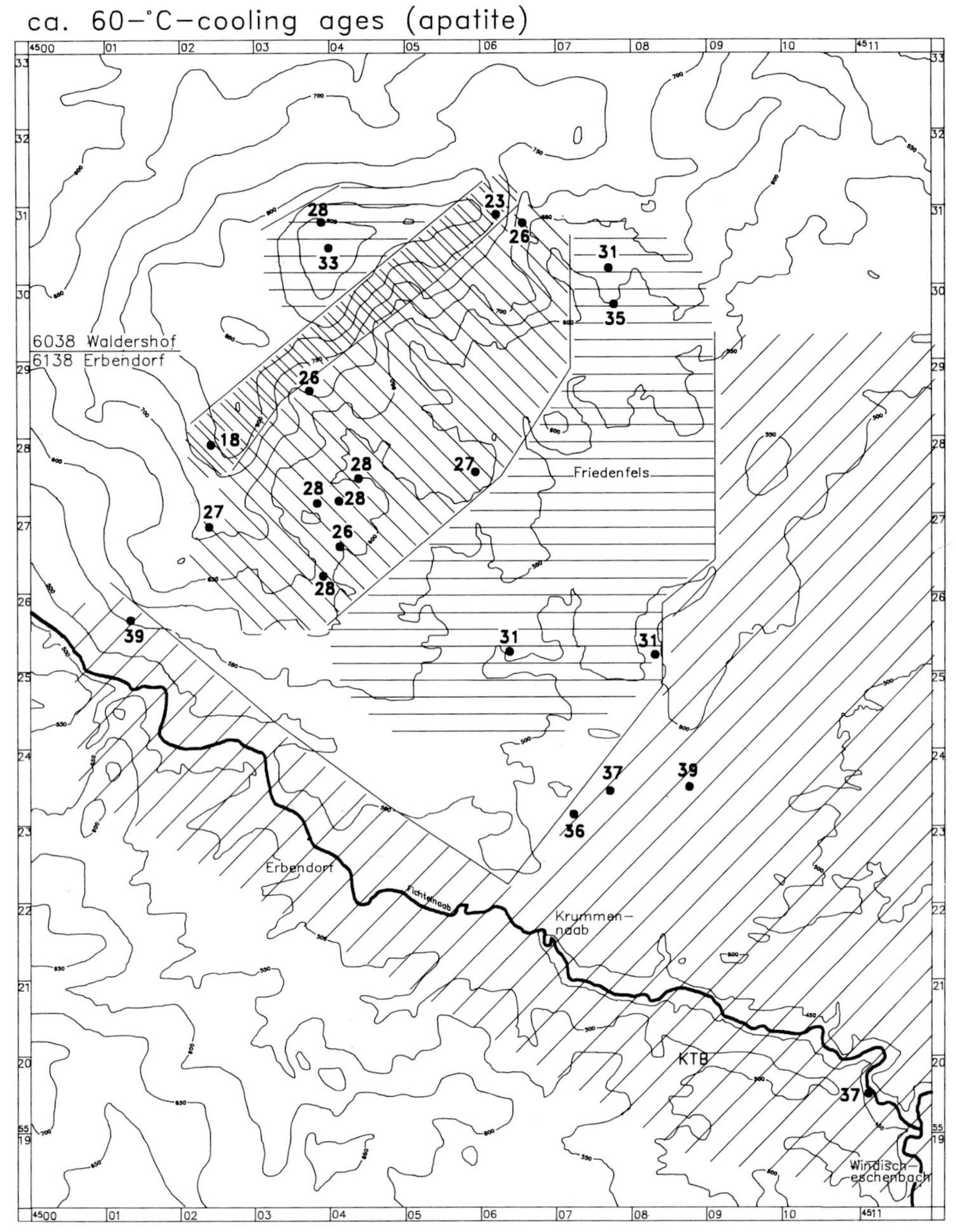

Fig. 7. Regional distribution of the ca. 60 °C cooling ages (Ma). The Steinwald region can be separated into 3 tectonic blocks (35–31 Ma [=], 28–26 Ma [\\\] and 23–18 [\\ \\]); the ages decrease from southeast to northwest. The main fault strike has now shifted to south-west-northeast.

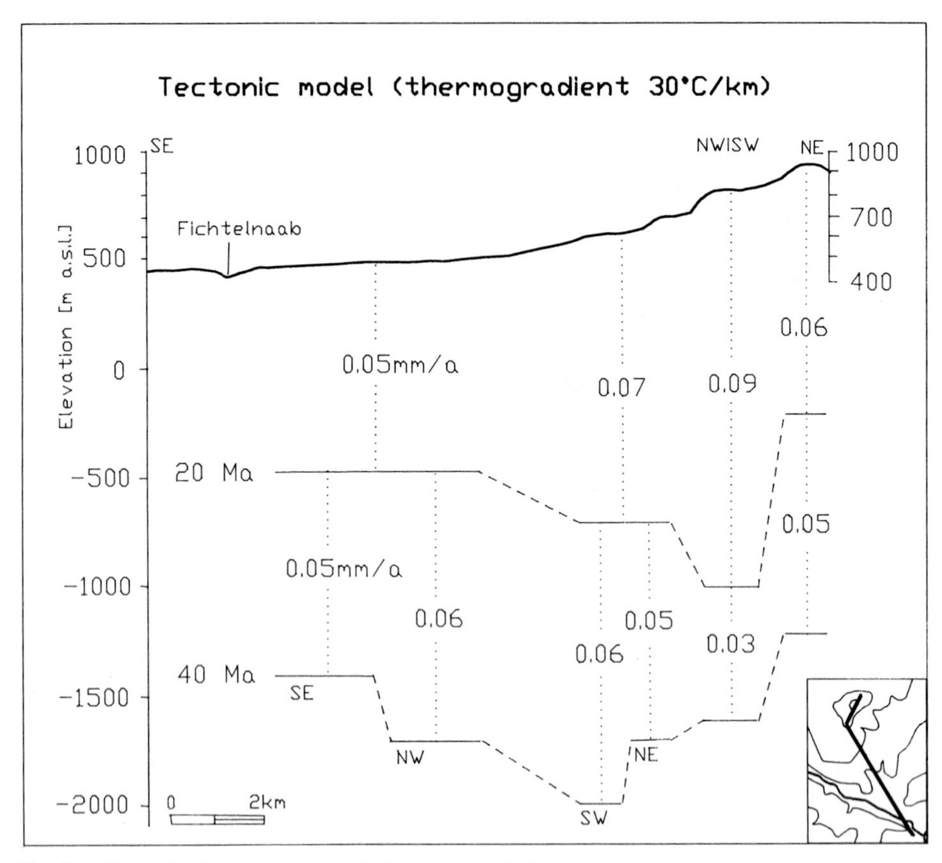

Fig. 8. Tectonic interpretation of the apatite fission-track ages along a profile from the Fichtelnaab to the top of the Steinwald (vertical scale exaggerated 4 times). The tectonic model (constant geothermal gradient: 30 °C/km) shows the palaeoposition of the current surface 20 and 40 Ma ago with corresponding uplift/denudation rates.

The combination of t_m- and t_f-ages allows the reconstruction of cooling paths for each sample, which permit conclusions about the cooling of the rocks – slow – constant – fast. If one applies for this interpretation a simplifying, constant geothermal gradient (tectonic model, cf. chapter 4), then the rates of uplift can be calculated. As demonstrated by the profile Fichtelnaab – Steinwald ridge (Fig. 8), the entire Palaeogene is characterized by an almost constant rate of uplift of 0.05 mm/a. This changes decisively in the Lower Miocene: whereas the ±500 m planation surface continues to be constantly uplifted, the surfaces on the flanks of the Steinwald ridge are indicating a distinctly accelerated uplift of 0.07 mm/a (samples from the 600/700 m surfaces), and of 0.09 mm/a for the 800/850 m surfaces. The crest of the Steinwald shows only a slightly higher rate of uplift of 0.06 mm/a. The area of the Steinwald was therefore uplifted considerably stronger than its surroundings in the course of the last 20 Ma, and was additionally subdivided into several fault blocks. One remark concerning the assumption of a constant geothermal gradient in this model: the application of a higher or variable gradient in this interpretation model would only cause a gradual shift, but not a change of the trend, because of the spatial limitation of the study area.

6 *Conclusion*

Already the ca. 100 °C ages demonstrate that the study area did not develop homogeneously during the Caenozoic, as was postulated by other authors because of cogent morphologic reasons (Büdel 1957, Wirthmann 1961, Désiré-Marchand & Klein 1987). With all necessary caution, which is appropriate for the interpretation of the ca. 60 °C cooling ages (t_f-ages) because of their low precision (1 σ ca. 10–15%), the results confirm the late Tertiary horst-like uplift of the Steinwald, as described by Louis (1984) on the basis of geomorphologic results. In addition, the disintegration of the Steinwald into fault blocks is evident. One should not expect, however, with regard to a geomorphologic analysis that the fission-track analysis will allow a precise dating of planation surfaces, whose elevations are only 100 m apart. Here, we meet the limitations of this method, as the systematic age differences of samples from the respective surfaces become smaller than the statistical error of the fission-track method. Nevertheless, because of dense sampling, we can state that the area of the Steinwald possesses a fault-scarp topography, which can be well correlated with the tectonic activities of the Eger graben, because of their similar ages. The rates of uplift calculated by means of the tectonic model, however, seem to be too large for the Neogene.

As a conclusion of this study, the statement can be made that the apatite fission-track analysis is a very useful tool for the solution of problems concerning the influence of tectonics on the development of landforms. Provided there are certain prerequisites, as e.g. rocks with accessory apatite-content, strong relief, supposed faults with larger displacement, the application of this method will be more than justified.

Acknowledgements

We would like to express our thanks to the Deutsche Forschungsgemeinschaft for the financial support and to many colleagues for their assistance. Above all to Director Dr. G. Stettner of the Geological Survey of Bavaria, to Prof. Dr. Lippolt, Laboratorium für Geochronologie der Universität, Heidelberg, and to Dr. E. Hejl, Max-Planck-Institut für Kernphysik, Heidelberg.

References

BISCHOFF, R. (1991): Geomorphologie und Spaltspurdatierung: Erste Untersuchungsergebnisse aus dem Umfeld der Kontinentalen Tiefbohrung bei Windischeschenbach/Oberpfalz. – Freiburger Geogr. H. **33**: 131–144; Freiburg.

BISCHOFF, R., A. SEMMEL & G. A. WAGNER (1990): Jüngere Tektonik und Reliefentwicklung im Umfeld der KTB – Geomorphologie und Spaltspurdatierung. – KTB-Report 90-4: 525; Gießen.

BISCHOFF, R., E. HEJL, A. SEMMEL & G. A. WAGNER (in print): Geomorphological interpretation of fission-track data – results from the Steinwald region, NE-Bavaria. – Nucl. Tracks Radiat. Meas.; Oxford.

BÜDEL, J. (1957): Grundzüge der klimamorphologischen Entwicklung Frankens. – Würzburger Geogr. Arb. **4/5**: 5–46; Würzburg.

DAVIS, W. M. (1912): Die erklärende Beschreibung der Landformen. – XVIII a. 565 p., B. G. Teubner, Berlin.

DÉSIRÉ-MARCHAND, J. & C. KLEIN (1987): Fichtelgebirge, Böhmerwald, Bayerischer Wald – Contribution à l'étude du problème des Piedmonttreppen. – Z. Geomorph. N. F., Suppl.-Bd. **65**: 101–138; Berlin, Stuttgart.

EMMERMANN, R. & H. RISCHMÜLLER (1990): Das Kontinentale Tiefbohrprogramm der Bundesrepublik Deutschland (KTB). – Die Geowiss., **8**: 241–257, Weinheim.

FLEISCHER, R. L., P. B. PRICE & R. M. WALKER (1965): The ion explosion spike mechanism for formation of charged particel tracks in solids. – J. Appl. Phys. **36**: 3645–3652.

––– (1975): Nuclear tracks in solids – Principles and applications. – Berkeley, Los Angeles, London.

FUCHS, K., K. v. GEHLEN, H. MÄLZER, H. MURAWSKI & A. SEMMEL (1983): Plateau uplift. – 411 p., Springer, Berlin, Heidelberg, New York, Tokio.

GLEADOW, A. J. W., I. R. DUDDY, P. F. GREEN & J. F. LOVERING (1986): Confined fission track lengths in apatite: a diagnostic tool for thermal history analysis. – Contrib. Mineral. Petrol. **94**: 405–415; Berlin, Heidelberg, New York.

GLEADOW, A. J. W. & J. F. LOVERING (1974): The effect of weathering on fission track dating. – Earth. Planet. Sci. Lett. **22**: 163–168; Amsterdam.

HEJL, E. & G. A. WAGNER (1990): Geothermische und tektonische Interpretation von Spaltspurdaten am Beispiel der Kontinentalen Tiefbohrung in der Oberpfalz. – Naturwissenschaften **77**: 202–213; Berlin, Heidelberg, New York.

HIRSCHMANN, G. (1992): Das Bruchstörungsmuster im KTB-Umfeld. – KTB-Report **92-3**: 85–124; Gießen.

HURFORD, A. J. (1990): Standardization of fission track dating calibration: Recommendation by Fission Track Working Group of the I.U.G.S. Subcommission on Geochronology. – Chemical Geology (Isotope Geoscience Section) **80**: 171–178; Amsterdam.

LOUIS, H. (1984): Zur Reliefentwicklung der Oberpfalz. – Relief, Boden, Paläoklima **3**: 1–66, Berlin, Stuttgart.

PRICE, P. B. & R. M. WALKER (1962): Chemical etching of charged particles tracks. – J. Appl. Phys. **33**: 3407–3412.

RICHTER, P. & G. STETTNER (1987): Die Granite des Steinwaldes (Nordost-Bayern) – ihre petrographische und geochemische Differenzierung. – Geol. Jb. **D 86**: 3–31; Hannover.

SEMMEL, A. (1979): Geomorphological criteria for recent tectonic – A discussion of examples from the north Upper Rhine area. – Allg. Vermessungs-Nachricht **86**: 370–374; Karlsruhe.

– (1984): Geomorphologische Kriterien für junge Krustenbewegungen in Mittelgebirgen. – Z. Geomorph. N. F., Suppl.-Bd. **50**: 79–90; Berlin, Stuttgart.

– (1991): Neotectonics and geomorphology in the Rhenish Massif and the Hessian Basin. – Tectonophysics, **195**: 291–297; Amsterdam.

STETTNER, G. (1989): Geologische Kartierungen im engeren und weiteren Umfeld der Tiefbohrung, Beispiele von deren Auswertung in Beziehung zum Profil der Vorbohrung und das Vorhaben einer geologischen KTB-Umfeldkarte 1:10 000. – KTB-Report 89-3: 10–23; Gießen.

WAGNER, G. A. (1968): Fission track dating of apatites. – Earth Planet. Sci. Lett. **4**: 411–415; Amsterdam.

– (1969): Spuren der spontanen Kernspaltung des Uran-238 als Mittel zur Datierung von Apatiten und ein Beitrag zur Geochronologie des Odenwaldes. – N. Jb. Miner. Abh. **110**: 252–286; Heidelberg.

– (1972): The geological interpretation of fission-track-ages. – Trans. Amer. Nucl. Soc. **15**: 117.

– (1988): Apatit fission-track geochrono-thermometer to 60 °C: Projected length studies. – Chemical Geology (Isotope Geoscience Section) **72**: 145–153; Amsterdam.

– (1990): Apatite fission-track dating of the crystalline basement of Middle Europe: Concepts and results. – Nucl. Tracks Radiat. Meas. **17**: 277–282; Oxford.

WAGNER, G. A., I. MICHALSKI & ZAUN (1989): Apatit fission track dating of the central european basement: Postvariscan thermo-tectonic evolution. – In: EMMERMANN, R. & J. WOHLENBERG (Hrsg.): The German Continental Deep Drilling Program (KTB). – Springer, Berlin, Heidelberg, New York, p. 481–500.

WAGNER, G. A. & P. VAN DEN HAUTE (1992): Fission-track dating. – 285 p., Enke, Stuttgart.

WIRTHMANN, A. (1961): Zur Geomorphologie der nördlichen Oberpfälzer Senke. – Würzburger Geogr. Arb. **9**: 41 p.; Würzburg.

Addresses of the authors: R. BISCHOFF, Max-Planck-Institut für Kernphysik, Postfach 103980, D-6900 Heidelberg, Prof. Dr. Dr. h. c. A. SEMMEL, Theodor-Körner-Str. 6, D-6238 Hofheim a. Ts., Prof. Dr. G. A. WAGNER, Max-Planck-Institut für Kernphysik, Postfach 103980, D-6900 Heidelberg.

Z. Geomorph. N.F.	Suppl.-Bd. 92	145–157	Berlin · Stuttgart	Mai 1993

Geomorphological and Geoecological Processes in the Mountain Forest Steppe of Northern Mongolia

by

CHRISTIAN OPP, Leipzig, and HEINER BARSCH, Potsdam

with 4 photos, 3 figures and 1 table

Summary. In the course of the experiment GEOMON 89/90 forest steppe ecosystems in North Mongolia were investigated. Results showed that spectral signatures of the vegetation reflect the spatial moisture distribution and by that also the structure of soil and relief. All investigated sites showed degradation features whose main reasons are overpasturing, faulty irrigation and a resulting soil compaction and soil salification.

Résumé. Dans le cadre des expériences GEOMON 89/90 on a reconnu des ecosystèms pour les forêts-steppes au nord de la Mongolie. Il en résulte que les signatures spectrals de la végétation dans des régions arides figurant le régime spatial des pluies et encore les structures du sol et du relief. Tours les emplacements qui sont enquêtés montrent des dégradation à cause de la surpaturation et l'arrosage incorrect. Enfin se montre une compression du sol et la salification du sol.

Zusammenfassung. Im Rahmen des Experiments GEOMON 89/90 wurden Waldsteppen-Ökosysteme in der Nordmongolei erkundet. Dabei ergab sich, daß in ariden Gebieten spektrale Signaturen der Vegetation die räumliche Feuchteverteilung und damit die Strukturen von Boden und Relief abbilden. Alle untersuchten Standorte wiesen Degradationserscheinungen auf, als deren Ursachen Überweidung, fehlerhafte Beregnung sowie die nachfolgende Bodenverdichtung und Bodenversalzung anzusehen sind.

1 *Questions and themes*

In 1990, German geographers were included in the international remote-sensing experiment GEOMON 89/90 in Mongolia. Within this framework the investigations made had the aim:

- to check a combination of terrestrial investigations tested in Central Europe, East Europe, and in the Caribbean (BARSCH, MAREK, WEICHELT & GEBHARDT 1990) and to examine remote-sensing methods for their applicability under the conditions of the extremely continental steppe-climate of northern Central Asia,
- to map site-characteristics – especially features of relief, soil and vegetation – of Central Asian ecosystems, their condition and spatial differentiation in an area of about 1.200 km^2,

0044-2798/93/0092-0145 $ 3.50

- to say sth. about the influence of agricultural use and possible changes in the climate on the ecosystmes of northern Mongolia and, resulting from that,
- to show possibilities and limits of the present as well as an extended agricultural use of northern Mongolia.

2 *Test region and test areas*

The test region "Zagaan Tolgoj" in the north of Mongolia is situated on the south side of the SW-NE-striking Selenga mountain range between the rivers Selenga and Orchon (cf. Fig. 1). It includes the intramontane Enchtal-basin, its flood-plain and the bordering pediments.

The Selenga mountains are a low mountain range with maximal altitudes of 2,000 m NN. Like other transbaikalian mountains it has repeatedly been ruptured and imbricated in the collisional belt between the North-Asian and the East-Asian plate since the proterocoicum. It received its today's morphostructure from ruptural-tectonic processes and from denudation processes in the mesocoicum and cenocoicum, which set in simultaneously. Old denudation forms were exposed to an extensive weathering and the production of periglacial debris during the pleistocene and holocene. In addition, there were winderosional and accumulation processes of the fine material and intensive linear and/or sheet erosion processes.

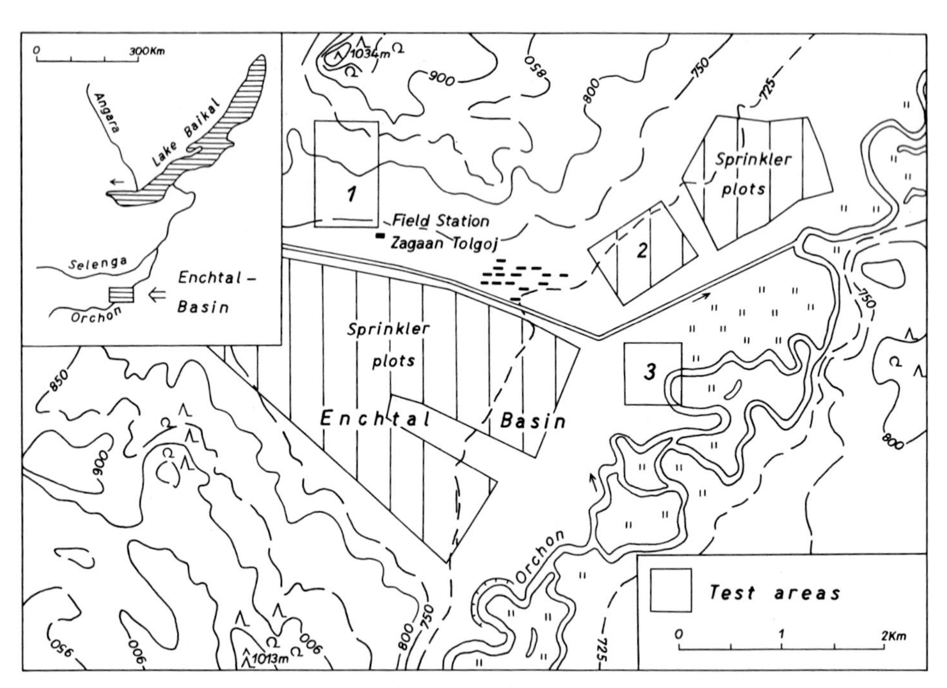

Fig. 1. Test region Zagaan Tolgoj.

Table 1 Mean values of selected climatic data of the Station Barun-Chara (Mongolia) after 40 years of measuring:

Mean annual temperature	− 1.9 °C
Mean July temperature	+19.0 °C
Mean January temperature	−26.0 °C
July maximum	+40.0 °C
January minimum	−47.2 °C
Days free of frost	98
Annual amount of precipitation	296.0 mm

The extremely continental climate of the present shows big daily and annual amplitudes of the air temperature, a low atmospheric humidity, low annual precipitations, long cold winters with little snow and short hot summers.

In Table 1 there are some data from the meteorological station Barun-Chara (807 m NN), which lies nearest to the test-region.

Under these climatic conditions the cultivation of agricultural crops, e.g. grain, is possible only because about 90% of the annual precipitation falls in periods of baric troughs over East Siberia during the warmest months (May to September), supplying 95% of the annual Orchon-runoff. Nevertheless, in dry years, evaporation is high and has to be made up for by additional irrigation.

In this region people used to farm land of lamaseries already in the past. Evidences of irrigation by canals go back to at least the 17th century. But in all this time the land has repeatedly been cultivated and fallen waste again. Additional irrigation has been attempted since the beginning of the 1950s. But the 50s with a lot of precipitation were followed by the 60s with low precipitation and sprinklers had partly become (Opp 1992). For this reason most agriculturally used areas in the test-region Zagaan Tolgoj had to be given up in 1962. From 1976 to 1978 the Russian irrigation system "Fregatt" was installed on the state farm Zagaan Tolgoj. It made agriculture possible by taking water from the river Orchon.

After extensive preliminary explorations in the test region, 7 test-areas along a transsect of the river Orchon up to the water shed of the Selenga mountains were chosen. Three of them are to be introduced here (cf. Fig. 1):

1: Proximal pediment
   incline: 4–9 °
   height: 760/790–840 m NN
2: Distal pediment
   incline: 1–2 °
   height: 710–745 m NN
3: Lowland terrace and flood plains of the Orchon height: 698–705 m NN

3 *Investigation methods*

Within the test-areas a detailed landscape analysis was carried out. Besides the usual ways of the marking of the vegetation- and soil cover, soil samples were taken to determine soil standard data (grain size, humus content, C/N, CEC, and pH) and selected material contents. On some test-sites undisturbed soil samples were taken with the help of 250 cm^3-cylinders in six parallels to determine specific soil-physic parameters (pore volume, pore-size distribution, soil density, saturated vertical water conductivity) which, e.g., allow statements about natural and man-made soil degradation processes. The penetration resistance of all investigated soil horizons was measured with the help of a pocket-penetrometer. On some of the test-sites the current soil temperatures were measured.

On 15 test-sites calibration values for the interpretation of remote sensing data were terrestrially measured with the help of a multi-channel-radiometer which registers signals in the VIS (visible light), NIR (near infra-red), SWIR (shortwave infra-red) and the TIR (thermic infra-red) spectral ranges. Moreover, planes flew on 7 tracks with a relative altitude of 200 m (in the Enchtal-basin) up to 300 m (in the mountains). The relation between the spectral radiation intensity in the red range of the visible light and in the near infra-red reflected the normalized vegetation index (NDVI). The water index could be derived from the relation between the values of the near and the short-wave infra-red. Unfortunately, there were only multispectral photographs available for the following extrapolation from these data on all test-sites (Barsch, Itzerott, Schwarzkopf & Loeper 1992). For this reason the digital image processing and the interpretation of the results had to be concentrated on the vegetation index. It shows the vitality of the plant cover and indirectly also the soil moisture, as in spring the soil moisture influences the development of the plants. Chlorophyll pigments of vital plant stocks absorb the red part of the visible light by far more than those of plant stocks under stress. The same applies to the reflection of the leaves in the near infra-red. In high turgor it is by far more intensive than in a low one.

4 *Results*

Test-area 1 lies in a proximal pediment with concave-convex slopes. Its inner structuring is clearly reflected by the spatial arrangement of the spectral signatures. Extremely dry ramps in the upper part (Photo 1, above) that are nearly free of vegetation and have a very low vegetation index (black) stand out from the moderately some wetter furrows with a closer vegetation cover (white). They lead in alluvial fans forming one wider accumulation plain. It is only little wet, to be seen in the mean vegetation indexes (middle and light grey). All in all, NDVI-values are very low as dry grass covers at most 30% of the soil, in the upper part of the pediment even less than 10%. Radiation temperatures in the thermic infra-red of 55 to 65 °C (measured on 25 June, 1990 between 12 a.m. and 1 p.m.) show an extraordinarily little flux of latent heat on the soil surface.

The spatial pattern of the ramps, furrows and accumulation plains proves that periglacial slope denudation processes as well as postglacial sheet erosion and ditch erosion (caused by heavy falls of rain) formed the relief. The loamy substratum does not consist of erosion material only, but it receives its silt also from aeolian material, which, with its fine pores, causes a relatively slow vertical flux of moisture and hereby furthers soil erosion. Reduction of the hydraulic pressure on the side of the

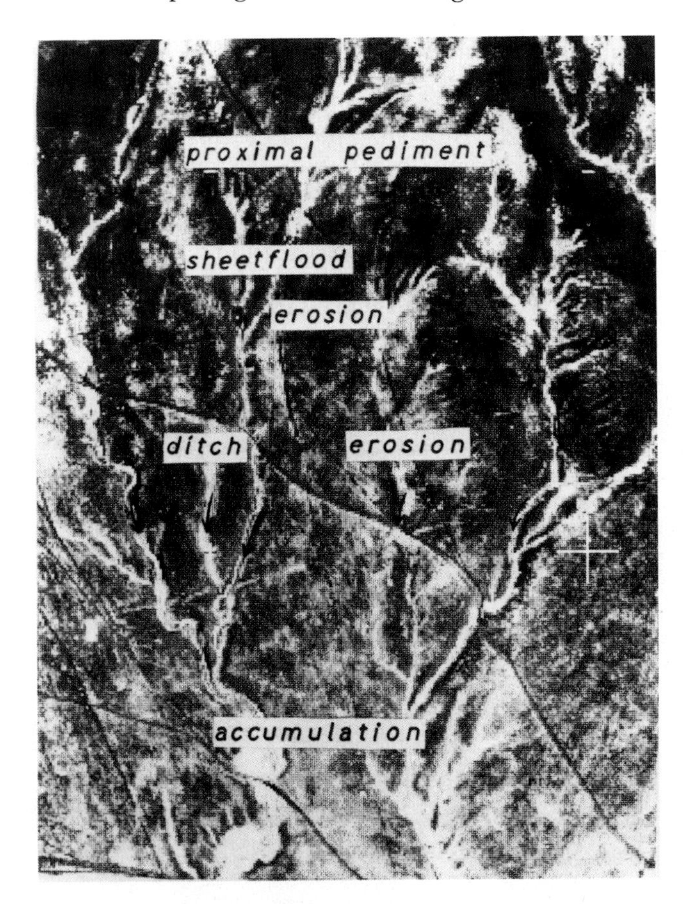

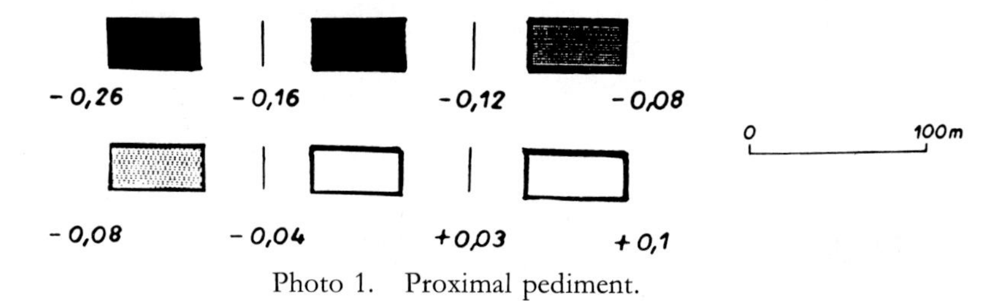

Photo 1. Proximal pediment.

furrows and on the alluvial fans lead to discharches of water. For this reason the canopy of the dry grass is closer there.

Dark Castanozeme soils of little thickness occur in the area of the proximal pediments (OPP & TULGAA 1992). The soils of the furrows differ from the ones of the ramps only in the deeper humus penetration. The superficial flush of fine material during heavy rain and especially during thaw leads to an relative enrichment of the upper soil with sand. The resulting little elasticity of the upper soils causes an enormous soil strength due to physical stress. Natural stress, caused by the splash effect of the raindrops, as well as stress caused by man, e.g. the grazing of the lifestock, are responsible for the degradation of the soil and vegetation cover.

Photo 2. Distal pediment.

During the investigation period the upper soils of these test-sites were compressed to such a high extent that made cylinder samples as well as the penetration of the pocket-penetrometer into the soil impossible.

In contrast to that there was hardly any or only little soil compaction on the agriculturally used Castanocemes in Test-area 2 on the distal pediment (Photo 2). Striking macromorphological differences in the structure are more a result of the decalcification of the upper soil than of the technical influence. On the margin of sprinkler plots the irrigation splash causes a differentiation in the texture, esp. a shift of silt particles. Areas which had not been irrigated for some time and tracks made by the wheels of the sprinklers often showed crustifications and following salifications in the upper soil (Opp 1992).

Wheat, partly also lucern, is cultivated. But the bad condition of the sprinklers is clearly to be seen in the spatial distribution of the vegetation indexes (Photo 2). Sprinkler circle 6 does hardly stand out from the not-irrigated surroundings, the same is true for sprinkler circle 5. In both cases NDVI attains only relatively low values (in the picture dark grey and black), which are similar to those of the dry grass in Test-area 1.

The sprinklers in the circles 2, 3 and 4 are in a better working order. Here vegetation indexes are relatively high, similar to the wet meadows of the neighbouring valley flood plain. The radiation temperature in the thermic infra-red measured at 12 a.m., on 23 June 1991, fluctuates between 25 and 50 °C depending on the density of the canopies. An estimation of the biomass derived from spot checks on test-sites in the sprinkler circles comes to results between 9.1 decitonnes/ha in circle 6 and 17.1 decitonnes/ha in circle 1.

This shows that the inner structuring of these morphologically homogenous accumulation plains is determined today by the more or less functioning irrigation. It influences the distribution of wet areas in the loam layer, which covers the underlying middle terrace debris. Towards the Orchon there first follows the flood plain of a little tributory. Only then we come to the low terraces and the flood plain of the Orchon itself (Test area 3). In this area pediment and fluvial deposits are partly dovetailing. Nevertheless, a clear distinction between low terraces and flood plains on the one side and distal pediments on the other side is possible. The younger forms were produced by the meandering river. That the river changed its course again and again with the rising water-level in summer and during the ice-drift in spring is reflected in a sequence of old river-branches, bank barriers (natural levees), low terraces and flood plains that stand out from the plant cover (Photo 3). The wet meadows in the flood plains show high vegetation indexes (white) in comparison to the bank barriers in dune formation (light grey). There Achnatheretum splendens covers at most 50% of the soil surface. Behind the bank barriers there is an alluvial plain which lies dry today and is locally covered with salt grass. The canopy is accordingly thin, to be seen in the low vegetation index (dark to black). In the direct vicinity of the river one has to look at the same NDVI-value differently. Here they mark flood areas on a slide slope (also dark grey to black). Narrow bank barriers, whose grass juts out of the water, are marked out (light).

The different density of the canopy again is combined with different radiation temperatures. During the flight on 27 June, 1990, temperatures in the micro-wave range rise above the Orchon from 160 °K to 210 °K in the inundation belt of the

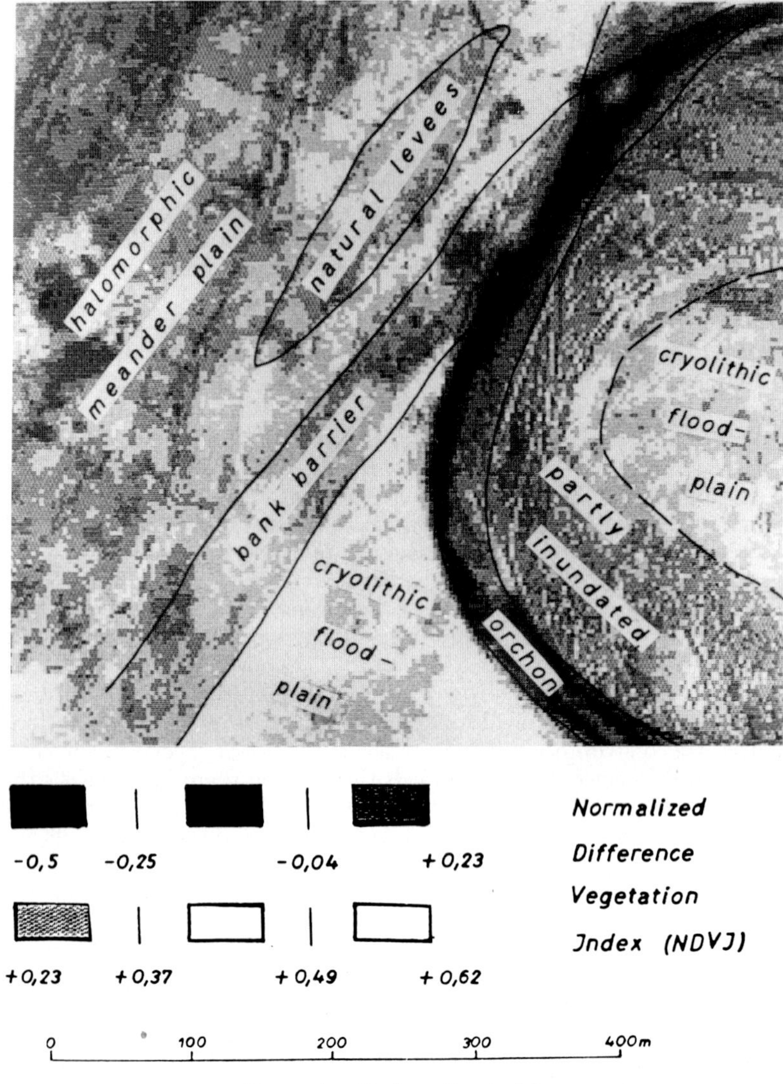

Photo 3. Orchon flood-plain.

slide slope, to 260 °K on the wet meadows and to 280 °K on dry grass. This shows how much soil and vegetation moisture in the flood plain of the Orchon vary.

Soils outside locally encountered salt crusts are Solonchaks. By far the biggest part of the areas between the irrigated fields and the Orchon (cf. Photo 3), which were first cultivated and then given up again, is mostly covered today with *Elymus*

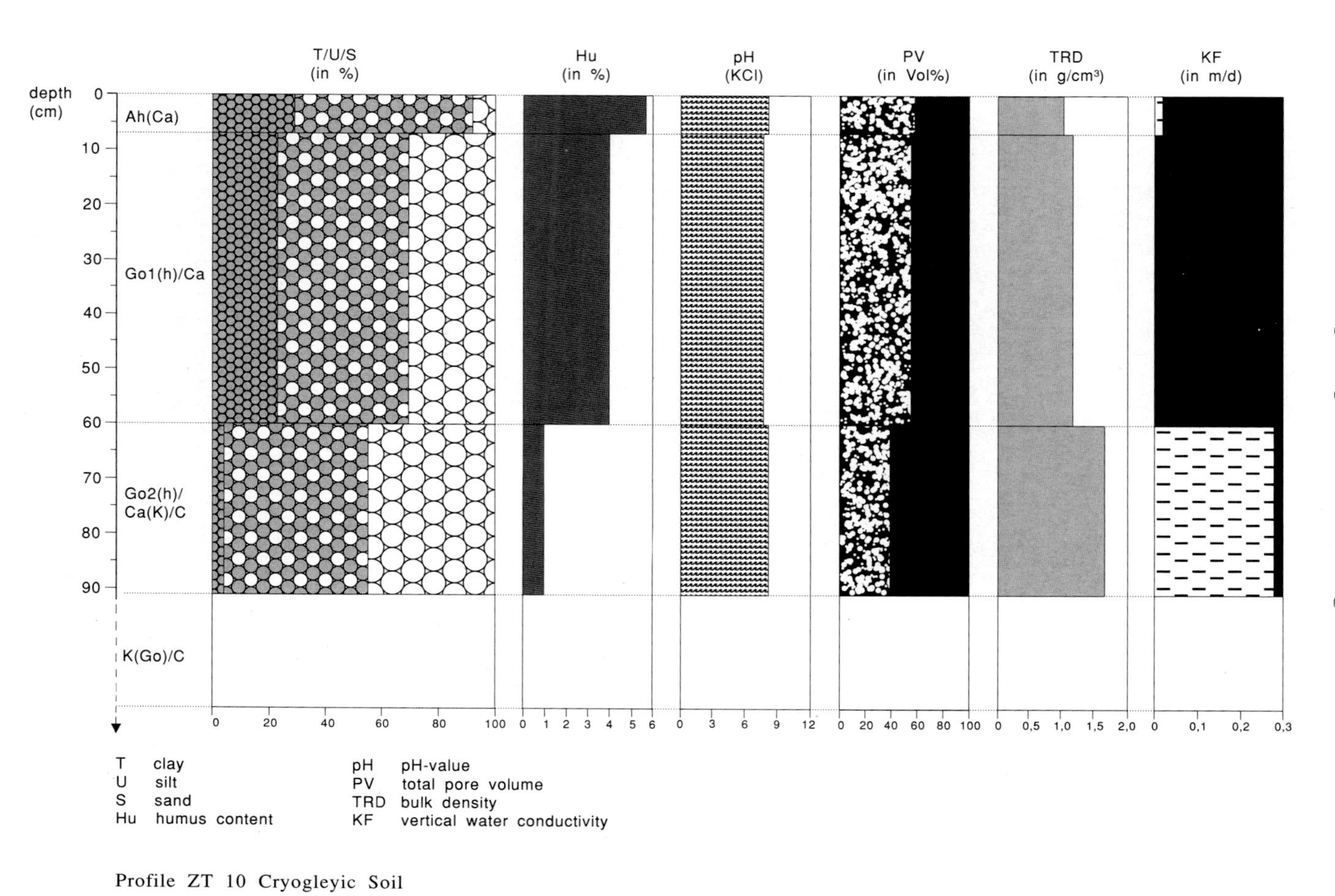

Profile ZT 10 Cryogleyic Soil

Fig. 2. Soil standard data and soil physical data of a Cryogleyic Soil at Zagaan Tolgoj, Northern Mongolia.

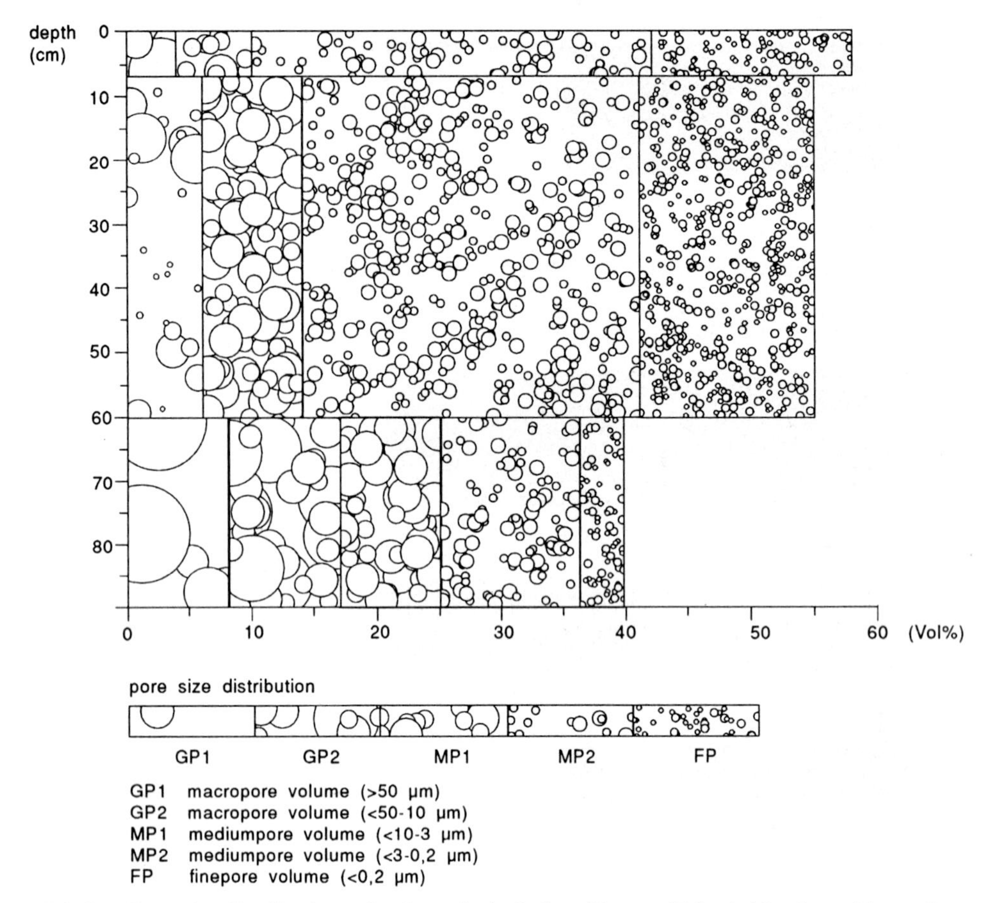

Fig. 3. Pore size distribution of a Cryogleyic Soil at Zagaan Tolgoj, Northern Mongolia.

chinensis. It is episodically used as grazing land. The typical soil of these sites is a Solonized Castanoceme.

The wet meadows near the Orchon are used for cutting grass. Though the pore volume in the upper soils of these sites is bigger than 50% (cf. Fig. 2), it is striking that their macro pore volume content is very small. The percentage of quickly draining coarse pores is zero (cf. Fig. 3). One reason for this seems to lie in the permafrost in the subsoil (OPP 1991). Measurements of the soil temperature on 24 July, 1990 showed that, whereas there were 16 °C on the surface, temperatures in 90 cm depth were below 0 °C. The permafrost-table at that time lay in a depth of more than 90 cm. Photo 4 shows a Cryogleyic Fluvisol which is a typical soil of these sites. The glacially pushed cryoturbate curving-ups of the layers in 70–30 cm depth are distinctable. They are in all likelihood also responsible for the deformation of the

Photo 4. Cryogleyic Soil within the flood-plain of the river Orchon, Northern Mongolia.

upper soil. Systems of polygonal ice wedges were observed near the wet meadows. A differenciated distribution of the soil moisture of the ice wedges – wetter margin of the polygon, drier core of the polygon – is reflected in the different colouring of the grass. Besides that ice-wedge gaps of a width of 2 up to 20 cm are typical for a soil surface which was deformed by ice push and shrinking. The biggest ice-wedge gaps were found in the direct proximity of the Orchon.

In the digitalized aerial photograph (Photo 3) the meandering course of the Orchon, which mainly bases on an intensive side erosion and is additionally increased by the perma-frost, is clearly to be seen. As a result of the streaming of the Orchon against the opposite slope during the summer months, the ice in the soil there melts, causing the opposite slope lose part of its stability and slide into the river. This "fluvio-cryo-thermoerosion" of the river bank is a characteristic phenomenon of many rivers in northern Central Asia, as permafrost south of the perma-frost border of Siberia and out of periglacial belt of high mountains is mostly only found in the wet lowlands and floodplains of the river valleys.

5 *Conclusions*

Under the conditions of the extremely continental climate in north Central Asia, a combination of terrestrial and remote-sensing methods is, from a methodical point of view, a suitable way of getting geoecological as well as geomorphological information about causal connections between the spatial pattern of the vegetation, soil moisture, substratum, and relief, which are reflected in the vegetation index. The distribution of moisture is in this case the limiting factor to the growth. Where irrigation is not possible, growth depends on the position of the test-site in the relief and on the character of the sediment or denudation cover there. I.e., the vegetation cover, which makes the interpretation of remote-sensing data nearly impossible for humid areas, is in arid areas a key to the extrapolation of single geomorphological data on a whole region. Irrigated areas only show the man-made landscape patterns. Geoecological conclusions can in all these cases be drawn from digital remote-sensing data.

Under a regional aspect it must be emphasized that all test-sites, also those which are only extensively used, show a number of degradation features in the soil and plant cover. It is true that in comparison to fields in Central Europe agriculturally used soils in Mongolia are not hardly compacted, but there are, on the other hand, superficial crustifications (Test-area 2), caused by not-functioning sprinklers or by wrong irrigation times. These crusts decrease the yield as the infiltration of the irrigation water or rain is hindered, causing this way surface salifications.

Over-pasturing is one of the biggest problems on the pediments (Test-area 1). As these sites are – because of the little plant cover – much eroded, an elastic soil cover is not likely to develop there. Since the 1950s, temperatures in Mongolia have steadily risen, increasing aridity in this area. This is reflected in increased evaporation, a further decline in the density of canopies, a reduction of the soil fauna and an increase in erosional processes (OPP 1991).

The relief in Test-area 3 has no eroding effect on the sites there, but the productivity of the meadows is limited, as it depends very much on the water supply. Fig. 3 shows a soil structure with no or hardly any macro pores, making the drainage of the water impossible. The Solonchaks and the Solonized Castanocemes between fanglomerat formations and the Orchon terrace-residuals are especially impermeable and enormously compacted. On the whole, it has become obvious that problems of the use of geoecologically little investigated areas like Mongolia can only be solved with the help of a combination of terrestrial and remote sensing methods. International and interdisciplinary work teams are of great use there. For further investigations, the global climatic change has to be considered by these work teams as well as the present changes in the Mongolian economic system, to provide – on a longerlasting and technically better basis – the geoscientivic basis for a more effective use of the steppe-ecosystems.

Acknowledgements

The authors are especially grateful to their colleagues Dr. BÖTTCHER (Magdeburg), Dr. ITZEROTT, Ing. LOEPER, Dr. SCHWARZKOPF (Potsdam), Dr. TULGAA, Dipl.-Ing. ENKTUVSHIN (Ulan-Bator), who were substantially included in the field-research and in the digital image processing.

References

BARSCH, H., K.-H. MAREK, H. WEICHELT & A. GEBHARDT (1990): Yield Prognosis by the Productivity Criteria Using Spectral Signatures in the VIS, NIR and TIR ranges. – In: SHAHROKI, F., N. JASENTULIYANA & N. TARABZOUNI (Eds.): Space commercialization. Satellite technology. – Progress in astronautics and aeronautics **128**: 214–226.

BARSCH, H., S. ITZEROTT, U. SCHWARZKOPF & H. LOEPER (1992): Maßstabswechsel bei der Kennzeichnung der Vegetation in Trockengebieten nach Fernerkundungsdaten. Ergebnisse und Probleme des Interkosmos-Experiments GEOMON 89/90. – Peterm. Geogr. Mitt. **136,1**: 5–16.

OPP, CH. (1991): Erste Ergebnisse bodenphysikalischer, bodenchemischer und landschaftsökologischer Untersuchungen in der Mongolei. – Mitt. Dt. Bodenkundl. Ges. **66 I**: 197–200.

– (1992): Investigations of natural and land-use related processes of land and soil degradation on the lower Orchon river. – In: 2. Internationales Symposium „Erforschung biologischer Ressourcen der Mongolei" in Deutschland vom 25. 3.–30. 3. 1992. Thesen zu den wissenschaftlichen Beiträgen. 99–100; Halle.

OPP, Ch. & CH. TULGAA (1992): Results of pedological and landscape-ecological investigations in Mongolia. - In: 2. Internationales Symposium „Erforschung biologischer Ressourcen der Mongolei" in Deutschland vom 25. 3.–30. 3. 1992. Thesen zu den wissenschaftlichen Beiträgen. 101–102; Halle.

Addresses of the authors: Dr. CHRISTIAN OPP, KAI e.V., Funkenburgstraße 24, D-O-7010 Leipzig, Germany. Prof. Dr. HEINER BARSCH, Department of Physical Geography and Landscape Ecology, University of Potsdam, PF753, D-O-1574 Golm, Germany.

Z. Geomorph. N.F.	Suppl.-Bd. 92	159–172	Berlin · Stuttgart	Mai 1993

Monitoring of Geomorphological Processes for a Sustainable Range Management in Kenya

by

RÜDIGER MÄCKEL, Freiburg i. Br., and DIERK WALTHER, Netphen

with 4 figures and 1 table

Summary. The investigation in the southwestern part of the Marsabit District of Kenya considers different stages of degradation due to human impact and drought. The monitoring of geomorphological processes for a sustainable range management is based upon the comparative observation of several test sites from 1980 to 1991. The main effect of aquatic and aeolian erosion processes is the loss of top soil and nutrients and the drying up of the habitat. Subsequently, grazing plants disappear and less palatable species invade the areas (invaders) and/or resistant and avoided plants increase (increasers). The reduction of the protective basal vegetation and the replacement of perennial plants by annuals accelerates soil erosion.

Recommendations for the management of the range areas include an erosion handbook in order to recognize and to avoid erosion damages as well as rehabilitation measurements and range control.

Zusammenfassung. Die Untersuchungen im Südwesten des Distrikts Marsabit, Kenia, zieht verschiedene Stadien der Landdegradierung in Betracht, die auf menschliche Einwirkung und auf Dürre zurückzuführen sind. Die langfristige Erfassung und Bewertung (monitoring) der geomorphologischen Prozesse für eine nachhaltige Weidebewirtschaftung (range management) stützt sich auf vergleichende Beobachtungen zahlreicher Testflächen zwischen 1980 und 1991. Die Erosionsprozesse durch Wasser und Wind verursachen vor allem einen Verlust an Boden und Nährstoffen sowie eine Austrocknung des Wuchsortes. Als Folge verschwinden immer mehr Weidepflanzen, während weniger schmackhafte Arten eindringen, beziehungsweise widerständige und gemiedene Pflanzen zunehmen. Der Verlust der schützenden Vegetationsdecke am Boden und der Ersatz perenner Arten durch annuelle beschleunigt die Bodenerosion. Die Empfehlungen für das Management der Weidegebiete enthalten ein Erosionshandbuch, mit dessen Hilfe Erosionsschäden erkannt und vermieden werden sollen, ferner Maßnahmen zur Rehabilitation degradierter Flächen und zur geregelten Beweidung.

1 *Introduction*

The implications of landforms and geomorphological processes for range management were discussed in different investigations on the semi-arid and arid lands of Kenya (MÄCKEL & WALTHER 1984, 1988 and 1992, MÄCKEL et al. 1989, TOUBER 1991 and HERLOCKER 1992). The reason for including geomorphology in range

0044-2798/93/0092-0159 $ 3.50

research is the fact that range potential is directly influenced by relief, or is at least interrelated with geomorphological factors. There is, for example, an effective influence of relief and soil erosion on the water regime (infiltration, storage capacity) and on soil fertility, on vegetation cover and composition and on recuperation. Other examples show the dependence of accessibility and vegetation use on relief and stoniness. However, most research work on these topics has been restricted to relatively short periods of observation. In order to classify range conditions for management purposes long-term observations are necessary. This paper discusses the research results of a 12-year observation period at the same test sites. The results are used for a monitoring programme in connection with the Range Management Handbook of Kenya (Schwartz et al. 1991).

The test sites for monitoring the geomorphological processes were selected in the nomadic pastoral areas of the Marsabit District of Kenya (Fig. 1). These areas are characterized by a moderately hot, predominantly arid tropical climate with two very short, mainly subhumid seasons. The mean annual rainfall is 240 mm for the climate station at Korr. About two thirds of this amount fall in the first (or large) rainy season between March and May, one third falls in the second (or small) rainy season between October and December. The high variability of precipitation is shown by the different amounts of rainfall and by its temporal and regional distribution. Quite often the rainfall is concentrated on short, heavy rains. There may also be a shift or total lack of rainfall. The heavy rains are not favourable to vegetation growth because of the resulting rapid surface flow and low depth of penetration. However, they give effect to various geomorphological processes such as sheet wash or linear erosion with subsequent impact on the vegetation cover and composition and land use.

The vegetation of the research area consists mainly of open thorn bush – dwarf shrub associations with annual or perennial grasses and herbs (Herlocker 1979, Schultka 1991). Where surface water is retained for a while, a sumptuous vegetation cover appears, as shown at Ballah 30 km north of Korr, with dense Sorghum arundinaceum stands observed after the rainy season in 1987.

The test sites were first studied by the authors in 1980 and 1981 (dry and wet season) and revisited several times up to 1992. A multistage remote sensing and field research method was used to document and evaluate the range potential and land degradation due to overutilization and/or climatic hazards. The research results of the first campaigns in the Marsabit District (1980 to 1987) were published in several papers and books (Mäckel & Walther 1984, Mäckel 1986, Mäckel et al. 1986; Walther 1987, Mäckel & Walther 1988, Dreiser et al. 1989).

This paper covers the continuation of research in 1991 and the comparison with the observations of the former years. The studies put an emphasis on geomorphological processes and their influence on soils and vegetation changes and present recommendations for the rehabilitation of degraded sites.

2 *Geomorphological Processes in the Southwestern Part of the Marsabit District*

Seventy test sites were documented in the 1980, 1981 and 1987 campaigns. Their sizes range between 50 and 100 ha and they are more or less homogeneous consid-

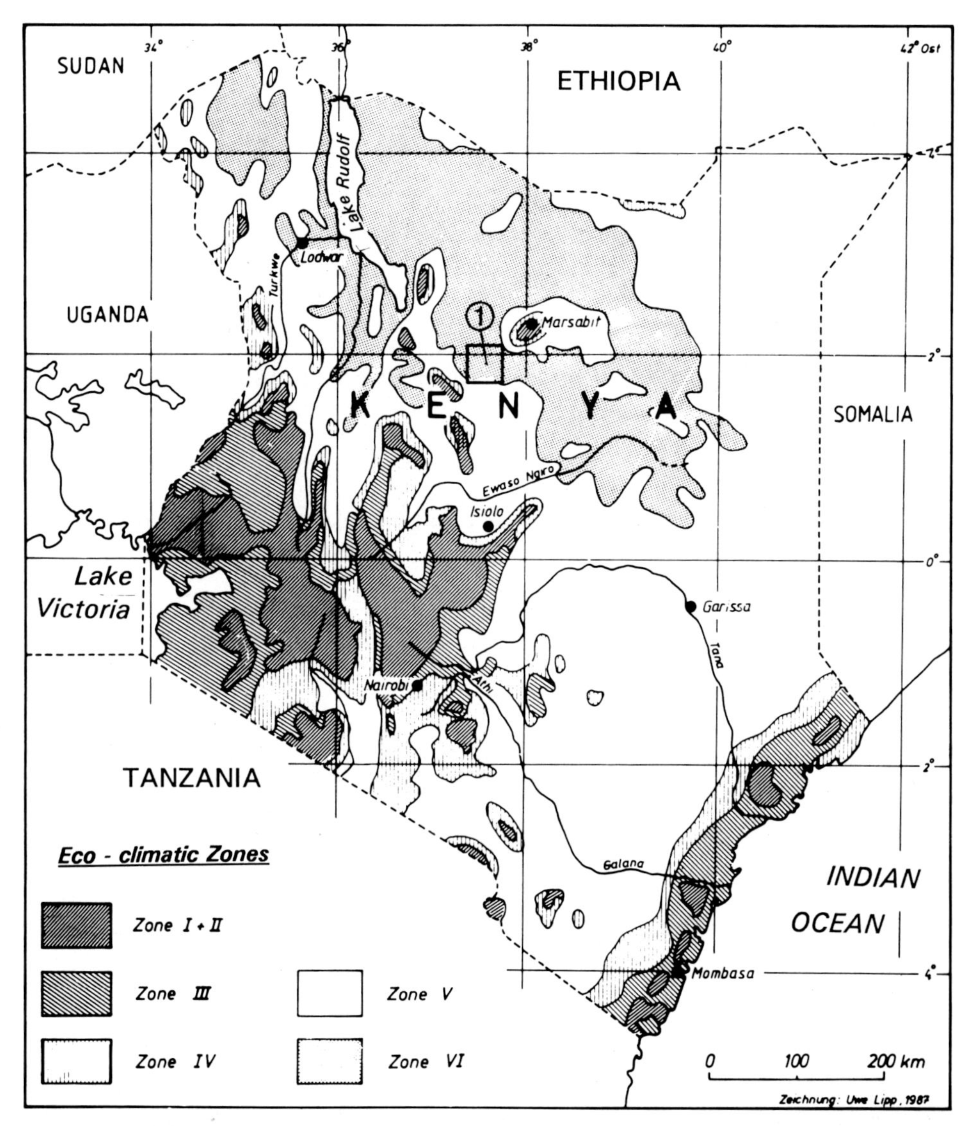

Fig. 1. Location of the study area in Kenya based on the map of eco-climatic zones by PRATT & GWYNNE (1978). 1: Korr area, southwestern Marsabit District. Eco-climatic zones I+II: Altitudinal zones and humid to dry-humid with forest and woodland. III: Dry-sub-humid with (semi-)evergreen bushland and savanna. IV: Semi-arid with dry savanna and Acacia woodland. V: Arid with thorn bushland and thicket. VI: Very arid with desert thorn scrub.

ering relief, soil type and vegetation composition. The test sites belong to different landscape types, which were defined according to their respective relief conditions and morphodynamics, substratum and soils, and vegetation. The test sites revisited in 1991 represent critical areas of fluvial and aeolian erosion, as they are in the vicinity of waterholes or permanent settlements such as Korr (MÄCKEL & WALTHER 1984).

The test sites belong to the following landscape types:

1. Stony hill slopes with open *Commiphora thorn bush and Indigofera* dwarf shrub (test site 53),
2. Hillfoot plains with open *Commiphora* thorn bush and *Duosperma* dwarf shrub (test site 29),
3. Plains with *Indigofera* dwarf shrub and Dactyloctenium grassland (test site 54),
4. Interfluve plains with *Indigofera-Heliotropium* dwarf shrubs (test site 38).

2.1 *Hillfoot Plains with Open Commiphora Bush and Duosperma Dwarf Shrub*

The hill foot plains of the Korr area are depositional plains. The sediments, mainly sands, were washed down from the adjacent hill slopes or brought into the foot

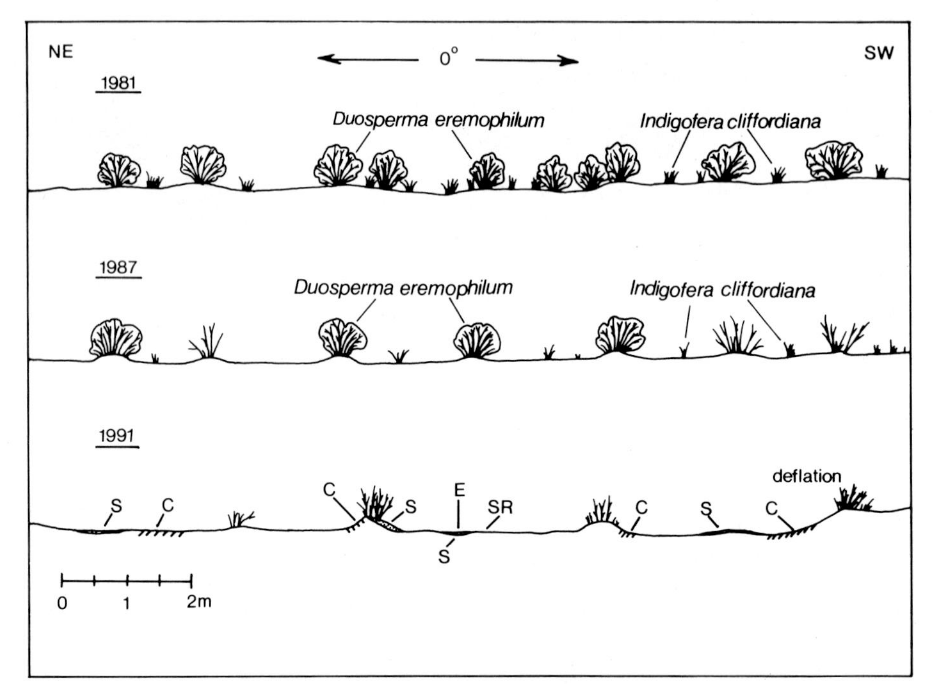

Fig. 2. Different stages of degradation of a dwarfshrub formation (*Duosperma eremophilum – Indigofera cliffordiana*) about 4 km south of Korr between 1981 and 1991 (C = Soil crust, E = Erosion, S = Sand, SR = Sand ripple marks).

plains by discontinuous laggas (seasonal water courses). During dry phases the aeolian activities increased and led to dune development on top of the fluvial and colluvial sediments. In a wetter phase, finally, the dunes were flattened by erosion and stabilized by a thorn bush – dwarf shrub vegetation.

The selected test site 29 represents an advanced stage of impairness (moderately to heavily impaired). The intensive sheet wash is noticeable in the height of the growth heaps of *Duosperma eremophilum*, which rise 15 to 20 cm above the present-day surface (Fig. 2). Another indicator of intensive erosion are the exposed roots of *Indigofera spinosa*. The rate of down-wash can be estimated by young invaders such as *Jatropha dichtar*. Their roots are already exposed 10 to 30 cm above the surface within the last five years. This implies a down-wash rate of at least 2 cm, at the most 6 cm per year according to the geomorphodynamics of the different test sites. However, taking into consideration the dry years of 1984, 1986 and 1991 down wash must have been considerably higher in the rainy seasons of the remaining years. In 1991 daily rainfall was ranging between 0.3 and 9.3 mm in the first rainy season (altogether 14.5 mm) and between 0.2 and 7.2 mm in the second rainy season (altogether 14.0 mm). This means that no erosion effective rainfall occurred.

Compared with the first studies in 1980 and 1981 there is an increase in aeolian activities. More wind ripple marks and small dunes behind dwarf shrubs were observed. Especially near the lagga the reactivation of dunes is remarkable. Here the Indigofera dwarf shrubs rise with their roots 10 cm above the deflation surface.

The geomorphological processes, especially sheet flood and deflation, influence the ecological conditions of the habitat. Dwarf shrubs with exposed roots do not survive, especially during prolonged dry seasons. These plants usually have a lateral root system near the surface in order to use the sparse rains which only penetrate the upper 10 to 20 cm of the soil. Once exposed, there is no chance for them to adapt their root system in order to get moisture. The plants then dry out and die.

Duosperma eremophilum seems to be a resistant plant concerning sand accumulation by wind. At test site 29, however, the vitality of *Duosperma eremophilum* on the growth heaps was weakened. In 1987 about 60% of the plants were already dead because of dryness. In 1991 *Duosperma eremophilum* had partly recovered (Table 1).

Quite obvious is the change in the composition of species in the tree and bush stratum. Most of the original bushes and trees were only represented by a few remaining specimen in 1987 and had totally disappeared by 1991 such as *Salvadora persica* or *Cordia sinensis* (Table 1). In contrast to the decline of the original species there was an increase of invaders, which are avoided by domestic livestock, for example *Jatropha parvifolia* and *Euphorbia cuneata* (which did not exist in 1981), *Acacia nubica* and, on the sandy lagga floor *Calotropis procera* (Table 1). In 1991 there was a remarkable decrease of these invaders. The reason might be the use of some of the shrubs (for example *Jatropha* species) under grazing pressure and/or the destruction of the plants by trampling. The site is only 4 km south of Korr and many animals pass the area to reach the permanent water holes. *Acacia reficiens*, however, shows an increase, in particular in vital regeneration growth.

Striking changes took place in the dwarf shrub layer concerning the rapid increase and decrease of *Duosperma eremophilum* and *Indigofera* species (Table 1). In the grass/herb layer there was an increase of annual species and weeds in place of permanent grasses between 1980 and 1987. In 1991 no grasses and herbs were found.

Table 1 Changing composition of species at four test sites in the Korr area. The surveys were carried out in 1981, 1987 and 1991 (rainy season). There is a decrease of browsing/grazing plants and an increase of avoided plants (grazing weeds), which either derive from the original vegetation type ("increasers") such as *Acacia reficiens* or which have invaded from drier areas ("invaders") such as the *Jatropha* species. There is also an increase of annuals at the cost of perennial plants caused by heavy use and subsequent soil erosion.

Landscape type/test site	2/53			5/29			6/54			7/38		
Year of survey	1981	1987	1991	1981	1987	1991	1981	1987	1991	1981	1987	1991
Trees/shrubs	5.4	4.1	n	2.7	6.0	2.1	2.2	6.0	1.2	0.2	0.2	0.2
Acacia nubica	–	–		0.2	1.0	0.9	0.2	1.2	0.0	–	–	–
Acacia reficiens	–	–		–	0.2	3.6	0.2	0.6	0.1	–	–	–
Acacia senegal	0.3	0.3		–	–	–	–	–	–	–	–	–
Acacia tortilis	0.4	0.4		–	–	–	0.1	0.3	0.0	–	–	–
Boscia coriacea	0.3	0.2										
Boswellia rivae	0.3	0.3										
Cadaba farinosa	–	–		–	–	–	0.2	–	–	–	–	–
Calotropis procera	–	–		–	0.6	–						
Commiphora flaviflora	0.5	–		} 0.6	–	–	} 1.0	0.5	1.1	0.2	0.2	0.2
Commiphora paolii	1.2	0.2										
Commiphora p/y	0.4	0.2										
Commiphora rivae	0.2	–										
Commiphora samharensis	0.3	0.3										
Cordia sinensis	0.5	0.3		0.3	0.1	–	–	–	–	–	–	–
Euphorbia cuneata	–	–		–	2.0	0.1	–	2.4	–	–	–	–
Grewia bicolor	0.5	0.3										
Jatropha dichtar	–	–		1.1	2.0	0.2	–	0.2	–	–	–	–
Jatropha parvifolia	0.3	1.0		–	–	0.1	–	–	–	–	–	–
Maerua crassifolia	0.3	0.3		–	–	–	} 0.4	0.2	0.0	–	–	–
Maerua spec.	–	–		0.2	–	–						
Lycium europaeum				–	–	0.9						
Salvadora persica	–	–		0.3	0.1	–	0.1	–	–	–	–	–
Dwarf shrubs	2.2	1.3		14.2	6.5	30.8	21.0	10.5	13.6	6.0	17.4	16.8
Duosperma eremophilum	0.3	0.3		12.8	4.2	8.5	3.6	2.0	0.2	0.3	0.2	2.0
Heliotropium steudneri	–	–		–	–		–	–	–	3.7	3.0	0.0
Heliotropium albohispidum	–	–		–	–	1.5	–	3.0	5.4	–	0.5	7.7
Heliotropium spec.						0.1						
Indigofera cliffordiana	0.5	0.2		1.4	1.0	8.3	10.4	3.5	0.0	0.5	2.0	1.3
Indigofera spinosa	1.0	–		–	–	11.8	7.0	1.5	5.8	2.0	11.7	5.6
Sericocomopsis hildebrandtii	0.5	0.3		–	–	1.6	–	–	2.1	0.6	0.0	0.0
Solanum incanum	–	0.5		–	–		–	0.5	0.1	–	–	–
Barleria spec.								–	–	–	–	0.2
Grasses/herbs	n	0.9		n	2.5	–	31.0	11.0	1.3	n	5.1	0.6
Aerva javonica		–			0.2	–	–	–	–		0.7	0.0
Aristida adscensionis		–			} 2.2	–	1.0	9.0	0.2		2.7	0.1
Aristida mutabilis		0.3									–	–
Oropetium minimum											–	0.1
Dactyloctenium aegyptium		–			–	–	} 30.0	1.0	0.2		–	–
Dactyloctenium bogdanii		–			–	–					–	–
Enneapogon spec		–			–	–	–	1.0	–		–	–
Papalia lapacea		0.2			0.1	–	–	–	–		1.7	0.0
Grass, unknown									0.2		–	0.4
Herb, unknown									0.7			

n = no observation, values in % cover.

This fact again indicates overgrazing of this area, which damaged the grazing area especially in the poor rainy season (April/May 1991).

The reason for this down trend may be again the drastic effects of the geomorphological processes in connection with the climatic conditions and growing grazing pressure in the environs of Korr. The climatic conditions are characterized by short heavy rains which are of low use for the plant growth but effective for erosion allowing only a small number of camels to browse for up to 14 days.

As a result of the changes in plant composition the usable biomass dropped from 1100 kg/ha in 1980/81 to 550 kg/ha in 1987. Despite the increase of dwarf shrub species the biomass production was not higher in 1991 because of lack of grasses and herbs. For a possible recovery of the area (vegetation and soils) range management recommends control of the number of grazing days and animals.

2.2 *The Sedimentary Plains with Indigofera Dwarf Shrub and Dactyloctenium Grassland*

The level plains (below 1 ° inclination) are built up by fluvial and colluvial sediments and covered by flattened dunes. Accordingly, the top soil is composed of sand or loamy sand. With increasing depth the substratum changes to sandy loam. Test site 54 is located 1 km east of Hafare lagga, 11 km north of Korr. The first surveys in 1980 and 1981 showed a *Dactyloctenium* grassland with a coverage of 30% to 40% and a tree/bush-stratum composed of *Commiphora flaviflora, Maerua* spec. and *Acacia tortilis*. The dwarf shrubs are composed of *Indigofera cliffordiana* and *I. spinosa*. *Duosperma eremophilum* only occurs in wetter wash flows and shallow swales.

The vegetation surveys (Table 1) show the heavy impairness of the vegetation since 1980. The reason is the continuity of settlement in the Hafare area. Overgrazing, trampling and wood cutting took place and were harmful in the droughts of 1984 and 1991. Many plants died and the unprotected sites were subject to heavy wind erosion, as the wind blows along continuous wind passages (Hafare gap), and because of the frequent occurrence of dust devils. As a result special aeolian forms developed, such as deflation swales (depressions) and crusts, small dunes behind dwarf shrubs and ripple marks. Flattened and formerly stabilized dunes were reactivated. Heavy sheet wash works between the very open dwarf shrub stands and leads to a lowering of the surface and to an exposition of growth heaps (Fig. 2). Even at level places linear erosion forms develop, along which sandy material is transported in a discontinuous flow. The concentration on linear flow supports a further exposition and drying-out of the growth heaps until the protective vegetation on top dies of the lack of water. The reaction of the dwarf shrubs to this process is varied. While *Indigofera cliffordiana* and *Duosperma eremophilum* suffer or die, *Sericocomopsis hildebrandtii* and *Indigofera spinosa* seem to be quite vital. Finally, the intensified geomorphodynamics by water and wind favours the growth of resistant plants, which are frequently of a minor fodder value than the former plants. The important grass species of *Dactyloctenium* disappeared almost completely (Table 1). Among the dwarf shrubs there was an increase in *Heliotropium albohispidum* and *Sericocomopsis hildebrandtii* and also a recovery of *Indigofera spinosa*. With regard to the bushes only an increase in *Acacia refisciens* was observed (in 1987 and less in 1991), *Acacia nubica* was frequent at neighbouring test sites.

The usable biomass for test site 54 in 1981 was 1000 kg/ha (dry season) and 2900 kg/ha (wet season). In 1987 and in 1991 the biomass was below 500 kg/ha. The degradation of the grazing ground might be connected with poor rainfall. This would be the case, if the composition of the species were similar, even if the percentage of cover were less than during good rainfall seasons. However, the disappearance of certain species, especially of permanent grasses, point to high grazing pressure, which – together with unfavourable climatic conditions and intensive geomorphological processes – have caused the heavy impairness of the site. Regeneration may be possible, if the area is closed off completely allowing seeds to sprout and the new plants to produce more seeds. This should continue for more than one year and – provided that the amount of rainfall is sufficient – the vegetation may recover. Once recovered, a strict grazing management must be observed to avoid further degradation.

2.3 *Stony Hill Slopes with Open Commiphora Thorn Bush and Indigofera Dwarf Shrub*

The weakly to moderately inclined slopes of the Korr hills (2 ° to 7 ° inclination) are characterized by stony soils. The fine material, mainly loamy sand, originates from weathering products of the quartz enriched parent rock (Precambrian gneisses) and from aeolian sediments. The proportion of stones at the surface is extremely high (up to 60%). Along the course of sheet wash and deflation the stoniness increases, which locally leads to the development of stone pavements. Linear erosion forms (rills) are confined to less rocky soils in the middle and lower part of the slopes. Test site 53 represents this landscape type.

The vegetation consists of open *Commiphora* thorn bush and *Indigofera* dwarf shrub under less impaired conditions. In 1981 *Commiphora species* such as *C. flaviflora*, *C. paolii* and *C. samharensis* were common. However, in the following years the number of specimen was drastically reduced through cutting (for fences and firewood). On the other hand indicator plants invaded, for example *Jatropha parvifolia* and *Euphorbia cuneata* (Table 1). In 1991 only very few *Commiphora* specimen existed, while the amount of *Euphorbia cuneata* had increased. Due to the high grazing pressure near the permanent water holes a protective basal cover cannot grow. As a result of effective sheet wash the residual enrichment with stones and boulders proceeds at many places of this landscape type.

Although regeneration seems to be arduous, at these sites abundant regrowth of *Indigofera* species (mainly *Indigofera spinosa*) was observed. The seeds are protected by the stones – so are the young seedlings. The continuous grazing pressure results in the loss of young plants. Therefore a strict closure, mainly of the sloping parts is needed. *Jatropha parvifolia* should be uprooted to allow an undisturbed growth of dwarf shrubs.

2.4 *Interfluve Plains with Indigofera-Heliotropium Dwarf Shrub*

The level to very weakly inclined interfluves between the drainage bands and laggas in the inland plains south of Korr are built up predominantly by fluvial and colluvial material (sands and loamy sands). Test site 38 represents this landscape type.

The area is characterized by strong sheet erosion between the dwarf shrubs, which causes a continuous down-wash of the surface. As a result, the bases of the protective dwarf shrubs rise in 15 to 20 cm high heaps above the surface. When isolated and exposed, the growing conditions become unfavourable because of the increasing dryness and the plants die. Without vegetative protection the heaps can be easily flattened by water (rain and sheet flood) or wind (Fig. 1). Patches bare of vegetation become larger and extend to 50 m wide areas with a characteristic sheet wash pattern.

At some places wash flows (20 cm wide) with low incision occur on level surfaces. In 1987 the growth heaps rose up to 10 cm above the surface and were at a distance of 1 to 2 m from each other. However, in 1991 the growth heaps were up to 30 cm high and the distance between them was already 3 to 8 m, indicating an alarming degree of destruction.

In 1981 some dwarf shrubs showed a weakened vitality (for example *Heliotropium steudneri* and *Indigofera spinosa*), which was caused by extreme dryness in combination with overgrazing. Perennial grazing plants were replaced by annual grasses (*Aristida* spec.) and weeds (i.e. *Heliotropium albohispidum*, *Aerva javanica*, *Papalia lapacea*) or avoided shrubs (*Jatropha parvifolia*).

Test site 38 was already moderately impaired in 1981, especially by sheet wash and wind erosion. The biomass was 1024 kg/ha. After the destruction of the nearby water pump the nomads did not use the area any more because of the long distance to the next water place. Therefore a regeneration of site 38 could begin. In 1987 a regrowth of *Indigofera spinosa* and *I. cliffordiana* from seeds was observed and the biomass increased to 2695 kg/ha. Altogether the area seems to recover and to develop into a favourable grazing ground for camels and goats, as grazing did not take place for at least two rainy seasons. In 1991, however, three formerly abandoned animal enclosures (fora) stood in the centre of the site marking the strong use of the area. The result was a decrease in fodder plants and an increase in patches free of vegetation (see above and Fig. 2). The biomass was estimated to be only 1700 kg/ha. The dwarf shrubs were not green and had no flowers compared to 1987. Some were even dead, as there was only a minor rainfall of 5.6 mm (!) on seven days during the main rainy season. No further remarks on range management are needed, as possibilities of regeneration are explained above.

2.5 *Degradation of Lagga Sides*

The inland plains between the Korr hill country and the Ndoto mountain area are characterized by a net of seasonal drainage systems. They show various different forms from steep sided laggas to shallow flow bands without a marked channel (compare part 2.4). Under slightly impaired conditions the laggas are accompanied by a riparian woodland. The tree and bush stratum mainly consists of *Acacia tortilis* and a few other species such as *Maerua crassifolia*, *Cordia gharaf* and *Grewia* species (in some places a total of 20 to 50% cover and a growth height up to 8 m). The dwarf shrub layer is composed of *Duosperma eremophilum*, which may reach 80% cover. In addition, some perennial or annual grasses and herbs occur. The uncovered surface amounts to less than 10% of the site.

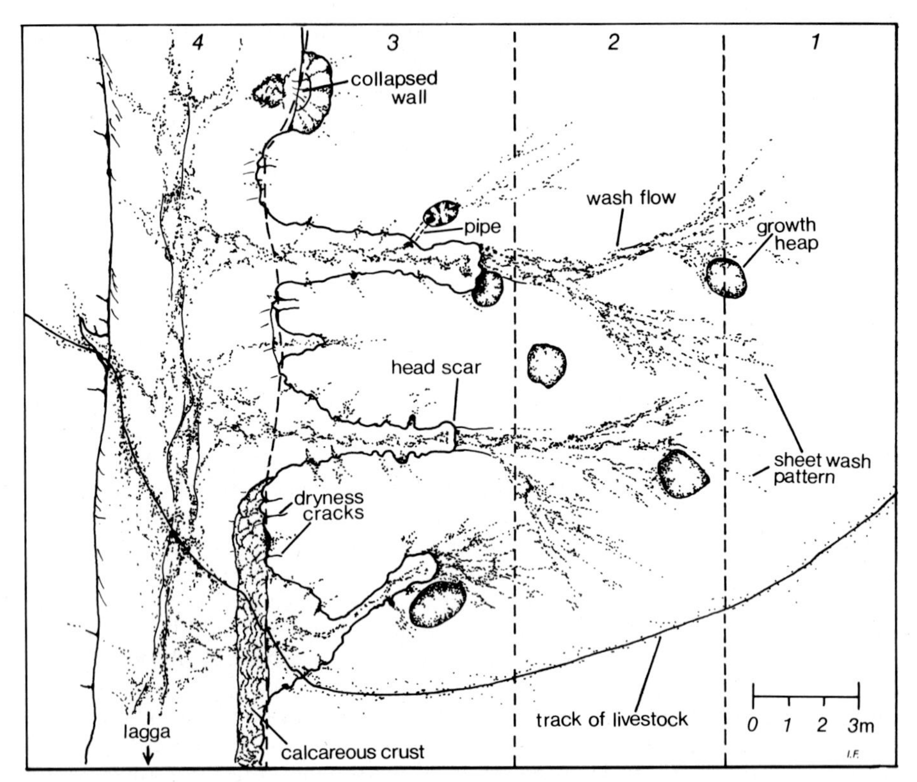

Fig. 3. Zonation of impaired areas along a lagga. 1: Sheet wash with minor growth heaps above the surface. 2: Heavy sheet wash with linear wash flow and minor head scars and growth heaps above the surface. 3: Linear incision with deep head scars and piping features. 4: Lagga with seasonal water flow.

The rate of stream flow varies even in neighbouring laggas. In some, highwater marks of one and a half metre are recognized, while in others nearby the channel floor is covered by dwarf shrub stands, which indicate a lack of water flow for several seasons. Usually the lagga beds are covered with transported sand and locally by fanlike accumulations of stones and grus. Flood waves after heavy rains are able to transport big stones and even boulders. Lateral erosion frequently expose calcareous layers, which form the hardened resistant duricrust steps along the lagga bed (Fig. 3). Many lagga walls up to 5 m high show different sediment layers and buried soil horizons. Organic findings such as charcoal and shells could be dated with the ^{14}C-method. Their ages range from Early Holocene to historic times and can be attributed to climatic changes and/or human activities with their resulting influence on erosion and accumulation, or geomorphological stability allowing for soil formation (MÄCKEL & WALTHER 1989).

The drainage systems with riparian woodland and dense dwarf shrub and grass cover form valuable ecosystems with regard to the regeneration potential, water regime and erosion protection. During the period of observation the riparian woodlands declined due to human impact. Different stages of degradation were observed in the Ngurunit and Saangani Valley (MÄCKEL 1986, MÄCKEL & WALTHER 1992). Here *Duosperma*, widespread where felling and fire has removed the *Acacia tortillis* stands, was probably not part of the original vegetation. Being an intolerant plant it invades disturbed places along the lagga. Without the basal vegetation cover composed of dwarf shrubs, grasses and herbs there exists no effective protection against intensive soil erosion. In the first observation years of 1980 and 1981 steep-sided gullies with active head scars and lateral extensions were found under a nearly closed canopy of *Acacia tortilis*. However, the basal cover had been destroyed by livestock of the pastoralists.

The lagga sides with an open thorn tree/bush and dwarf shrub cover are characterized by intensive sheet wash. The continuous lowering of the surface is illustrated by growth mounds of the dwarf shrubs, which reach up to 30 cm above the present-day sheet flow level. At overutilized sites the dwarf shrubs are destroyed by trampling and browsing or grazing. Finally, more and more bare patches develop, which again are subject to increased soil erosion. Transported sand and grus alternating with exposed soil crust form a characteristic sheet wash pattern. Between the dwarf shrubs the surface is lowered along shallow spillways (wash bands) which show no scars. However, with increasing livestock pressure there is a change form sheet wash to linear erosion, starting with headscars at rills and discontinuous gullies along trampling paths and tracks. Heavily used lagga sides show a dense net of steep lateral gullies extending from the main channel (Fig. 3). They often start with a marked headscar and a plunge pool at the base. In addition to the overland flow, which causes a deepening and widening of the gullies, various features of subsurface erosion occur. They are horizontal and vertical pipes and tunnels, soil bridges and depressions caused by collapsed top layers over erosion tunnels. These piping features are mainly found in clayey soils such as Vertisols or Fluvisols with vertic properties or Luvisols composed of clay loam.

Comparative observations of lagga sides show the increase of lateral erosion in connection with the destruction of vegetation. Concerning the rehabilitation there is a need of detailed mapping. Such maps should mark the different zones of degradation along the laggas and define the zones which must be protected for recuperation of the vegetation. Intensively eroded lagga areas with steep sided gullies need special anti-erosion work such as the building of stone walls or tree bars in order to prevent further erosion and to favour accumulation and regrowth. However, the costs have to be considered.

3 *Conclusion*

The monitoring of geomorphological processes for a sustainable range management is based upon the comparative observation of test sites from 1980 to 1991. The test sites belong to different landscape types and represent areas of severe aquatic and aeolian erosion. The erosional forms differ according to inclination, substratum and

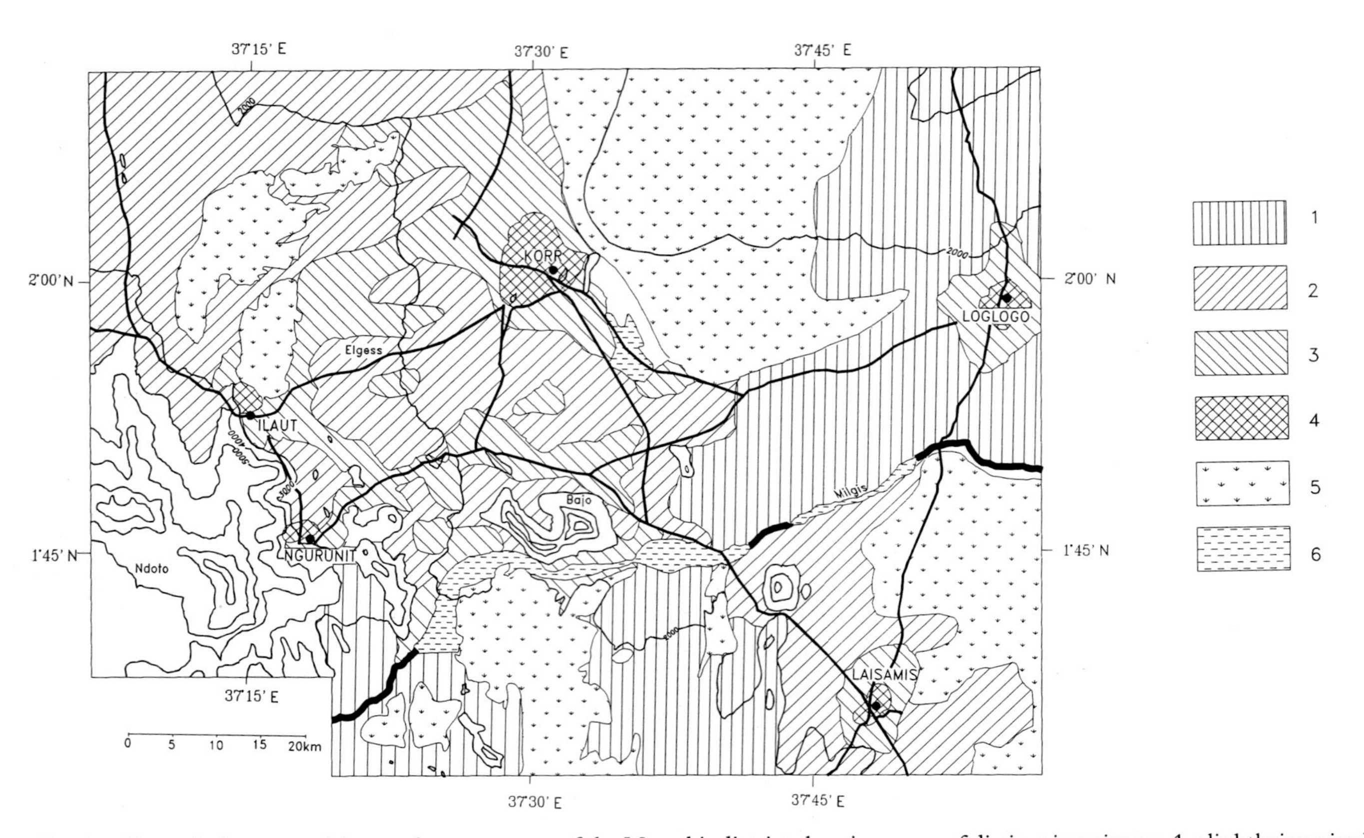

Fig. 4. Degradation map of the southwestern part of the Marsabit district showing areas of distinct impairness. 1. slightly impaired, 2. moderately impaired, 3. heavily impaired, 4. very heavily impaired, 5. lava plateau, 6. seasonally flooded area.

rainfall conditions of the landscape type and the intensity of land use. The resulting types of geomorphological processes indicate the stage of impairness of the individual grazing area (Fig. 4). The main effect of geomorphological processes such as sheet wash and linear erosion is the loss of top soil and nutrients. Secondly, the lowering of the surface leads to a drying up of the area. Grazing plants die and less palatable plants – accustomed to drier habitats – invade the area, or resistant or avoided plants increase. The loss of the protective basal vegetation (herbs and grasses) and the replacement of perennial plants by annuals accelerate soil erosion. Consequently, geomorphological indicator plants are important in connection with geomorphological processes and forms.

The management of range areas in the southwestern part of the Marsabit District has to follow three main considerations:

1. The costs of technologically oriented rehabilitation do not pay against the benefits from an increase in grazing/browsing animal production.
2. However, a low cost (low technology) oriented rehabilitation is possible by encouraging the population to "invest" into their closest neighbourhood by planting trees, shrubs and dwarf shrubs beneficial to humans and animals. Rehabilitation of range areas can also be supported by the common recognition of chiefs, elders and warriors responsible for camels etc. who agree on a management system which allows the vegetation to recover and to spread out again. This means control of the number and type of animals and the duration of use in a given area.
3. Soil erosion processes should be fought against, preferably in the initial stages (stage I and II in Fig. 3) to reduce intensive runoff, which – if not prevented – results in accelerated flow causing severe linear erosion (stages III and IV in Fig. 3).

In order to use the information collected by the monitoring programme for practical usage a handbook (erosion guide) is being prepared by the authors. It shows typical processes and forms of erosion at different stages of degradation and the subsequent vegetation conditions in connection with rainfall data and land use pressure. In addition to this, maps are necessary to illustrate the different stages of impaired areas and erosion risks (Figs. 2 & 4). They form the basis for protective measurements to regenerate degraded rangelands.

Acknowledgements

The field work of R. Mäckel was supported by the German Research Council during the first phase and by the award of a Volkswagen Foundation grant during the second phase.

D. Walther was an associate expert on landuse/ecology with Unesco/IPAL from 1979 to 1981 and coordinator of the Range Management Handbook of Kenya (Ministry of Livestock Development) from 1986 to 1991. Both projects were financed by the Government of the Federal Republic of Germany.

The authors wish to thank Dr. D. Herlocker, formerly Unesco/IPAL, at present coordinator of the Range Management Handbook of Kenya, for stimulating discussions in the field and many help for more than 12 years.

References

DREISER, C., R. MÄCKEL, D. WALTHER & R. WINTER (1989): Abschätzungen des Weidepotentials in ariden Gebieten Kenias mit Hilfe von LANDSAT-TM-Daten. – Geogr. Rundschau **41**: 690–695.

EIDEN, G., C. DREISER, G. GESELL & T. KÖNIG (1991): Large Scale Monitoring of Rangeland Vegetation Using NOAA/1 AVHRR LAC Data. – Range Management Handbook of Kenya Vol. III.4, 92 p. and 49 col. plates. Nairobi/Oberpfaffenhofen.

GÖTTING, R. & R. MÄCKEL (1986): Spatial Analysis of the Dynamics of an Ecosystem by Multistage Remote Sensing in Kenya. – 1986 IEEE International Geoscience and Remote Sensing Symposium (IGARRS '86), Zürich: 53–56.

HERLOCKER, D. (1979): Vegetation of southwestern Marsabit District, Kenya. – IPAL-Techn. Rep. No. D-1; Nairobi.

– (1992). A Survey Method for Classification of Range Condition. – Range Monitoring Series No. 1, Range Management Handbook of Kenya Vol. III.7 (ed. S. B. SHABAANI & D. WALTHER), 31 p.; Nairobi.

MÄCKEL, R. (1986): Oberflächenformung in den Trockengebieten Nordkenias. – Relief, Boden, Paläoklima **4**: 85–225.

– (1989): Die Bewertung des Weidepotentials in den Trockengebieten Kenias mit Hilfe der kombinierten Erderkundung. – DFD/DFVLR-Mitt. **89-04**: 47–51

MÄCKEL, R., G. MENZ & D. WALTHER (1989): Weidepotential and Landdegradierung in den Trockengebieten Kenias, dargestellt an Testflächen im Samburu-Distrikt. – Erdkunde **43**: 253–267.

MÄCKEL, R. & D. WALTHER (1984): Change of Vegetation Cover and Morphodynamics – A Study in Applied Geomorphology in the Semi-arid Lands of Northern Kenya. – Z. Geomorph. N. F. Suppl.-Bd. **59**: 77–93.

– – (1988): Die Bedeutung der Vegetationsveränderung und Geomorphodynamik für die Subsistenzweidewirtschaft in den Trockengebieten Kenias. – Abh. Dt. Geogr. Tag München 1987: 540–546.

– – (1989): Die Reliefentwicklung der Trockengebiete Nordkenias im Jungquartär. – Z. Geomorph. N. F. Suppl. – Bd. **74**: 71–81.

– – (1992): Naturpotential und Landdegradierung in den Trockengebieten Kenias. – Erdkundl. Wissen, 235 S. (im Druck).

MÄCKEL, R., R. WINTER & D. WALTHER (1986): Vegetation and Landscape Classification of the Dry Savana in Eastern Africa Combining Field Work and Digitally Processed LANDSAT-MSS Imagery. – Fernerkundung und Raumanalyse – Klimatologische und Landschaftsökologische Auswertung von Fernerkundungsdaten: 99–127.

PRATT, D. J. & M. D. GWYNNE (1978): Rangeland Management and Ecology in East Africa. – 310 p., London.

SCHULTKA, W. (1991): Vegetation Types. – Range Management Handbook of Kenya Vol. II.1: 25–52.

SCHWARTZ, H. J., S. B. SHABAANI & D. WALTHER (eds.) (1991): Marsabit District. – Range Management Handbook of Kenya Vol. II.1. 164 p.; Nairobi.

TOUBER, L. (1991): Landforms and Soils. – Range Management Handbook of Kenya Vol. II.1: 5–24.

WALTHER, D. (1987): Landnutzung und Landschaftsbeeinträchtigung in den Rendille-Weidegebieten des Marsabit-Distriktes, Nordkenia. – Freiburger Geogr. H. **28**: 330 S.; Freiburg i. Br.

Addresses of the authors: Prof. Dr. R. MÄCKEL, Institut für Physische Geographie der Albert-Ludwigs-Universität, Werderring 4, D-7800 Freiburg i. Br., Dr. D. WALTHER, Kiefernweg 13, D-5902 Netphen 3.

Z. Geomorph. N.F.	Suppl.-Bd. 92	173–188	Berlin · Stuttgart	Mai 1993

A Stratigraphy of Slope Deposits and Soils in the Northeastern Great Basin and its Vicinity

by

ARNO KLEBER, Bayreuth

with 4 figures and 1 table

Summary. In and near the northeastern Great Basin, there is a recurring record of at least three major phases where slope deposits – loess and especially loess-mixed layers – were laid down. Each was succeeded by soil formation. Relations to dated deposits and relief allow to constrain their ages: the youngest layer is ca. 13,000 years old. It bears a soil of intermediate maturity. Two other layers were deposited after the penultimate glaciation but prior to the last glacial maximum. They are separated by a disconformity of probably early Wisconsin age. Both display mature soils with argillic and mostly calcic horizons, with the upper soil usually welded to the lower one.

Zusammenfassung. Im nordöstlichen Großen Becken, USA, und in den angrenzenden Gebirgen können drei weit verbreitete Deckschichten ausgegliedert werden, die aus Hangschutt und Löß bestehen. Jeder Ablagerungsphase einer Deckschicht folgte eine Phase der Bodenbildung. Gut datierte Sedimente und Oberflächenformen in den Becken und in den Gebirgen erlauben, das Alter der Sedimente und Böden einzugrenzen: Die jüngste Lage ist ca. 13 000 Jahre alt. In ihr ist ein mittel entwickelter Boden ausgebildet. Die zwei älteren Deckschichten sind nach der vorletzten Kaltzeit, aber vor dem Maximum der letzten Kaltzeit entstanden. Sie werden von einer Diskordanz getrennt, die wahrscheinlich frühletztkaltzeitlichen Alters ist. In beiden sind jeweils Tonverlagerungshorizonte ausgebildet. Damit sind meist ausgeprägte basale Kalkanreicherungshorizonte vergesellschaftet, deren oberer in der Regel den darunter liegenden Boden überprägt hat.

1 *Introduction*

With the work of RICHMOND (1962), MORRISON (1964, 1965b), or REIDER (1975, 1977), a pedostratigraphic concept of the Great Basin and of the Rocky Mountains was proposed – and applied to landscape outside of stratigraphic control by the latter author. The stratigraphic and some of the pedologic knowledge of that time, however, especially the potential of absolute dating, was less developed than today. Therefore, a discrepancy arose between that work and the evolving stratigraphic framework independent from soils. This may have led many contributors to debase the stratigraphic relevance of soils: soils are thought not to form during distinct soil forming intervals due to climatic control. Instead, soil formation is considered a function of the lapse of time during which more or less the same tendencies of

0044-2798/93/0092-0173 $ 4.00

pedogenesis prevail; climate, especially climatic change, is not believed to critically influence soil-formation (cf. BIRKELAND 1984 for the surficial soils of the Rocky Mountains, HARDEN 1982 for those of the central Californian Ranges, or SCOTT et al. 1983 for paleosols in the Great Basin; recent state of discussion in BIRKELAND 1990 and MCFADDEN & KNUEPFER 1990). This view is somewhat inconsistent with humid areas where climatic fluctuations appear to be decisive factors of soil-formation (e.g. SEMMEL 1977, also BIRKELAND 1984). Essential for this discussion is whether or not a statigraphic relation of distinct soils exists, i.e. whether they are bound to distinct periods of time, or occur randomly dispersed in time. This study contributes to this discussion as it reports some sites from the Great Basin and the Rocky Mountains to reveal indications of the existence of a pedostratigraphic framework.

Such a framework was developed in Germany (SCHILLING & WIEFEL 1962, SEMMEL 1964, recent summary in KLEBER 1991 a and b). The emphasis was placed on the nature of the soil's parent materials, not just the soils and their underlying bedrock. It was found that the soils on the slopes are not formed from bedrock by weathering alone. In fact, there are several layers of creeped material (mainly by solifluction and sometimes additional hillwash), some of which are intermixed with loess, which are not random regarding their occurrence and the vertical sequences they display; KLEBER (1990) proposes the English term "*cover-bed*" for such surficial layers. This study attempts to outline such a concept of a genetic and chronologic relationship between cover-beds and soils in parts of the western United States based on more than 100 measured profiles, ca. 40 auger cores, 540 soil samples, and ca. 300 heavy mineral analyses to evaluate soil inhomogeneity. Some of this material is reported in this study.

2 *Paleo-lakes of the Great Basin*

Several rises and falls of the lake-level in the northeastern Great Basin, in the so-called Bonneville Basin, during the lake-history are known. During the deep-lake cycles, the lake exceeded the modern level of Great Salt Lake (1280 m a.s.l.) by 100 m or more. The lake fluctuations during the last ca. 150,000 yr are sketched in Fig. 1.

This study deals with both of the last major lake-cycles: The Little Valley- and the Bonneville-cycles. The former cycle occurred ca. 140,000 B.P. (MCCALPIN 1986, MCCOY 1981). Its deposits are usually overlain by Bonneville deposits. Both are often separated by a paleosol, the Promontory soil (MORRISON 1965 a). SCOTT et al. (1983: 278–279) debase the use of this soil for stratigraphic purposes because the soil is considered to vary too much regionally, soils of different age are said to be indistinguishable from each other, and the soils are never continuously traceable for greater distances. A minor lake-advance between Little Valley and Bonneville is TL-dated at ca. 80,000 B.P. (MCCALPIN 1986, ROBISON & MCCALPIN 1987). At the Bear River, north of Great Salt Lake, a similar lake deposit, the Cutler Dam deposit (OVIATT et al. 1987), yields an aminostratigraphic age of >60,000 yr (OVIATT, pers. comm. 1990).

The Bonneville deposits accumulated during the last deep-lake cycle. Besides lake deposits, several shorelines formed during this highstand (cf. Fig. 1 for age ranges, approximate heights, and designations of the major such shorelines after

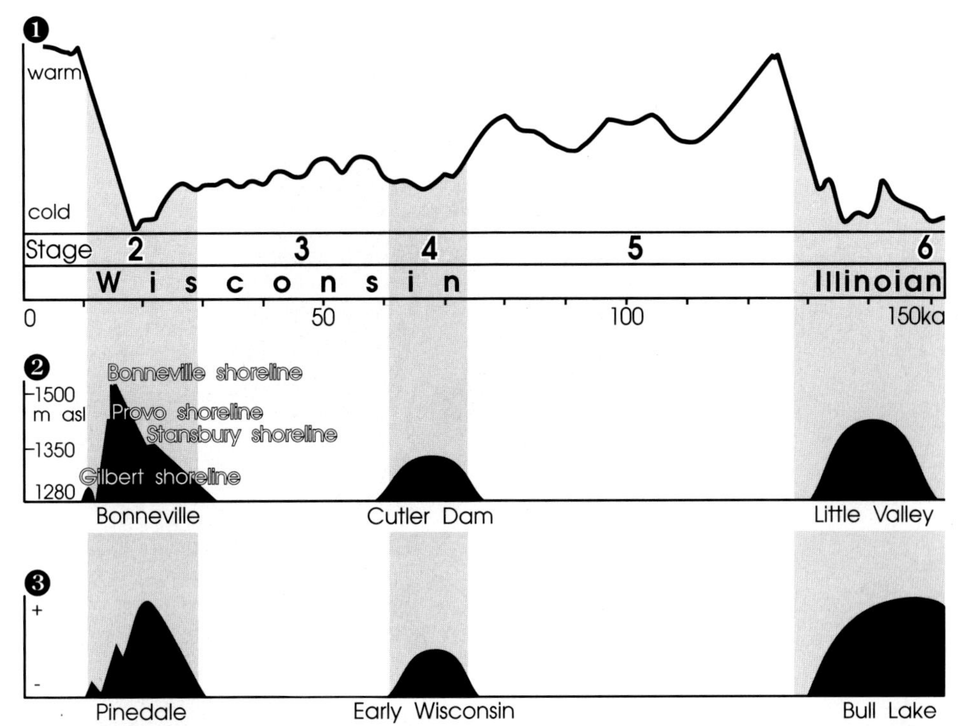

Fig. 1. Lake cycles in the Bonneville Basin (2) and glacial stages in the Rocky Mountains (3) since ca. 150,000 yr compared to the marine oxygene-isotope record/stages and to the stratigraphic divisions (1): Sources of information listed in text. Shaded areas indicate cold stages. Scale for lake levels is approximate, due to regionally varying isostatic rebound of several lake terraces.

BENSON et al. 1990, CURREY & BURR 1988, CURREY 1990, CURREY et al. 1984, OVIATT 1987, SACK 1989, SCOTT et al. 1983).

A lake-complex of similar magnitude, Lake Lahontan, existed several times during the Quaternary in the northwestern Great Basin. The chronology of this lake is similar to that of Lake Bonneville (BENSON et al. 1990). The lake deposits of the last and of the penultimate major highstand (Sehoo and Eetza, respectively) are separated by a mature paleosol as in Lake Bonneville, the Churchill soil of MORRISON (1964).

3 *The Promontory soil and its parent materials*

The Promontory soil was studied at several locations (Fig. 2) where age constraint by dated (MCCOY 1981, 1987, SCOTT et al. 1982, 1983), or at least inferred (SCOTT & SHROBA 1980), overlying and underlying lake deposits was available. All these

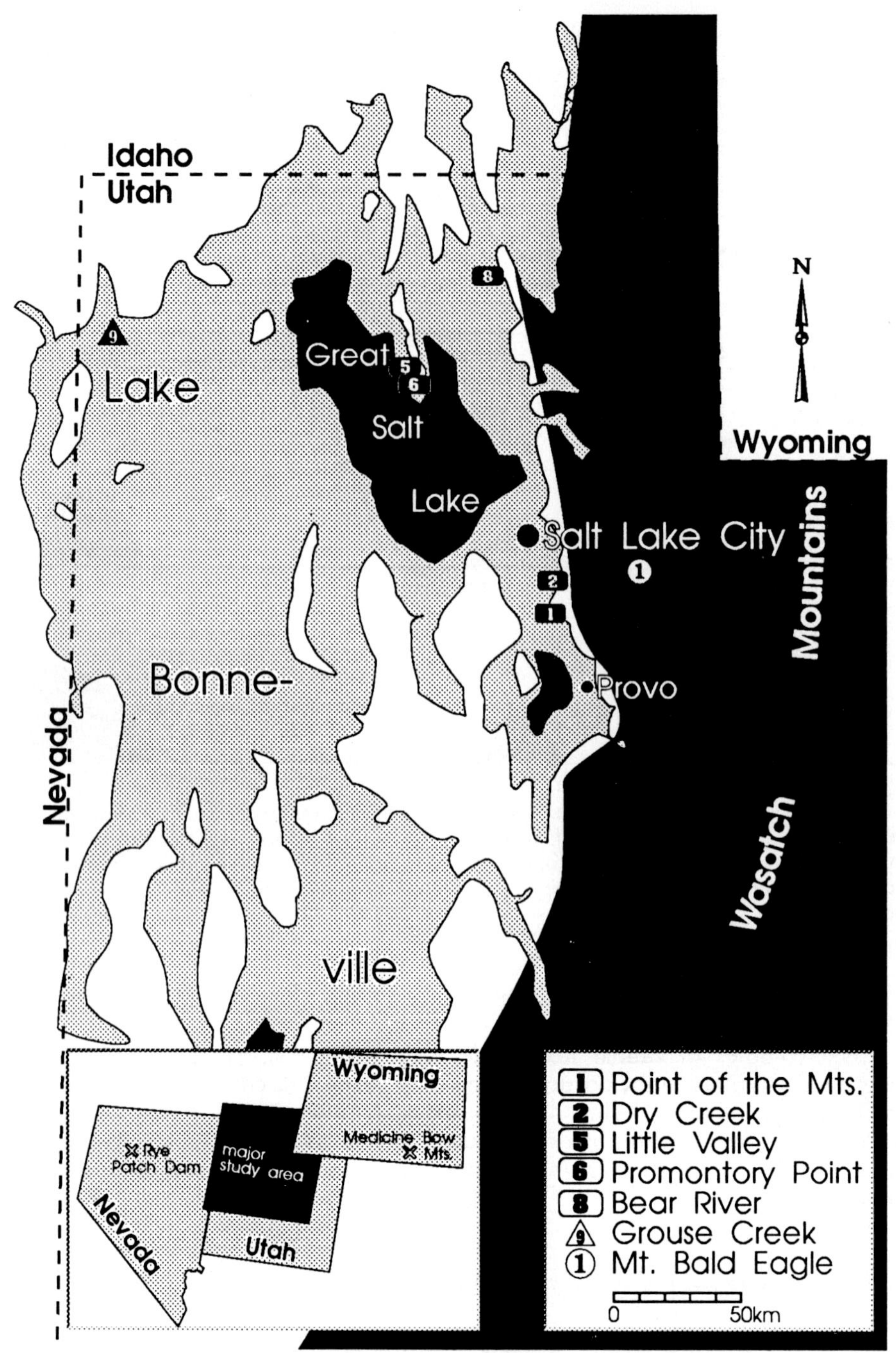

Fig. 2. Former extent of Lake Bonneville and location of mentioned sites in its vicinity. Approximate location of the other sites is found on the insert map.

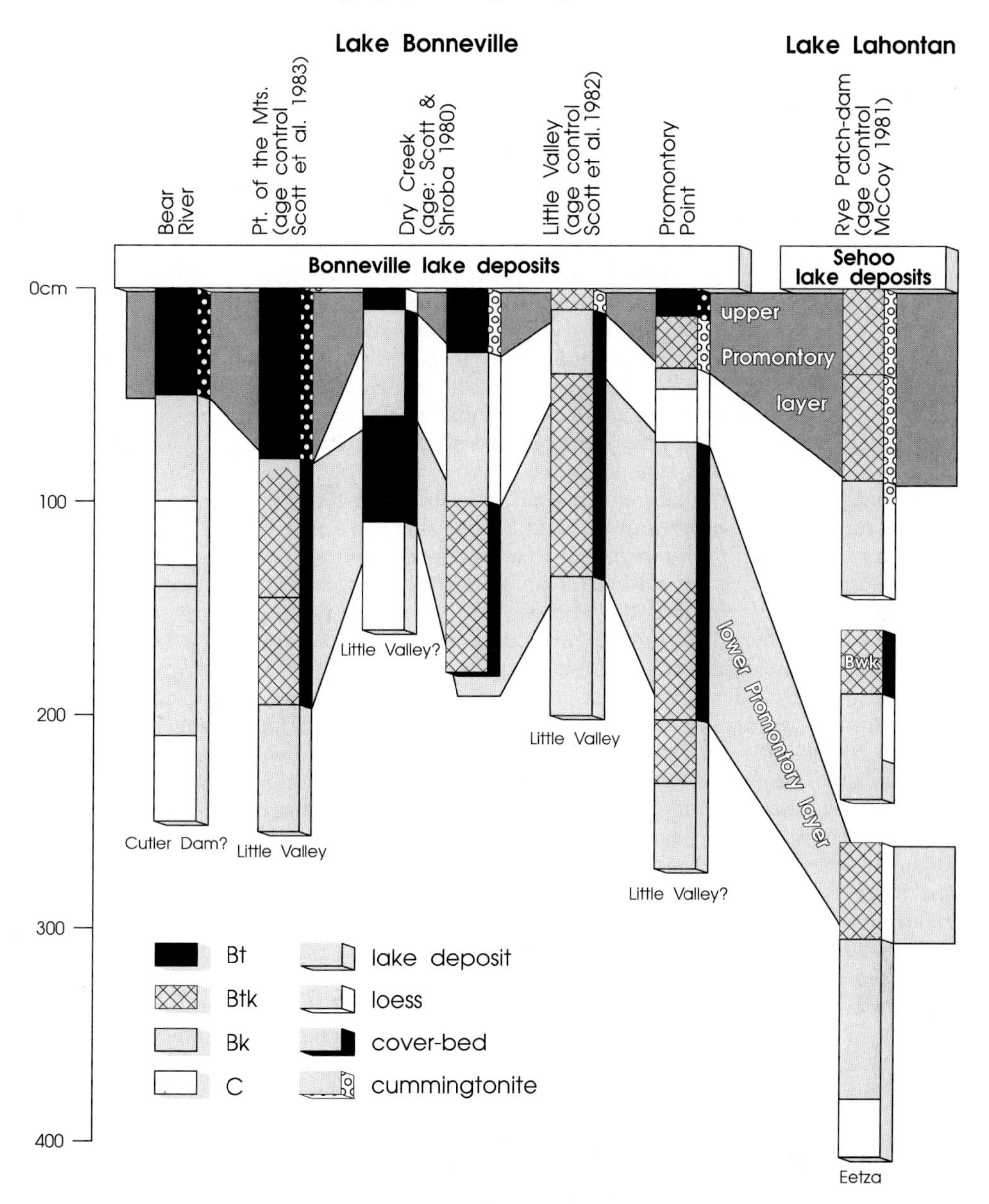

Fig. 3. Correlation of Promontory soil-profiles. Columns indicate horizons (front-facing), parent material (3rd dimension), and layers that yield traces of cummingtonite (dotted).

incarnations of the soil are developed in loess-derived layers as their content of coarse silty particles and of allochthonous heavy minerals suggests. There are always similar sequences of horizons (cf. Fig. 3): the soils are topped by an argillic (Bt) horizon essentially free of carbonate, which is delimited by distinct disconformities. The layer below displays intense carbonate enrichment (Bk horizon) that abates downwards. Clay-films then may be observed as in the upper Bt horizon. Following another disconformity, usually already in the underlying lake beds, all profiles but one again display intense carbonate enrichment. This sequence is not easy to explain pedogenetically. The lower Bk horizon probably reflects the carbonate depletion and succeeding clay-translocation in the horizon(s) above it, but there is another area of carbonate enrichment higher in the profile that shares its horizon with argillic properties. Both properties cannot have formed simultaneously since carbonate enrichment and clay-translocation are mutually exclusive in one horizon at the same time (e.g. Birkeland 1984); argillic horizons usually form only when most of the carbonate is depleted from the horizon. To explain the Promontory soil's horizonation, it is thus necessary to distinguish phases where clay-translocation took place from phases with carbonate enrichment, engulfing an older argillic horizon in the case of the upper Bk horizon. One could argue that the upper Bk horizon represents a phase of less leaching intensity, so that the enrichment occurred higher in the soil rather than in a more intense earlier phase. However, the question then arises as to the source of the carbonate *after* the depletion that was needed to form the argillic properties. The fact that engulfment occurs just below the upper Bt horizon, especially where no significant change in texture could apply for abruptly reduced percolation, precludes addition of carbonate to the soil, composed as it is, after the formation of the argillic properties. A conclusive explanation of the horizon sequence is that two composite soils are considered: The lower part of the profile belongs to an older soil with Bt-Bk-C horizonation. This soil shall be referred to as the *lower Promontory soil*; its parent material – as far as it consists of a cover-bed and/or of loess – is referred to as the *lower Promontory layer*. The upper part consists of a younger soil developed in loess and loess-derived cover-beds, deposited on top of the older soil. This soil shall be called *upper Promontory soil*; its parent material is referred to as the *upper Promontory layer*. This soil tended to have a Bt-Bk-C horizonation as well; since, however, the thickness of its parent material usually was equal to or less than the potential depth of the soil, the carbonate enrichment reached the soil buried immediately below. This buried soil probably had a higher bulk density, due to the soil-formation it already had undergone, reducing percolation. Hence, a composite Bk-/Bt horizon developed.

If this is the case, examples should exist where the upper layer was thick enough to prevent both soils from welding. At Promontory Point, in a neighboring pit of Little Valley, such an example was observed (cf. Fig. 3). Field evidence, texture, and mineralogy of this soil are essentially the same as at Little Valley, while the only older paleosol observed there to display a similar composite horizonation differs in all this respect. Compared to the other instances of the Promontory soils, an additional loess-layer occurs in this profile, which intersects between the upper argillic horizon and the compound calcic and argillic horizon. This layer is enriched with carbonate, too, but only to a lesser degree so that, with regard to the experience from the other profiles, clay-films should be preserved if any had existed. This is not the case.

Table 1. Heavy mineral contents of selected profiles and horizons. Transparent grains from the fraction 100–200 μm were counted until ca. 200 grains were reached. Samples with a sum of significantly less than 200 are poor in heavy minerals due to small contents in the original deposit or due to weathering or veiling effects.

Promontory Point	*2Bt*	*3CBk*	*4Bk*	*4Btk*	*5Btk*	*6Bk*
augite	40.5	49.3	39.9	43.4	42	11
br. hornblende	17.5	12.3	16.6	13.6	12	5
cummingtonite	0.5					
epidote	0.5				3	3
garnet	1.0	0.5	1.8	0.5	1	
gr. hornblende	31.5	29.6	32.3	27.8	32	74
hypersthene	3.5	4.9	4.0	7.1	4	1
mica	1.0		0.9	2.5	2	2
titanite		1.5		0.5		
topas	0.5		1.8	1.0		
turmaline	1.0	1.0	0.9	1.0	2	2
others	2.5	1.0	1.7	2.5	3	2
total	200	203	223	198	136	105

Grouse Creek area	*Bw*	*2C*
augite	29.3	21.6
chlorite	4.5	2.6
cummingtonite	1.7	
disthene	2.3	2.0
epidote	3.9	0.7
garnet	11.3	10.6
br.hornblende	1.1	2.0
gr.hornblende	26.0	22.5
hypersthene	1.7	
sulfate	13.0	31.1
titanaugite	1.1	0.7
zircon	1.1	0.7
others	3.0	5.4
total	177	151

Medicine Bow Mts., Bull Lake	*2AB*	*3Bt*	*4Btk*	*5C*
augite		1.7		
cummingtonite	0.5	1.1		
chondrodite	1.0			
epidote	15.6	21.1	23.9	26.3
garnet	2.0	6.7	8.4	3.5
gr.hornblende	80.4	68.3	64.6	69.2
muscovite			1.3	1.0
olivine			0.4	
pumpellyite		0.6	0.4	
titanite			0.4	
turmaline		0.6		
zoisite & clinoz.	0.5		0.4	
total	199	180	226	198

Mt. Bald Eagle	*A*	*2E*	*3Bt*	*4Bt*	*fill*
allanite					93.2
andalusite		8	0.4	0.4	
augite	6.9	2		0.4	
cummingtonite	1.6				
diaspore	7.5	32	21.9	24.6	2.5
epidote	8.5	8	17.7	14.5	
garnet	1.0	3	2.3	6.9	
hornblende	65.3	40	54.2	45.6	2.5
mica	3.2	3	1.2	1.2	
staurolite	0.6	3	0.4		
titanite	1.9		1.5	3.6	
zircon	1.9		0.4	2.0	1.8
others	2.0	1	0.4	0.8	
total	308	60	260	248	161

Instead, clay infiltration from the overlying argillic horizon into this layer is visible in its uppermost part only, and abates quickly within the layer. This horizon is either the lowest part of the upper Promontory layer, which was so thick that the argillic horizon could not seize it entirely, or it is an overprinted relic of the lower Promontory soil's E horizon. Heavy mineral-analyses (Table 1) display strongly increasing augite, brown hornblende, and hypersthene in the Promontory layers compared to the lake deposit (6Bk) interpreted as eolian addition on these layers. The few occurrences of a very rare mineral, cummingtonite, are in accordance with several other analyses from the upper Promontory layer. This mineral is described in the western USA only from some ashes derived from Mt. St. Helens. Its occurrence may indicate a volcanic event more or less synchronous with the deposition of the upper Promontory layer, possibly correlative with layer C that is older than ca. 37,000 yr (MULLINEAUX et al. 1978).

The Churchill soil in the Lake Lahontan area, which intersects lake deposits broadly correlative to Little Valley and Bonneville, is time-equivalent to the Promontory soil (McCOY 1981). At Rye Patch Dam, Nevada, this soil bifurcates into three independent soils (MORRISON et al. 1965), interpreted as separated instances of the Promontory layers. Except for the intermediate soil, these are mature argillic/calcic soils (cf. Fig. 3), intercalated by subaerial layers and by a lake deposit that reflects a minor lake-advance (MORRISON & FRYE 1965).

The minor lake-advance could be correlative to the Cutler Dam in the Bonneville Basin. If this is the case, there is a chance to find the upper Promontory layer and its soil upon these lake deposits isolated from the lower Promontory soil. A possible instance exists at the Bear River, rejected to be correlative to the Cutler Dam by OVIATT et al. (1987, pers. comm. 1990) because of its gravelly character absent in the other deposits, and of its unusually mature soil. Since lake deposits in the basin generally vary in their composition (SCOTT et al. 1983), and a soil may likely preserve just above permeable ground with less surface run-off, it is possible that this exposure displays the Cutler Dam deposit and a post-Cutler Dam soil (this is considered possible by REIDER, co-author in OVIATT et al. 1987, pers. comm. 1990). Actually, the soil consists of only one argillic horizon developed in a cover-bed with basal carbonate enrichment into the underlying lake deposit – indications of a twofold soil development are absent (Fig. 3). This profile is therefore assumed to represent the upper Promontory layer covering a Cutler Dam deposit; the soil then represents the upper Promontory soil that is not welded to the lower Promontory soil in this case.

4 *Cover-beds and soils on lake terraces*

KLEBER (1990) utilizes the thoroughly dated shorelines of Lake Bonneville to provide age constraints for a surficial layer, which he describes to be widespread on some shorelines. It disconformably overlies bedrock or Bonneville lake deposits. The layer has a remarkable coarse silt content – often much more than the underlying lake deposit or bedrock; on slopes or where slopes are near, it often contains coarser rock-fragments than could be derived from the directly underlying substratum. Therefore, the layer must be addressed as a cover-bed with a mixed-in loess content. This layer is referred to as the *Provo layer* (named after the Provo-period of the lake).

The Provo layer occurs on the Bonneville and on the Provo shorelines; it is therefore younger than the latter, which began to form ca. 14,500 B.P. It was never found on the Stansbury shoreline. Since it is older than both Bonneville and Provo, the Stansbury shoreline must have been flooded by the lake when the cover-bed was deposited. This suggests the dating of this cover-bed into the Provo-period of Lake Bonneville. The layer formed before ca. 13,000 B.P. because the lake emerged at approximately this time from the Stansbury shoreline (SACK 1989, SCOTT et al. 1983). This layer may contain cummingtonite, which is always absent in the underlying material (example from Grouse Creek area, Table 1), indicating a correlation with a Mt. St. Helens-ash, preferably with layer S that dates ca. 13,000 B.P. (MULLINEAUX et al. 1978). All other ash layers, except for layer C that is too old to be correlative, were either not widespread, and/or did not contain abundant cummingtonite, or were not transported into southeastern directions (SMITH et al. 1977, MULLINEAUX et al. 1978, DAVIS 1985).

KLEBER (1990) describes the soil in the Provo layer, referred to as the *Provo soil*, in the eastern part of the Bonneville Basin to be strictly bound to the layer's occurrence: a decalcified A horizon in the Provo layer is succeeded by a weak Bk horizon below the layer. Further to the west, the Provo soil is not decalcified although a weak Bk horizon occurs as well; the color of the A horizons is paler there and their organic carbon content is smaller. In the driest parts of the Basin, a weak Bw horizon occurs in the Provo layer. In the extreme case, all horizons, i.e. including the Bk, concentrate in the Provo layer instead of extending into the underlying substratum (see example in Fig. 4).

There is no firm age constraint for the Provo soil. However, some suppositions may be drawn: Two instances of the Provo-soil were found buried by younger deposits. The radiocarbon dates register the time of burial, not that of the soil-formation itself, because organic carbon, especially humic acid, in soils renews rapidly (GAMPER 1987). The burial took place 1020 ± 115 (Hv17628, humic acid) and 1325 ± 105 B.P. (Hv17627, organic matter) in one instance, and 2785 ± 100 C^{14}-B.P. (Hv17630, humic acid) in the other. The latter yields an interesting age of 6575 ± 195 yr on the organic matter (Hv17629); this date is interpreted as an age-mixture between the time of origination of the soil and of its burial. The age of the Provo soil is then significantly higher than this date. A few published dates register the burial of soils that might be correlative to the Provo soil: slope deposits overlying a soil developed in loess-rich material in northeastern Utah are dated 9,000–11,000 B.P. (McCALPIN 1988), the Fallon deposit in the Lahontan Basin – a lake deposit that overlies a soil with an A-Bw-Bk horizonation (MORRISON 1964, MORRISON & FRYE 1965, MORRISON et al. 1965) – dates 10,000–11,000 B.P. (CURREY 1988 and pers. comm. 1992), and a buried A horizon in northwestern Nevada is dated at ca. 9000 B.P. (CURRY & MELHORN 1990).

5 *Moraines in the Medicine Bow Mts., Wyoming*

Basing on various relative age criteria (BIRKELAND et al. 1979), supported by some absolute dates, a stratigraphy of the glacial events in the Rocky Mountains' valleys is well established (RICHMOND 1986b). The oldest event to display a well defined

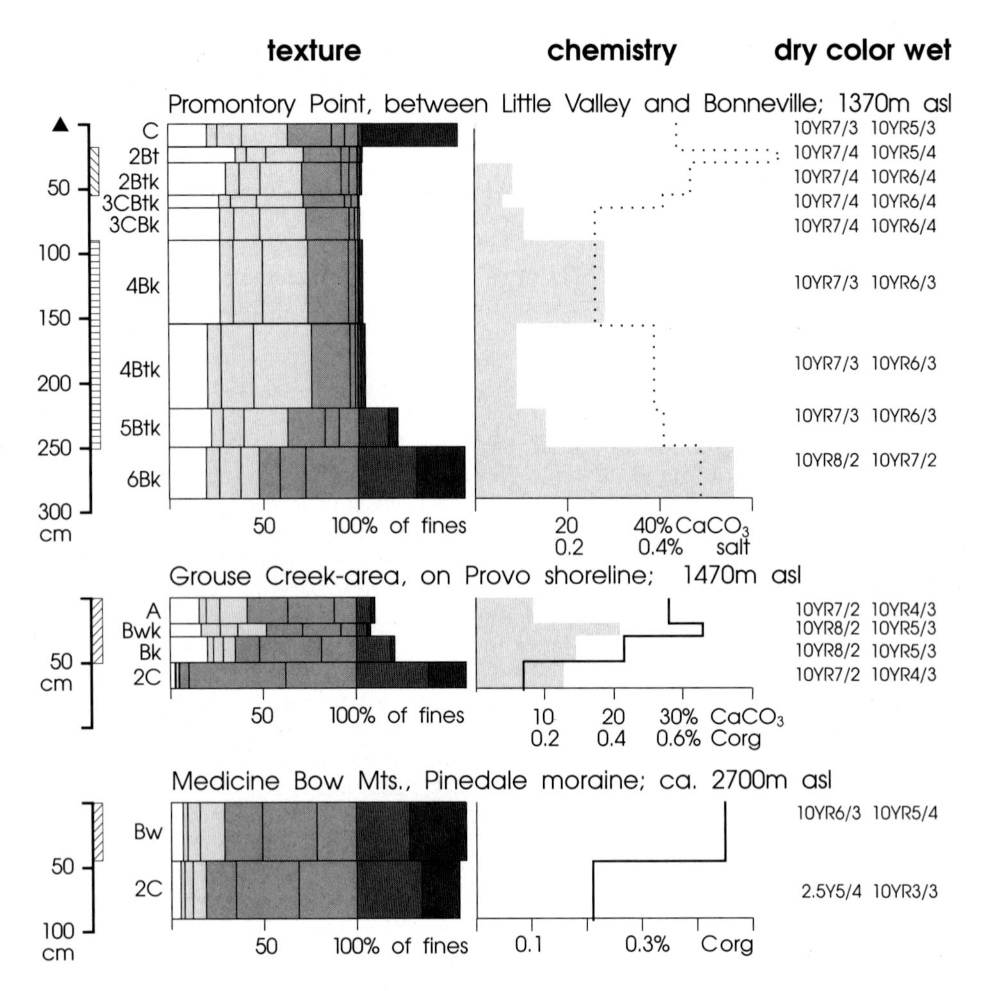

Fig. 4. Analytical characterization of the profiles discussed in text. Analyses are after German Institute for Standardization (DIN) nrs. 19682–19684 and after SCHLICHTING & BLUME (1966: 105–106; salt content). The leftmost column gives a scale of the horizons' depths. If it starts with an arrow, the exposure continues above the top of the measured section. To the right, the cover-beds are marked; a free space between them indicates cover-beds which are not undoubtedly attachable to one of them. The horizon-designations after Soil Survey Staff (1990) follow. Clay (<0.002 mm), silt (0.002–<0.063 mm), and sand (0.063–<2 mm) are specified in percent of the total fines; the latter two are divided into the fine, medium, and coarse fraction each. The clasts > 2 mm to 2 cm are specified in percent of the total sample. The clasts > 2 cm are volume-percent, estimated in the field. The following columns show some soil-chemical data (carbonate, salt, organic carbon) as far as derived, and the colors of the samples, dry and wet, determined in the laboratory.

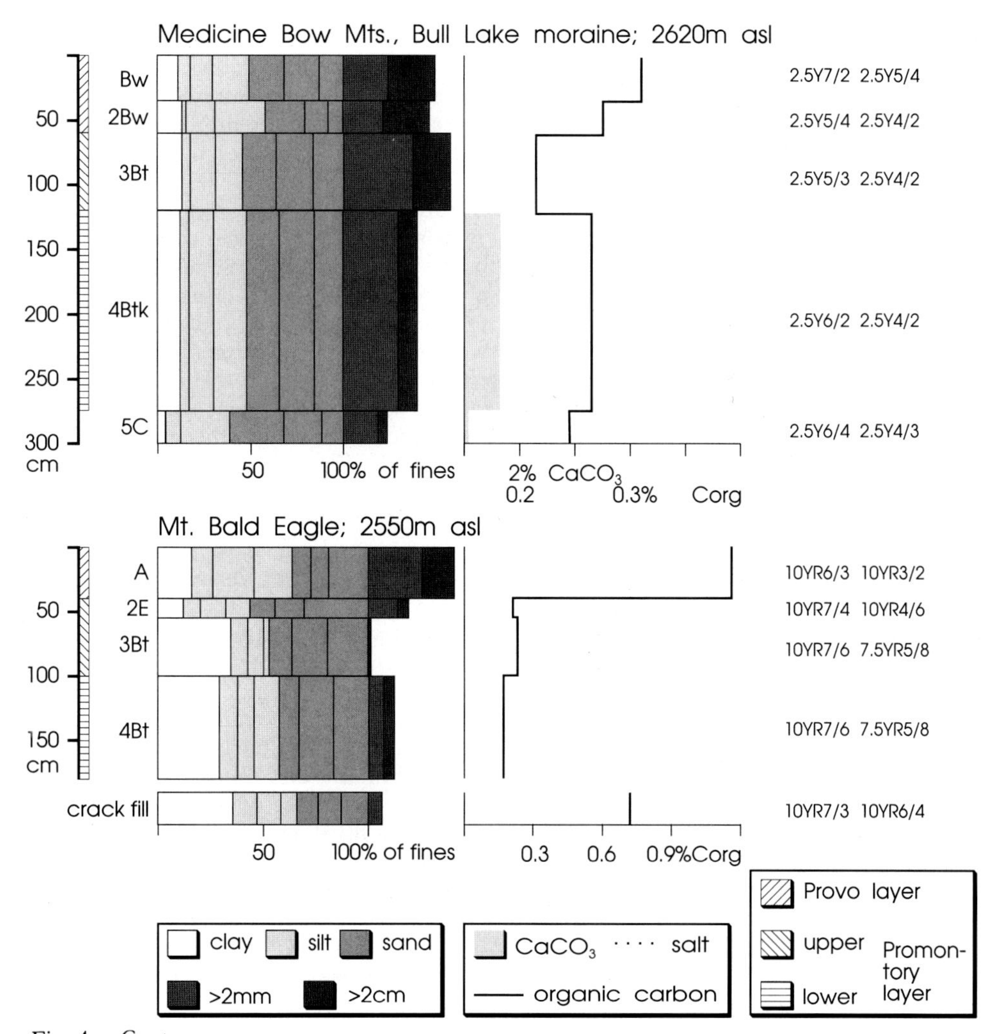

Fig. 4. Cont.

morainal relief is the Bull Lake at ca. 150,000 B.P. (Pierce et al. 1976, Pierce 1979). The latest major glacier advance is the Pinedale 22,000–18,000 B.P. with several later substages. There is evidence of further glacial advances of secondary order between the Bull Lake and the Pinedale (Richmond 1986a); most prominent is the early Wisconsin advance 70,000–55,000 B.P. (Pierce 1979, Colman & Pierce 1986).

The glacial record of the eastern flank of the Medicine Bow Mts. (Fig. 2 for location) follows the record evident throughout the Rocky Mountains: a Bull Lake and several Pinedale moraines may be distinguished (McCallum 1962, McCahon & Munn in review, Mears and Munn, pers. comm. 1990).

The soil on the Pinedale moraine is developed in a cover-bed with coarse clasts that are significantly more frequent than in the underlying moraine. The clasts are oriented downslope with their long axes, a feature absent in the moraine. Coarse silt and fine sand reflect eolian addition to the layer compared to the moraine. Immature pedogenesis in the upper layer is demonstrated by some organic carbon (Fig. 4) and by a weak subangular blocky structure.

The soil on the Bull Lake moraine starts from its top with two layers, subdivided by a minor disconformity, which have downslope oriented clasts, contrary to the lower layers. They contain more coarse silt and differ in their mineralogy from the other layers (Table 1). They likely correlate with the layer covering the Pinedale moraine. There is a disconformity between these suggested post-Pinedale layers and the underlying Bt horizon that separates two soils formed at different times: the Bt horizon must have been truncated, loosing its original E horizon that is absent now; later, new layers were deposited now carrying the weak surface-soil. Since soil development of this intensity is absent on the Pinedale moraine, the Bt horizon on the Bull Lake moraine is likely older than Pinedale. The soil contains only small amounts of carbonate. Abruptly below a major disconformity at 120 cm, indicated by textural change, carbonate enrichment sets in. This is unlikely explained by carbonate translocated from the thin uppermost layer, because there is no reason from the recent soil properties for the carbonate to precipitate abruptly just at that depth (cf. texture in Figure 4; bulk density is 1.43 and 1.33 in the horizons above and below, respectively, MCCAHON & MUNN in review). The carbonate more likely stems from the upper Bt horizon, indicating that this disconformity separates two soils of different ages as previously discussed. The heavy minerals in the various layers support the subdivision into different strata; the trace-minerals, especially, differ from layer to layer and are absent in the moraine (lowest horizon) entirely. The traces of cummingtonite in the upper layers are similarly distributed here as previously reported from the Great Basin profiles. The sequence of layers and soils is essentially the same as in the Bonneville Basin, except for the missing Bonneville lake deposits. Thus the post-Pindale layer and the uppermost layers on the Bull Lake moraine are suggested to correspond with the Provo layer, the parent materials of the upper and the lower Bt horizons with the upper and the lower Promontory layers, respectively.

Early Wisconsin moraines are scarce. A few studies exist, however, where soils on such moraines or on relief within their age range are described (BUSACCA 1987, COLMAN & PIERCE 1986, HARDEN 1982). Consistently, argillic horizons occur on such sites, and are roughly half as thick as on neighboring Bull Lake-relief. This suggests that the upper Promontory layer is younger than early Wisconsin, while the lower one is not; this is in accordance with the age suggestion for the layers previously reported from the Great Basin.

6 *Between former lakes and glaciers*

Processes leaving such clear traces in the high mountains as well as in the deepest basin – synchronously as far as the temporal resolution of the age control provided by moraines and lake deposits goes – were unlikely without effect in the areas between. In fact, similar sequences of cover-beds and soils were observed throughout the Rocky Mountains and the northern Great Basin. Examples are shown by KLEBER (1992) from the La Sal Mts., southeastern Utah. One profile from Mt. Bald Eagle, Wasatch Mts., may serve as an example here. The profile has an A horizon developed in a cover-bed with clasts oriented downslope. This overlies a thin layer with an E horizon. This horizon wedges out within a few meters from the measured section. Two Bt horizons follow, separated by a major disconformity. The lower one contains many cracks, probably through desiccation, which are filled with a material with more than 90% allanite (Table 1), a mineral that is absent in any other layer in this profile and in other nearby profiles. The source of this material is unknown. The filling, however, must have occurred when the cracks were open to the surface; since they do not trace into the overlying layer, the surface was close to the upper border of the lower Bt horizon. On the other hand, the crack fill appears not influenced by pedogenic alteration, it is therefore younger than the soil where it is developed. This demonstrates that both of the Bt horizons belong to different times of soil formation.

7 *Conclusion*

The stratigraphic framework for the soils and their parent materials, resulting from age-controlled studies in the Bonneville Basin and in the Rocky Mountains, is as follows:

age	name/*informal name*	process
10–11 ka	Gilbert shoreline	latest Pleistocene lake stand
	Provo soil	soil formation, properties depend on environment settings
ca. 13 ka	Stansbury shoreline set free	lake emerges the formerly flooded shoreline
	Provo layer	cover-bed with loess, some cummingtonite
13–14 ka	Mt. St. Helens layer S	eruption, possible source of cummingtonite
14.2–14.5	Provo shoreline	lake stand
14.5–15.3	Bounneville shoreline	highest lake stand
22 ka	Stansbury shoreline	first lake stand
25 ka	onset of the Bonneville cycle	major lake advance
	–	truncation of top-soil
	upper Promontory soil	argillic and moderate to strong calcic horizons, often welded to the lower Promontory soil
	upper Promontory layer	loess layer or cover-bed with mixed-in loess, some cummingtonite
<37 ka	Mt. St. Helens layer C	eruption, possible source of cummingtonite
~55–70 ka	early Wisconsin glaciation	minor glacial advance

age	name/*informal name*	process
~60–80 ka	Cutler Dam lake-cycle	minor lake advance
	–	truncation of top-soil
	lower Promontory soil	argillic hor. with moderate calcic hor. below
	lower Pormontory layer	cover-bed mixed with loess
~140 ka	Little Valley lake-cycle	major lake advance
~150 ka	Bull Lake glaciation	major glacial advance

The soils in the northern Great Basin and vicinity appear bound to distinct periods. The formation of each soil was enabled by the deposition of a cover-bed that acted as a new parent material for that soil. It is likely that the new soils would not have been able to overprint the older ones significantly if they had had to share the same parent material.

Acknowledgements

I am very grateful to DONALD R. CURREY, Salt Lake City, who introduced me into the area, to BRAINERD MEARS Jr. and LARRY C. MUNN, Laramie, who led me to the moraine sites in the Medicine Bow Mts., and to RICHARD W. ARNOLD, Washington D. C., CHARLES G. J. OVIATT, Manhattan, Kansas, and RICHARD G. REIDER, Laramie, for very helpful discussion. I especially thank RADU SCHILL who performed the analyses. I thank KEN FROESE who did the proof-reading on an earlier draft of this manuscript. The research was funded by the German Research Council.

References

BENSON, L. V., D. R. CURREY, R. I. DORN, K. R. LAJOIE, C. G. OVIATT, S. W. ROBINSON, G. I. SMITH & S. STINE (1990): Chronology of expansion and contraction of four Great Basin lake systems during the past 35,000 years. – Palaeogeography, Palaeoclimatology, Palaeoecology **78**: 241–286.

BIRKELAND, P. W. (1984): Soils and geomorphology. – New York.

– (1990): Soil-geomorphic research – a selective overview. – Geomorphology **3**: 207–224.

BIRKELAND, P. W., S. M. COLMAN, R. M. BURKE, R. R. SHROBA & T. C. MEIERDING (1979): Nomenclature of alpine glacial deposits, or, what's in a name. – Geology **7**: 532–536.

COLMAN, S. M. & K. L. PIERCE (1986): Glacial sequence near McCall, Idaho: Weathering rinds, soil development, morphology, and other relative-age criteria. – Quat. Res. **25**: 25–42

CURREY, D. R. (1988): Isochronism of final Pleistocene shallow lakes in the Great Salt Lake and Carson Desert regions of the Great Basin. – 10th Bienn. Meeting Amer. Quat. Assoc, Amherst, Mass., abstracts: 177.

– (1990): Quaternary palaeolakes in the evolution of semidesert basins, with special emphasis on Lake Bonneville and the Great Basin, U.S.A. – Palaeogeography, Palaeoclimatology, Palaeoecology **76**, 189–214; Amsterdam.

CURREY, D. R., G. ATWOOD & D. R. MABEY (1984): Major levels of Great Salt Lake and Lake Bonneville. – Utah Geolog. Mineral. Surv. Map **73**.

CURREY, D. R. & T. N. BURR (1988): Linear model of threshold-controlled shorelines of Lake Bonneville. – Utah Geol. Mineral. Survey Misc. Pub. **88-1**: 104–110; Denver, Colorado.

CURRY, B. B. & W. N. MELHORN (1990): Summit Lake landslide and geomorphic history of Summit Lake Basin, Northwestern Nevada. – Geomorphology **4**: 1–17.

DAVIS, J. O. (1985): Correlation of late Quaternary tephra layers in a long pluvial sequence near Summer Lake, Oregon. – Quat. Res. **23**: 38–53.

GAMPER, M. (1985): Morphochronologische Untersuchungen an Solifluktionszungen, Moränen und Schwemmkegeln in den Schweizer Alpen. – Physische Geogr. **17**, Zürich.

HARDEN, J. W. (1982): A quantitative index of soil development from field descriptions: Examples from a chronosequence in central California. – Geoderma **28**: 1–28.

KLEBER, A. (1990): Upper Quaternary sediments and soils in the Great Salt Lake-area, USA. – Z. Geomorph. N.F. **34**: 271–281.

– (1991 a): Die Gliederung der Schuttdecken am Beispiel einiger oberfränkischer Bodenprofile. – Bayreuther Bodenkund. Ber. **17**: 83–105.

– (1991 b): Gliederung und Eigenschaften der Hang-Schuttdecken und ihre Bedeutung für die Bodengenese. – Mitt. Dt. Bodenkundl. Ges. **66**: 807–810.

– (1992): Deckschichten und Böden in den nordwestlichen La Sal Mts., Utah, USA. – Bonner Geogr. Arbeiten: in press.

MCCAHON, T. J. & L. C. MUNN (in review): Morphology and genesis of Late Pleistocene till soils in the Medicine Bow Mountains, Wyoming.

MCCALLUM, M. E. (1962): Glaciation of Libby Creek Canyon, east flank of Medicine Bow Mountains, southeastern Wyoming. – Univ. Wyoming Contrib. Geol. **1**: 21–29; Laramie, Wyoming.

MCCALPIN, J. (1986): Thermoluminescence (TL) dating in seismic hazard evaluations: An example from the Bonneville Basin, Utah. – Proc. 22nd Sympos. on Engineering Geology and Soils Engineering: 156–176; Boise, Idaho.

– (1988): The history of Lake Bonneville in Cache Valley: updating G. K. Gilbert's observations. – Utah Geol. Mineral. Survey Misc. Pub. **88-1**: 111–116; Denver, Colorado.

MCCOY, W. D. (1981): Quaternary aminostratigraphy of the Bonneville and Lahontan Basins, Western U.S., with paleoclimatic implications. PhD-thesis, Univ. of Colorado; Boulder.

– (1987): Quaternary aminostratigraphy of the Bonneville Basin, western United States. – Geol. Soc. Amer. Bull. **98**: 99–112.

MCFADDEN, L. D. & P. L. K. KNUEPFER (1990): Soil geomorphology: the linkage of pedology and surficial processes. – Geomorphology **3**: 197–205.

MORRISON, R. B. (1964): Lake Lahontan: geology of southern Carson Desert, Nevada. – U. S. Geol. Surv. Prof. Pap. 401; Washington, D.C.

– (1965 a): New evidence on Lake Bonneville stratigraphy and history from southern Promontory Point, Utah. – U.S. Geol. Surv. Prof. Pap. **525-C**: C110–119; Washington, D.C.

– (1965 b): Quaternary Geology of the Great Basin. – In: WRIGHT, H. E. Jr. & D. G. FREY (eds.): The Quaternary of the United States: 265–285; Princeton, New Jersey.

MORRISON, R. B. & J. C. FRYE (1965): Correlation of the middle and late Quaternary successions of the Lake Lahontan, Lake Bonneville, Rocky Mountain (Wasatch Range), Southern Great Plains, and eastern Midwest Areas. – Nevada Bureau of Mines Rep. **9**; Reno, Nev.

MORRISON, R. B., M. D. MIFFLIN & M. M. WHEAT (1965): Rye Patch Dam. – Guidebook Field Conf. 1, Northern Great Basin and California, Internat. Assoc. Quat. Res. 7th Congr.: 29–33.

MULLINEAUX, D. R., R. E. WILCOX, W. F. EBAUGH, R. FRYXELL & M. RUBIN (1978): Age of the last major scabland flood of the Columbia Plateau in eastern Washington. – Quat. Res. **10**: 171–180.

OVIATT, C. G. (1987): Lake Bonneville stratigraphy at the Old River Bed, Utah. – Am. Jour. Sci. **287**: 383–398.
OVIATT, C. G., W. D. MCCOY & R. G. REIDER (1987): Evidence for a shallow early or middle Wisconsin-age lake in the Bonneville-Basin, Utah. – Quat. Res. **27**: 248–262.
PIERCE, K. L. (1979): History and dynamics of glaciation in the northern Yellowstone National Park area. – U. S. Geolog. Surv. Prof. Pap. **729-F**; Washington, D.C.
PIERCE, K. L., J. D. OBRADOVICH & I. FRIEDMAN (1976): Obsidian hydration dating and correlation of Bull Lake and Pinedale glaciations near West Yellowstone, Montana. – Geol. Soc. Amer. Bull. **87**: 703–710
REIDER, R. G. (1975): Morphology and genesis of soils on the Prairie Divide deposit (pre-Wisconsin), Front Range, Colorado. – Arctic and Alpine Res. **7**: 353–372.
– (1977): Geomorphic implications of pre-Wisconsin soils on the White River Plateau erosion surface of northwestern Colorado. – Catena **3**: 355–368.
RICHMOND, G. M. (1962): Quaternary stratigraphy of the La Sal Mountains, Utah. – U. S. Geol. Surv. Prof. Pap. **324**: Washington, D. C.
– (1986a): Stratigraphy and chronology of glaciations in Yellowstone National Park. – In: SIBRAVA, V., D. Q. BOWEN & G. M. RICHMOND (eds.): Quaternary glaciations in the northern hemisphere. – Quat. Sci. Rev. **5**: 83–98; Oxford.
– (1986b): Stratigraphy and correlation of glacial deposits of the Rocky Mountains, the Colorado Plateau and the ranges of the Great Basin. – In: SIBRAVA, V., D. Q. BOWEN & G. M. RICHMOND (eds.): Quaternary glaciations in the northern hemisphere. – Quat. Sci. Rev. **5**: 99–127; Oxford.
ROBISON, R. M. & J. P. MCCALPIN (1987): Surficial geology of Hansel Valley, Box Elder County, Utah. – Utah Geol. Assoc. Pub. **16**: 335–349.
SACK, D. (1989): Reconstructing the chronology of Lake Bonneville: an historical review. – In: TINKLER, K. J. (ed.): History of Geomorphology, The Binghampton Symposia in Geomorphology, Internat. Ser. **19**: 223–256; London, Unwin Hyman.
SCHILLING, W. & WIEFEL, H. (1962): Jungpleistozäne Periglazialbildungen und ihre regionale Differenzierung in einigen Teilen Thüringens und des Harzes. – Geologie **11**: 428–460.
SCHLICHTING, E. & H. P. BLUME (1966): Bodenkundliches Praktikum. – Hamburg, Berlin.
SCOTT, W. E. & R. R. SHROBA (1980): Stratigraphic significance and variability of soils buried by deposits of the last glacial cycle of Lake Bonneville. – Geol. Soc. Amer. Abstr. Progr. **12**: 304; Boulder, Co.
SCOTT, W. E., W. D. MCCOY, R. R. SHROBA & R. MEYER (1983): Reinterpretation of the exposed record of the last two cycles of Lake Bonneville, Western United States. – Quat. Res. **20**: 261–285.
SEMMEL, A. (1964): Junge Schuttdecken in hessischen Mittelgebirgen. – Notizbl. hess. L.-A. Bodenforsch. **92**: 275–285.
– (1977): Grundzüge der Bodengeographie. – Stuttgart.
SMITH, H. W., R. OKAZAKI & C. R. KNOWLES (1977): Electron microprobe data for tephra attributed to Glacier Peak, Washington. – Quat. Res. **7**: 197–206.
Soil Survey Staff (1990): Keys to Soil Taxonomy. – SMSS Techn. Monogr. **6**; Blacksburg, Virginia.

Address of the author: Universität Bayreuth, Lehrstuhl für Geomorphologie, Postfach 101251, D-8580 Bayreuth.

Z. Geomorph. N.F.	Suppl.-Bd. 92	189–200	Berlin · Stuttgart	Mai 1993

Etchplanation, Review and Comments of BÜDEL's Model

by

HANNA BREMER, Köln

with 8 figures

Summary. Concepts in tropical geomorphology are briefly outlined together with their basic observations and arguments. Due to morphogenesis there are many variations in landforms, weathering cover and processes, which render the models as abstractions explaining the major features only. There seem also to be some rules for the variations. These are mainly due to tectonic and climatic history. Therefore there is a strong plea for the morphogenetic approach which includes former and recent processes of weathering and denudation.

Zusammenfassung. Modelle der Reliefentwicklung in den Tropen werden kurz geschildert in Zusammenhang mit den grundlegenden Beobachtungen und Argumenten. Infolge der Morphogenese gibt es viele Variationen der Landformen, des Verwitterungsmantels und der Prozesse, die zeigen, daß die Modelle Abstraktionen sind, die nur die wesentlichsten Züge erklären. Anscheinend gibt es aber auch Regeln für die Änderungen. Diese hängen u.a. ab von der tektonischen und klimatischen Geschichte. Daher wird sehr stark für eine morphogenetische Untersuchungsweise plädiert, die frühere und rezente Prozesse der Verwitterung und Abtragung mit einschließt.

Research in tropical geomorphology has a very old tradition in Germany. A short historical review is given in a history of geomorphology in Germany (BREMER 1993). WIRTHMANN (1987) wrote the most recent geomorphology of the tropics in the German language where he confronted very ably the different concepts. There is a supplement volume of the "Zeitschrift für Geomorphologie" on "Geomorphology of the tropics" (GRUNERT 1992), where several colleagues reported their latest investigations. Therefore the following account is restricted more or less to the internationally well known model of BÜDEL.

1 *Concepts in tropical geomorphology*

The concept of climatic geomorphology was developed in central Europe some 100 years ago. Almost from the very beginning the different character of landforms and geomorphological processes in the tropics, as compared with temperate areas, was emphasised. Climatically controlled exogenic forces are most active due to the high temperatures and humidity, resulting in intensive weathering. It is interesting that

0044-2798/93/0092-0189 $ 3.00

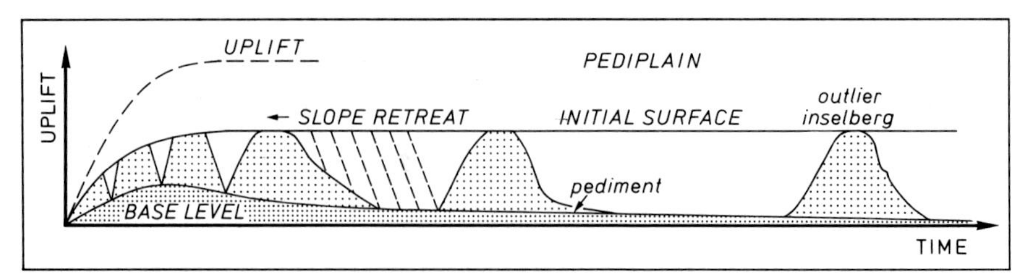

Fig. 1. The model of L. C. King (1949, 1962; drawing with reference to Thornes & Brunsden 1977, with changes) was developed in semiarid South Africa. The escarpment there shows more mechanical than intensive chemical weathering. As the drawing is a two-dimensional abstraction, scarp retreat in the areal context does not show. Therefore the main difficulties, i.e. the formation of inselbergs and the planation of divides, are not evident.

the main models of tropical geomorphology where derived from observations in Africa.

Two contrasting hypotheses are usually presented: L. C. King (1949, 1962) proposed pediplanation as the result of scarp retreat (Fig. 1), Büdel's (1957, 1977) model (Fig. 2) of double etch planation is based on intensive weathering and stripping processes. These are synchronous; as the surface is lowered the weathering front proceeds downwards. Weathering was already stressed by Wayland (1933). However, he was not explicit concerning the process of erosion, although he mentions peneplanation for surfaces with some relief, denudational stripping for extreme flatness.

There are three strong arguments for Büdel's model: 1. plains cutting across rocks of different resistance can be explained by intensive weathering. Only a uniform weathering mantle of several meters thickness can mask differences in rock hardness. 2. Denudational processes not decisively controlled by river erosion are responsible for planation even on the divides. A good indicator of these processes are planation passes crossing divides. 3. Inselbergs are not explained by slope retreat or as erosional remnants due to hard rock. Slope retreat starts from the rivers and is unlikely to occur in all four main directions. The distributional pattern of inselbergs does not suggest geological causes like narrower or wider spacing of faults or internal rock differentiation in crystalline rocks. These patterns should be repetitive and inselbergs are randomly distributed. Rather, the positions of inselbergs are to be understood from the morphogenesis, especially with respect to scarp evolution (Fig. 3, 4; Bremer 1971). The model of Büdel does not contradict explanations in terms of variable rock resistance but rather includes these, especially when the wash surface is close to the weathering front.

Slope retreat on the other hand is inferred by the areal succession from flat to steep slopes, from small to high gradients. The method of deducing an evolution from sequences is well established in geomorphology but is not in all cases unequivocal (Fig. 5). Different kinds of slopes also evolve with planational lowering. This is relevant to the explanation of outliers, relics of the higher plain. As slope retreat starts from the rivers it becomes difficult to explain why divides in front of scarps are planated to the level of the plain. Thus we must seek other criteria to evaluate

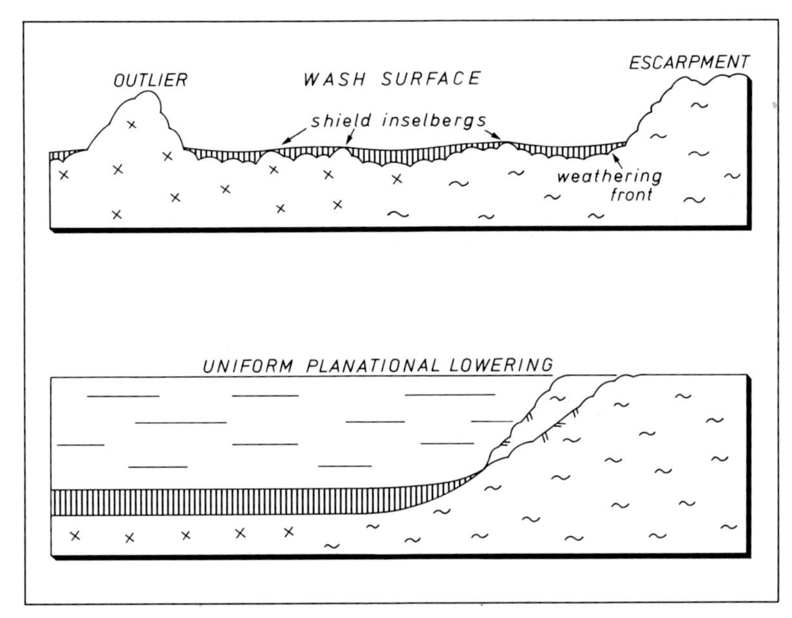

Fig. 2. The areal denudation of the surface and the contemporaneous lowering of the weathering front is called double planation or etch planation, as chemical weathering is the most important process. This concept of BÜDEL (1957, 1977) was deduced from the major landforms: plains with low overall gradient, slopes not necessarily adjusted to the rivers, inselbergs of different height (SI = shield inselbergs = initial stage of inselbergs which grow upward relatively by planational lowering of the surrounding), escarpments often crossed by rivers with waterfalls or rapids. The prerequisite for the development of these landforms is an intensive, rather uniform weathering mantle several meters thick. Presumably the differences of plains having more or less relief, few or many inselbergs are due to the thickness of the regolith at the beginning. The weathering front shows more (a) or less (b) relief and is more or less distant from the surface, due to intensity and differentiation in moisture conditions. While in a thick regolith under a perhumid climate there is little differentiation of the underground moisture and thus an even weathering front results, in the thinner regolith divergent weathering is initiated, leading to an uneven weathering front and eventually more relief at the surface (BREMER 1986).

the problem of slope retreat. Remnants of a volcanic flow or of a dated deposit in front of the escarpment prove a stationary scarp, but they are seldom to be found. BREMER (1971, 1981) proposed, from a variety of arguments, including the increase in number of inselbergs in front and on top of an escarpment and from the network of rivers, that slope evolution consists of initial steepening with retreat of the lower more intensely moistened slope, thus steepness increases. There is a relative stability of steep slopes and small shifts of the crest (Fig. 3, 4). Quite often, whale backs, very well rounded core stones, caves and karren in crystalline rock or sandstone and wide runnels provide further criteria for the recognition of slope stability and of weathering under humid tropical conditions, which is underlying this model.

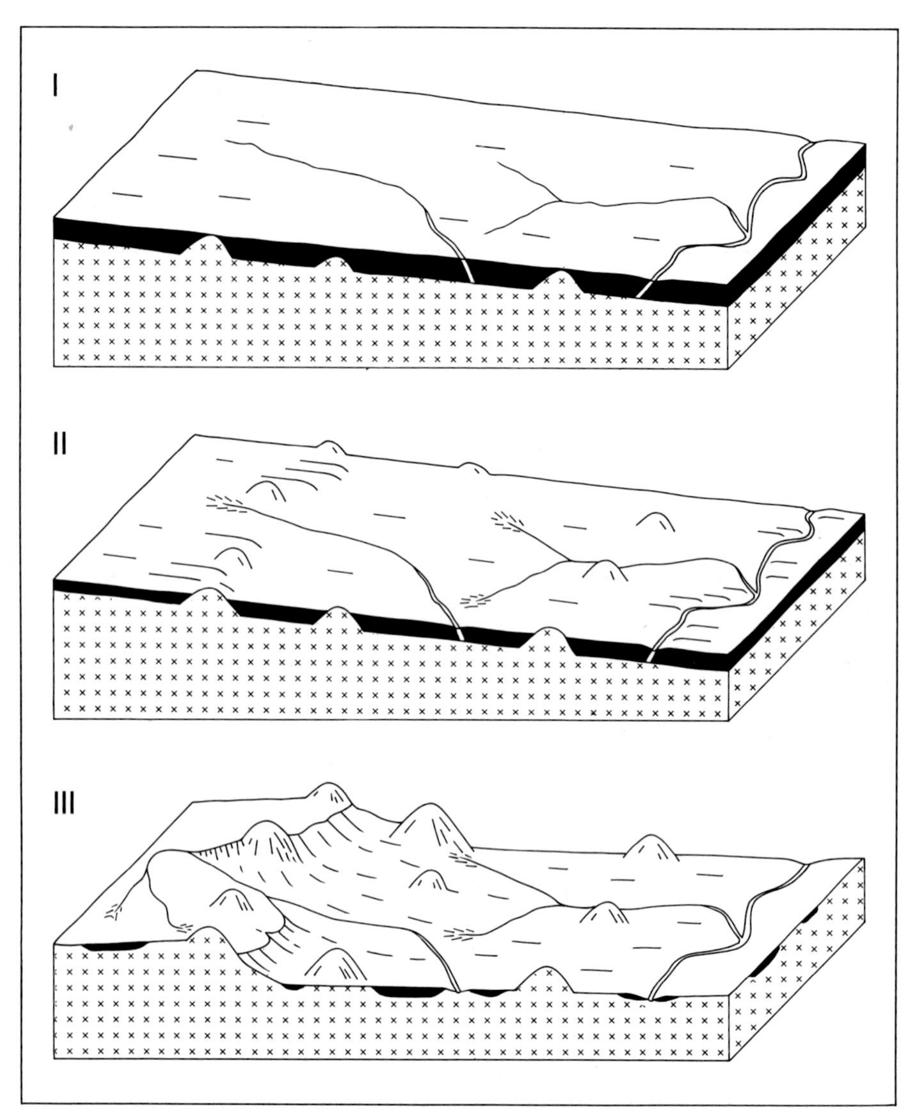

Fig. 3. Development of escarpment and inselbergs near Kabba, SW-Nigeria, simplified after BREMER (1971). During and after uplift planational lowering is more active near the base level (Niger R.) where soil moisture and therefore weathering and denudation are somewhat higher (stage 1). Once a rock surface is exposed, inselbergs and slight rises develop (stage 2). As the network of rivers is inherited, the rises have many wide openings. Rivers are attacking the rise from the rear, too. The proximal one may have been completely planated, (stage 3) and may only be deduced from a cluster of inselbergs. The distal rise evolves into an escarpment with inselbergs in front and on top and with wide indentations (triangular reentrant after BÜDEL 1977). The original weathering cover has been largely removed. The rivers today are slightly incised and accompanied by some alluvium. This is most likely due to climatic change.

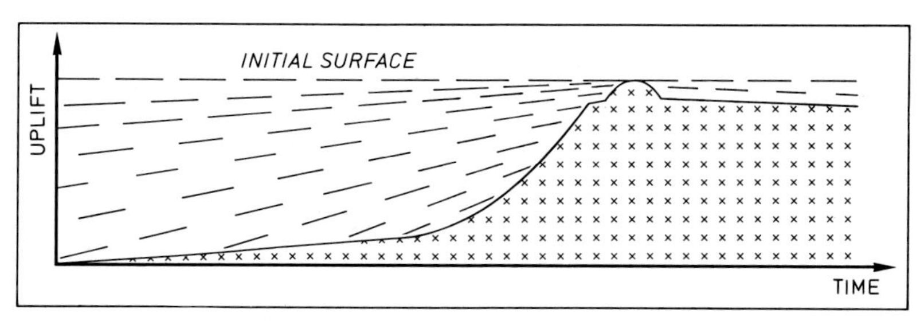

Fig. 4. Schematic presentation of the development of an escarpment. With uplift planational lowering is more intense near base level, as soil moisture there is more uniform. In time a band-like area with slightly higher gradient more or less parallel to base level evolves. Seepage is faster there, and the weathering front comes closer to the surface. Once rock has been exposed, inselbergs develop. With further difference in height above base level (about 100 m??) an escarpment evolves. Of course this is not a strict threshold but a figure with wide variations depending on climatic factors, e.g. the distribution of rainfall or the country rock attacked.

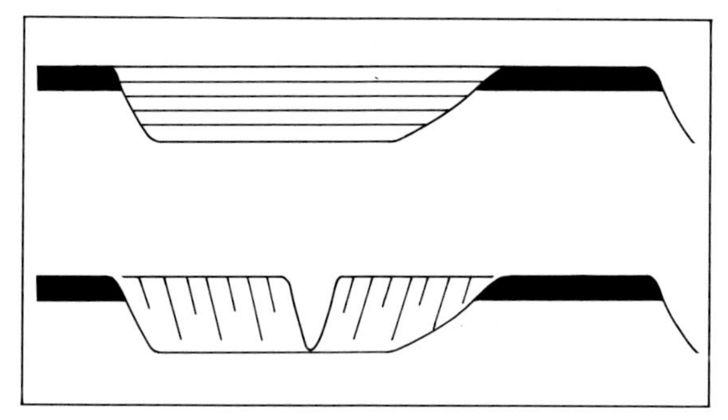

Fig. 5. Outliers in front of an escarpment, even if they are capped by basalt or dated sediments may be explained just as well by planational lowering as by scarp retreat.

The second argument for slope retreat, that is, the higher the gradient the greater the erosin, applies to extratropical denudational processes controlled mainly by gravity. In tropical regions control by friction is more important due to "divergent weathering" (BREMER 1971, 1981). Once a rock surface is exposed, rain water runs off very fast and there is almost no further weathering with decay rather hardening of the surface whilst in the damp and warm surrounding where the run-off soaks in there is intensive rotting and transformation of minerals. Thus even very steep rock surfaces are resistant and gradually grow relatively, whilst the surrounding area is lowered. Controlled by moisture conditions, a plain or a flat slope might show stronger erosion than a steep slope.

The model of BÜDEL explains the major landform features and provides a few basic principles: landforms in the tropics are unique and may not be explained by concepts evolved in the temperate zone. Only with intensive weathering do the differences in rock hardness become unimportant. Extensive plains are formed by planational lowering. It is a great advantage that the model of BÜDEL is an open system, open to changes of energy due to climatic change and/or uplift. It might be added that incision too is not completely comparable to fluvial erosion in the temperate zone (BREMER 1971). There are few tools in form of pebbles for mechanical erosion. Incision follows preweathered lines, where the fine material is washed out. This explains peculiar river nets (see p. 195).

2 *Morphogenesis in the tropics*

The huge variety of landforms in the tropics requires a genetic approach controlled by climatic change and/or uplift unique to each region. However, there seem to be certain general aspects. The first three concern red loam (oxisol, ultisol, latosol) formation, requiring a humid tropical climate of at least 1650 mm precipitation per year (SPÄTH 1981).

1. The highest parts of the landscape are usually the oldest ones – with, of course, the exception of volcanos or tectonic blocks. In most cases this oldest "relief generation" (BÜDEL 1977) has been shaped by etch planation and has been dissected into small relicts. These are often stripped of the regolith cover and only the roots of relict soils are preserved or new soils are developed. There is only slight or no further lowering of these planar relief elements but increasing dissection.

The lower plains are relatively younger. However, absolute altitude, as such is no indicator of a young age. One of the oldest surfaces appears to be the plain in the Rio Negro/Casiquiare region in South America (SCHNÜTGEN & BREMER 1985). Extreme flatness, few very high inselbergs, some very old soils, and the low ionic content in the rivers are indicative. There may be rapids in the rivers despite the low gradient. They are not due to renewed erosion caused by uplift but to exposure of the weathering front.

2. The younger relief generation i.e. the lower surfaces are intiated by uplift. The planational lowering is increasingly restricted to the areas of uniform weathering. These are controlled by more or less uniform moisture conditions, found near the main lines of discharge. These plains differ from plains of lateral erosion by rivers, or pediplanation by slope retreat, not only by their weathering cover but by special features like inselbergs, planation passes in the divides, broad planational embayments in escarpments (triangular reentrant, Fig. 6).

Probably the differences in detailed landform features, like number, height, form, and position of inselbergs, slope form and gradient of escarpments as well as their variously indented plan, overall inclination of the plain and detailed gradients from the divides to the rivers, might be explained in terms of the rate of uplift. It is proposed that with slow uplift there is no incision but rather homogenous weathering conditions in the whole area resulting in uniform landforms. Once the rates of denudation exceed the rates of uplift the thickness of the weathering mantle decreases and inselbergs, scarps and local gradients increase.

3. Incision and dissection are also controlled by the preceding chemical weathering. This explains the irregularity of valley width, gradient, and length of slopes, the

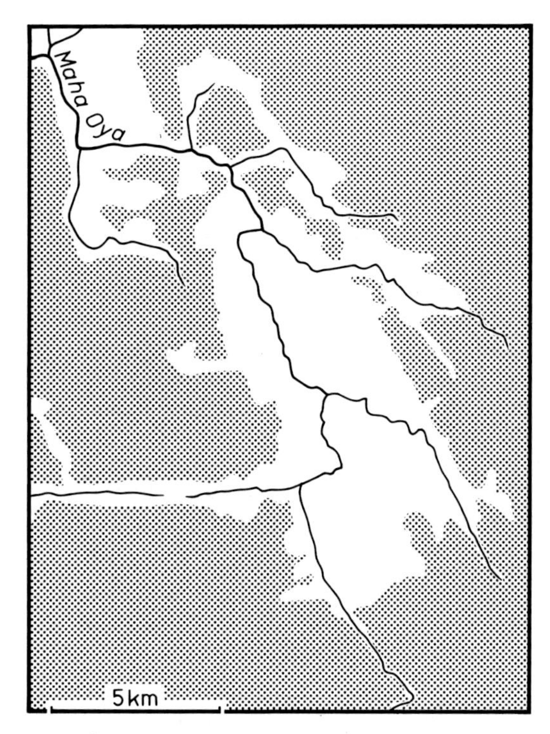

Fig. 6. An intramontane basin may have several outlets, as in the case of a basin at the western side of the highlands of Sri Lanka. Sometimes there is not even a river in these planation passes. The plan of the basin may be adapted to tectonic lines, but as was pointed out for the inselbergs, there are examples independent of rock hardness. From the position of inselbergs, the planated divides, the plan of the scarps, and their slope a planational lowering is deduced for the intramontane basin.

close connection of the network of rivers to fault lines in some places as well as the independence in others; incision starts on a thin regolith and proceeds along faults, or incision begins on a thick regolith and is superimposed. There are relatively few fluvial deposits, because the products of chemical weathering are carried away in suspension to a large extent. Little-transported lag deposits are quite often placers, but they are irregular in distribution. Large accumulations may of course be found in areas of tectonic subsidence or near the coast due to changes in sea level.

4. With the change to savannah climate in upper Tertiary or Pleistocene time there has been a great variety of overprinting. Generally, intensive weathering has become more restricted in area as have the planational processes. In the more humid savannah climates the palaeosoils have been widely removed. Brown, grey, and black soils, often rich in stones, of about 1 m thickness have developed. There is more wash on slopes and deposition in depressions. In the drier savannah climates the oxisols are more widely preserved but are often reworked. Either they are trans-

formed more or less in situ, whereby they become more clotted, more earthy in texture or they are transported and redeposited and undergo new soil development. Deposits at the foot of slopes are usually only a few meters thick and discontinuous. Along the lines of discharge either in dambos or in wide river plains 10–20 m of deposits may be found. Where thicknesses exceed this, tectonic subsidence should be suspected.

Morphogenesis is important in three regards: 1. The extrapolation of recent processes does not explain the landform assemblage. 2. A transfer and generalisation of observations and results from one area to another with similar landforms is not secure. For example, deep penetration of weathering along fault lines in plains may only be found in the older relief generations (often the second?), not in the younger ones. 3. Landforms and weathering relicts of the older relief generation have not only been reduced by dissection and erosion but also altered, to varying degree. Whilst relic soils in most cases are patchy in distribution, landforms, especially planar surfaces, are more widespread and less changed. There are features, though, which from the very beginning were patchy in distribution, like massive ironcrusts (plinthite) and deep gruss profiles. In these cases their present occurence is not indicative of a formerly extensive areal cover.

3 *Geomorphological processes in the tropics*

Divergent weathering tends to produce steep and flat but few intermediate slopes. The weathering products are clay size or medium size sand and coarser particles. Silt and fine sand are not absent but sometimes conspicuously less frequent than in extratropical samples.

There are three main denudational processes which are combined in different proportions due to climate and pre-existing soils and landforms. They are: 1. Subterraneous removal of material occurs in the form of solutions, gels, and fines. This seems to be more important in the humid tropics than elsewhere. 2. Wash seems to be the predominant surface process in the semi-humid to semi-arid tropics. Mainly fines are carried away and may be deposited in small areas where flow velocity is diminished. The transport takes place in many small steps of perhaps only centimeters, usually not more than a few decimeters per rainstorm event. 3. Clay suspension is more active on flat relief elements where the soil is wetter and water prevails for rather longer time on the surface. Rainsplash and bioturbation augment this process.

In the tropics subterraneous movement of water is not only more important but much faster than in the temperate zone. This also explains why small creeks respond faster and soils dry up faster after the rain. This appears to be attributable to more stable pores in the soil, well seen in thin sections (Fig. 7). There are several reasons for this. Kaolinite and aluminia oxides and hydroxides unlike three-layer minerals, have lower swelling and shrinking values. Iron and silica compounds are well crystallized in old soils. Sometimes they coat pores, thus stabilizing them.

The subterraneous movement has an important lateral component. The regular arrangement of soils from the divide to the low-lying area may be due to this process but also to differences in weathering due to laterally changing moisture conditions. In case of relic soils on the divides and alluvium near the discharge line surficial wash

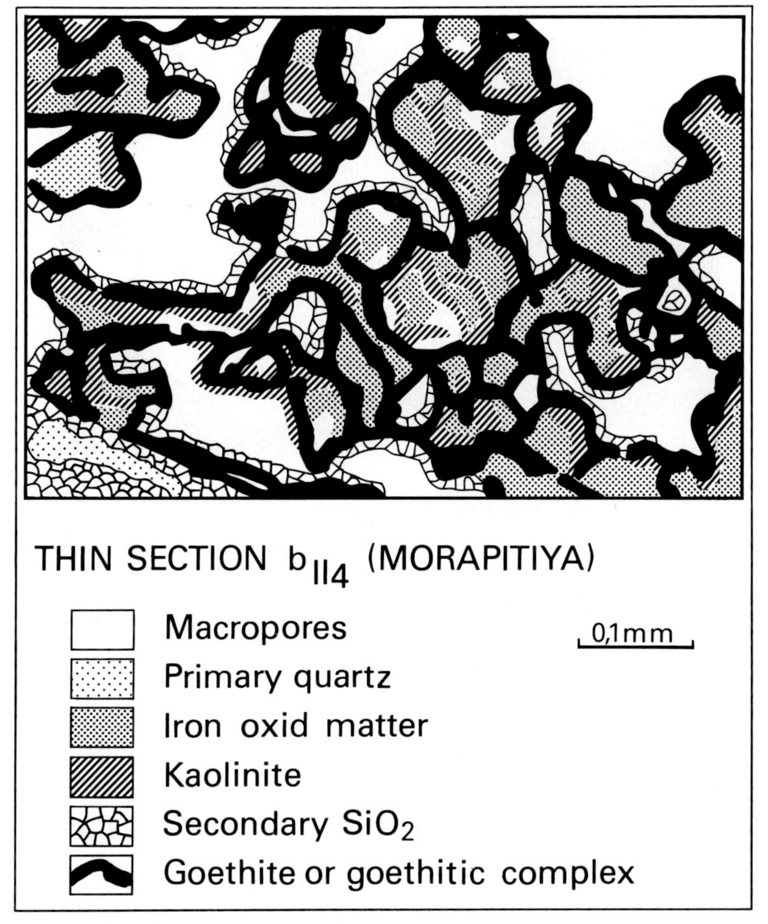

Fig. 7. With intensive weathering, in this example from near Colombo/Sri Lanka in the humid rain forest, nearly all minerals except for quartz are completely weathered. In the macropores there may be a crystallisation of secondary minerals, occasionally, as here, in successive stages (goethite, kaolinite, silica). Therefore the macropores are very stable, not shrinking or expanding in each rainy and dry season. Quite often solution features in associated quartz grains point to a longer weathering time as well. Drawing by SCHNÜTGEN and KUBELKE.

may be another reason. This Catena concept was developed in East Africa. The dryer the climate the more the subterraneous movement is restricted to broad linear bands with irregular widenings. Closed depressions and dambos are a visible expression of this. In combination with wash processes wide swales are formed. In the humid tropics, concentration of subterraneous water movement may also be attributable to tectonic uplift or fall of the sea level. Planar bands are formed.

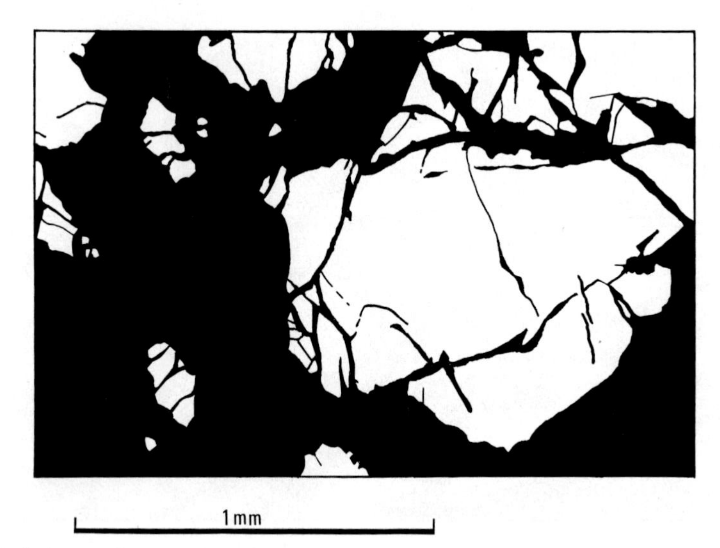

Fig. 8. Splitting and corrosion of quartz grains (white) by iron and clay complexes (black, including pores). The protusion of gels into the cleavages of the quartz grains is followed by the opening of pores probably in part due to crystallisation. From the different stages of splitting and corrosion a relative age may be deduced (after SCHNÜTGEN & SPÄTH 1983, Fig. 6). Weathering covers displaying these features have not been transported.

There seems also to be considerable subterraneous water movement on slopes, as may be deduced from the volume and rapid onset of storm run-off in small creeks. A change of density in the soil profile is the preferential pathway for the interflow. Initially this might be the lower limit of the main bioturbation activity, which is in the order of 1–1.50 m. In the process of self-enforcement a stoneline may develop (BREMER & SPÄTH 1989, MOEYERSONS 1989). Again these features are not unequivocal. Soils overlying stonelines cannot only be explained by accumulation on top of a stone pavement since stonelines also occur on divides. Most are, however, on slopes, with few in areas of accumulation. The areal extent of stonelines seems to be related to originally already well areated, i.e. porous soils. These are the above mentioned, aged red loams with macropores. In recent accumulation areas there is a higher proportion of swelling clays, especially in the black soils of the semi-humid zone. In these cases stonelines are most likely due to accumulation and in these horizons there is fast water movement whilst in topographically moist parts the soil or ground water may be almost stagnant. Due to high weathering rates a high ion content occurs in the latter. The faster moving water may be of good quality.

4 *Conclusions*

The models for relief development in the tropics are abstractions, explaining the major features of the landscape. As planation surfaces are usually very old landscapes they have undergone later changes. It is a generally accepted concept that younger plains are inset into older ones and surround these, due to general tectonic uplift and new planation. The stripping of the old plains, resulting from their higher position above base level and/or climatic change is more or less complete. This applies as well to the overprinting of the soils or their new formation. The landforms may also be altered by changes in the weathering cover. Mainly there is an increase in slope gradient and in minor relief features like inselbergs, small accumulation areas, rapids in rivers, etc.

These changes of the weathering cover and the landforms are due to tectonics and climatic change. Therefore individual landscapes evolve in uniquely different ways, such that they require investigation by the morphogenetic approach. The older relief elements leave their impact not only in relict forms and soils, but are also quite often a control for the recent processes.

Important parts in the explanation of landforms in the tropics are past and present moisture conditions. If this is accepted, the relief may provide feedback information with respect to the occurrence and movement of underground water. It is proposed that in old intensively weathered landscapes there is a greater potential for groundwater in fracture zones than in young landscapes. High inselbergs may be one indicator of an old development. Old weathering covers of the humid type contain more stable minerals of clay size and more permanent macropores. Water moves faster and has less ionic content than in younger soils, especially young ones in the semi-humid climate. Indicators of ancient weathering of the humid type are pisoliths, corroded quartz grains, the occurrence of haematite in the iron mineral suite and probably stone lines.

References

ALEXANDRE, J. & J.-J. SYMOENS (1989): Stone-lines. – Académie Royale des Sciences d'Outre-Mer. – Geo-Eco-Trop. 11 (1987, publ. 1989); Brüssel.

BREMER, H. (1971): Flüsse, Flächen- und Stufenbildung in den feuchten Tropen. – Würzburger Geogr. Arb. **35**.

– (1981): Reliefformen und reliefbildende Prozesse in Sri Lanka. – In: BREMER et al. (1981), 7–183.

– (1986): Geomorphologie in den Tropen – Beobachtungen, Prozesse, Modelle. – Geoökodynamik **7**: 89–112.

– (1993): History of German geomorphology. – Wiley. (in print).

BREMER, H. & H. SPÄTH (1989): Geomorphological observations concerning stone-lines. – In: ALEXANDRE & SYMOENS (1989), 185–195.

BREMER, H., A. SCHNÜTGEN & H. SPÄTH (1981): Zur Morphogenese in den freuchten Tropen. Verwitterung und Reliefbildung am Beispiel von Sri Lanka. – Relief, Boden, Paläoklima **1**; Stuttgart.

BÜDEL, J. (1957): Die „Doppelten Einebnungsflächen" in den feuchten Tropen. – Z. Geomorph. N. F. **1**: 201–228.

– (1977): Klima-Geomorphologie. – Berlin. Translated by L. FISCHER & D. BUSCHE (1982): Climatic geomorphology. – Princeton Univers. Press.

GRUNERT J. (ed., 1992): Geomorphology of the tropics with special reference to South Asia and Africa. – Z. Geomorph., Suppl. **91**.

KING, L. C. (1949): The pediment landform: some current problems. – Geol. Mag. **86**: 245–250.
– (1962): The morphology of the earth. – Edinburgh.
MOEYERSONS, J. (1989): The concentration of stones into stone-line, as a result from subsurface movements in fine and loose soils in the tropics. – In: ALEXANDRE & SYMOENS (1989), 11–22.
SCHNÜTGEN, A. & H. BREMER (1985): Die Entstehung von Decksanden im oberen Rio-Negro-Gebiet. – Z. Geomorph. Suppl. **56**: 55–67.
SPÄTH, H. (1981): Bodenbildung und Reliefentwicklung in Sri Lanka. – In: BREMER et al. (1981), 185–238.
THORNES S. B. & D. BRUNSDEN (1977): Geomorphology and time. – London.
WAYLAND, E. J. (1933): Peneplains and some other erosional platforms. – Protectorate of Uganda, Geol. Surv. Dept., Ann. Rep. Bull. Notes **1**: 376–377.
WIRTHMANN, A. (1987): Geomorphologie der Tropen. – Erträge der Forschung 248; Darmstadt.

Address of the author: Prof. Dr. HANNA BREMER, Geographisches Institut, Universität, D-5000 Köln 41.

Z. Geomorph. N.F.	Suppl.-Bd. 92	201–216	Berlin · Stuttgart	Mai 1993

Lateritic Crusts as Climate-Morphological Indicators for the Development of Planation Surfaces – Possibilities and Limits

by

JÜRGEN RUNGE, Paderborn

with 3 photos, 2 figures and 3 tables

Summary. In northern Togo investigations on parent rocks and the lateritic crusts developed above them were carried out. The question was as to whether or not laterites are suited, as BÜDEL maintains, to help in the reconstruction of extended planation levels. It will be shown that the genesis of lateritic crusts seems to be more dependant on the petrological conditions than had before been assumed. Furthermore lateritic crusts are often the result of redeposited material which can, therefore, not generally be used for the reconstruction of old erosion surfaces. Finally a comparison of the planation surfaces of Togo with the planation chronologies of West Africa existing will be carried out.

Zusammenfassung. In Nord-Togo wurden Ausgangsgesteine und darüber entwickelte Lateritkrusten vor dem Hintergrund untersucht, ob mit Hilfe der Lateritbildungen im Sinne BÜDEL's ausgedehnte Rumpfflächenniveaus rekonstruiert werden können. Es zeigte sich, daß die Entwicklung der Eisenkrusten in viel stärkerem Maße an die petrographischen Bedingungen gebunden ist, als bisher angenommen wurde. Darüber hinaus erweisen sich Lateritkrusten häufig auch als umgelagerte Bildungen und können somit zur Rekonstruktion von Altflächen nicht herangezogen werden. Abschließend wurde der Versuch unternommen, die in Togo beschriebenen Flächenreste in das in der Literatur bestehende System der Rumpfflächen in Westafrika einzuordnen.

1 *Introduction*

Since the "Düsseldorfer Geographische Vorträge" took place in 1927 (compare THORBECKE 1927 and others), German geomorphology has been divided into two different schools of thought: On the one hand there is the more structural-geomorphological line, especially represented by HERBERT LOUIS (1900–1985), and on the other hand there is the so-called climatic-geomorphology whose most popular representative was indubitably JULIUS BÜDEL (1903–1983). Germany's geomorphological research has been characterized by these two often conflicting lines of research. Especially during the decades after World War II and even still today structural- and climatic-geomorphology seem to contradict each other. In addition to the two schools of thought mentioned above, German tropical geomorphology has been

0044-2798/93/0092-0201 $ 4.00

very interested in the investigation of planation surfaces all over the globe. The sense and the intention of the intensive and often controversial discussions on planation surfaces in Germany are obviously not comprehensible to Geomorphologists from francophone and anglophone countries. This is expressed by a remark of OLLIER (1981: 152), who commenting on the scientific treatment of planation surfaces, says: "Most people who are not blind or stupid can tell when they are in an area of relatively flat country: they can recognize a plain when they see one . . .". Even though this comment is not very encouraging for further research on planation surfaces, the present study is nevertheless concerned with these problems. The causal conditions and interrelationships of the formation of lateritic crusts, their distribution and their importance for the development of planation surfaces are the topics of the following essay. With that a scientific statement of BÜDEL (1981) is taken up again, who was always convinced that there are causal correlations between the formation of laterite above latosols and the development of extended planation surfaces. "Lateritkrusten . . . greifen wie Rumpfflächen über die verschiedenen Gesteine hinweg . . . und bilden über diesen ein einheitlich hartes Deckgestein" (Lateritic crusts cut different rocks like planation surfaces and form a uniform resistant cover) (BÜDEL 1981: 107).

2 *Geological and geomorphological conditions of the study area*

The above mentioned conception of BÜDEL was examined on the planation surfaces and the lateritic crusts that are found in the northern part of Togo in West Africa (9°–11°N, 0°–1°30'E). Geologically this area is located in a palaeozoic panafrican orogenetic area called *Dahomeyiden* which strikes from SSW to NNE direction. The Precambrien Westafrican basement is found close to the panafrican orogen.

The palaeozoic folded rock-structures are interpreted as the result of a collision of continents and an overthrusting of continental plates. An easterly continental plate drifted from east to west and affiliated with the African craton (CRENN 1957, BESSOLES & TROMPETTE 1980, République Togolaise 1984). AFFATON (1975) distinguished and divided the *Dahomeyides* into structural units called *Buem, Atacora* and *Plaine de Bénin.* The outcropping geological structures determine considerably the very heterogeneously composed geomorphological landscape in northern Togo (Fig. 1, Fig. 2). The unit of *Buem* consists of sedimentary rocks of the Oti-Dapaong-Bombouaka-group. They cover and fill out the northeasterly continuation of the ghanesian Volta-basin. In the northern part these sand- and clay-stones strike out forming two marked monoclinal ridges (Fig. 2b, 2c). The relief of cuestas leads over to the Precambrian basement formation (*Birrimien*) that continues far into the state of Burkina Faso. Resistant quartzitic-schists and mica-schists of the structural unit *Atacora* are connected to the Oti-Volta-basin in the eastern and southeastern parts of the area studied. The dominant relief features of Togo are built up by these weathering resistant rocks forming steep escarpments and ridges (*Togo-Atacora-mountains*) (Fig. 1, Fig. 2a). Further east the structural unit *Plaine de Bénin* adjoins (Fig. 1). It is characterized by granites, gneisses and muskovite-quartzites forming an extended planation surface with a mostly overall low relief intensity. Locally inselbergs and groups of inselberg-mountains occur, an example being *Monts Kabyé.* These moun-

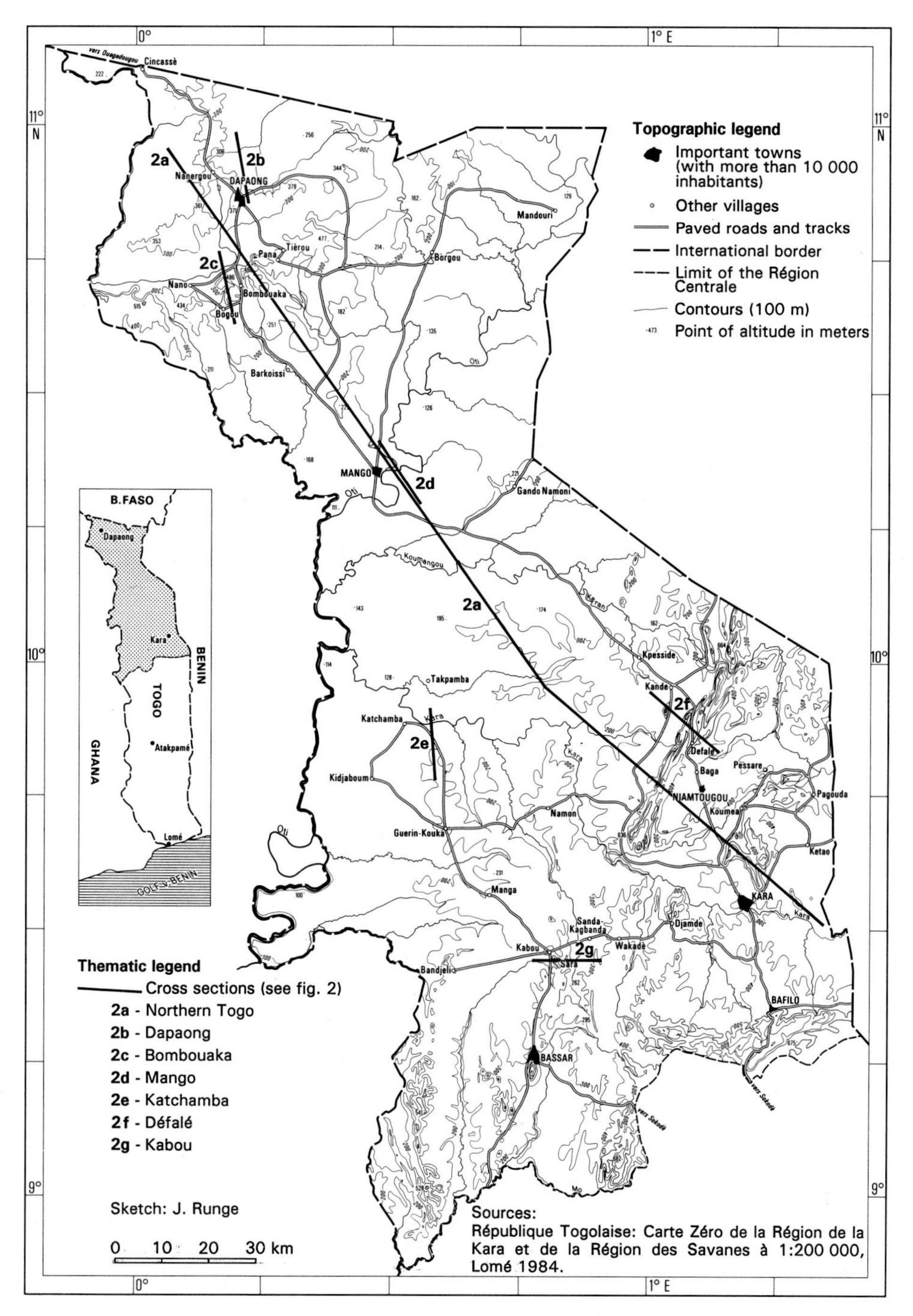

Fig. 1. Location map of Northern Togo and position of cross sections.

tains consist of basic and ultrabasic rocks that intruded into the upthrusted continental edge in the course of the panafrican orogeny (Fig. 2a). Therefore from the viewpoint of geomorphology, northern Togo can be described as an extended basin filled up by thick layers of sediment showing cuestas on the northern margins and by ridges and escarpments, planation surfaces interspersed with groups of inselbergs in the south and southeast (Fig. 1, 2a).

Inserted into the geomorphological variety of forms that is dependant on the heterogeneous geologic conditions a big number of erosion surfaces and planation levels can be observed. Some of them are covered by massive lateritic crusts.

The relationship between lateritic crusts and planation surfaces were investigated in six selected areas. Geologic and topographic cross sections were mapped and measured out, and the distribution of the different forms of laterites were studied in detail.

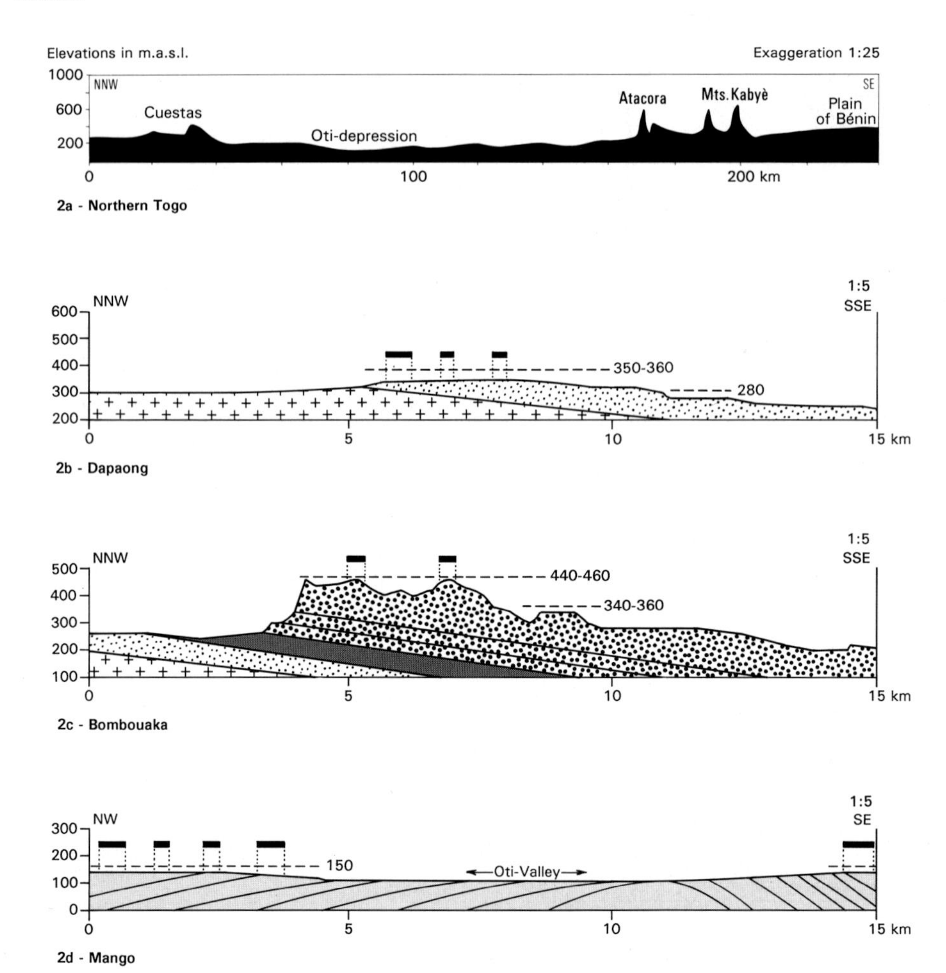

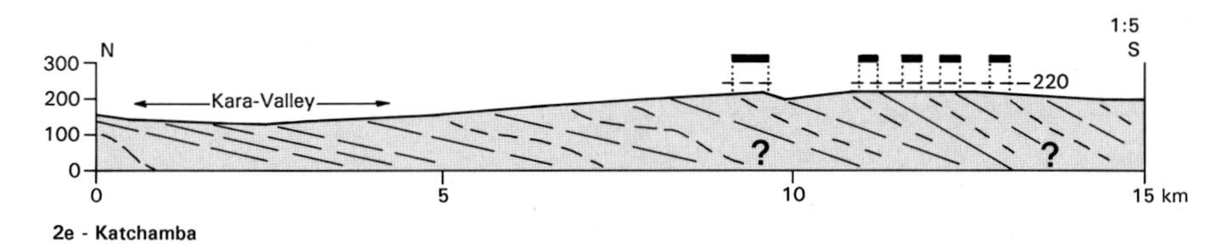

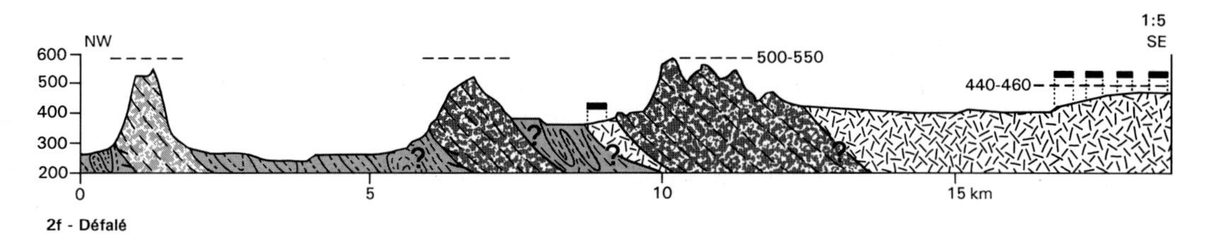

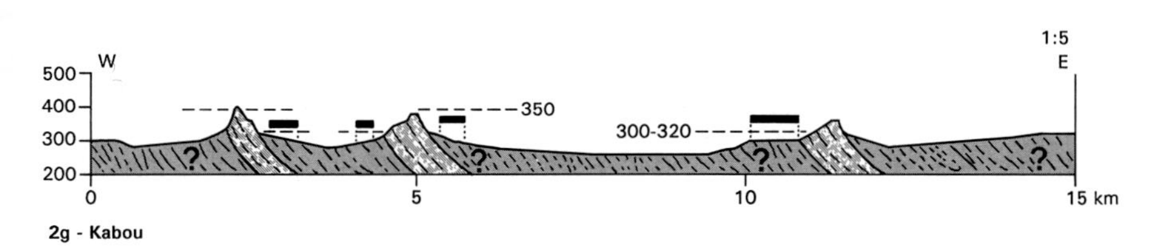

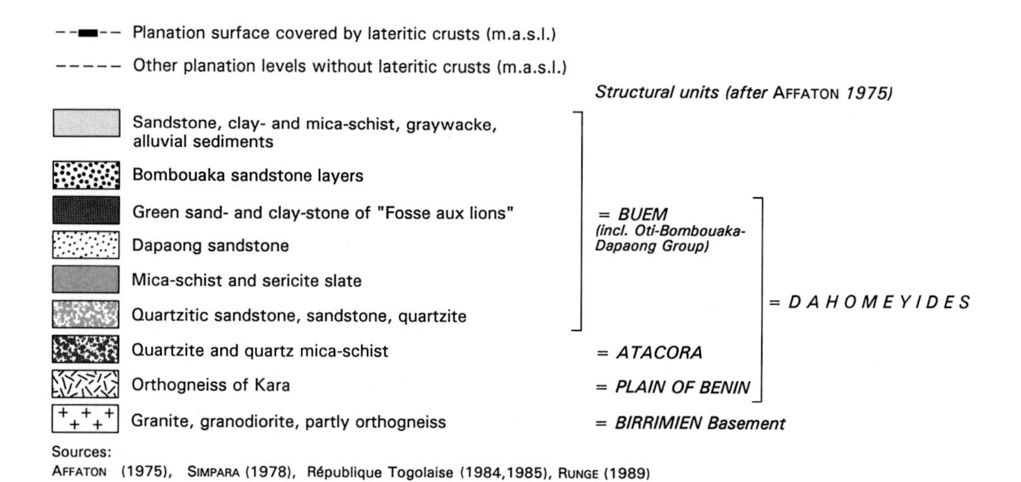

Fig. 2. Geologic cross sections and occurrence of lateritic crusts in Northern Togo (see Fig. 1).

3 *Distribution, occurrence and characteristics of lateritic crusts*

3.1 *Geological and geomorphological cross sections*

The examination of six cross sections located in northern Togo (Fig. 1, Fig. 2) resulted in the following observations:

Cross section Dapaong (Fig. 2b): Near the village Dapaong a single sandstone layer (*Grès de Dapaong*) covers discordantly the basement forming a monolithic monoclinal ridge. The ridge rises up only 20 to 30 m above the foreland consisting of granitic *Birrimien*-basement. The bottom rocks of the Dapaong sandstone layer consist of a heterogeneously composed base conglomerate within a yellowish, finegrained sandstone. The overlaying sandstone appears in reddish, pink to white colours and includes quartzitic weathering resistant material. The layering of the Dapaong sandstone is not uniform. Layers at a thickness of some centimetres to several meters occur. The dip of the sandstone directs to a SSE-direction with 4°–6°. In place of a clear scarp often a 30°–45° steep slope forms a kind of transitional ramp from the foreland to the sandstone plateau. Although the sandstone is widespread, it is formed only out of locally monoclinal ridges and scarps, presumably because of individual geologic structures effecting weathering resistance (compare Barth 1970). Lateritic crusts occur near to the escarpment in 320–350 m a.s.l., and they probably increase the morphological valence of the structure-affecting Dapaong-sandstone at this position (Fig. 2b). Poss & Rossi (1987: 41) interpreted these crusts above a planation surface as remnants of a fossil, late-pleistocene erosion surface. On the basement rocks north of the scarp, lateritic crusts are found only locally on *interfluves* with no plain-wide expansion. This granitic area is featured by a dendritic drainage net with flood areas which are sometimes extended.

Cross section Bombouaka (Fig. 2c): Near the village Bombouaka (Fig. 1), there is a clear sandstone escarpment of more than 200 m in height caused by the *Grès de Bombouaka* (Fig. 2c). These sandstones are characterized by a very massive structure. They show outcrops in banks of several meters of thickness. The quartz grains are light pink to light beige in colour and are sugar-like fine to partly arcose-like. The upper rocky and sometimes overhanging parts of the scarp transmit to the foreland by a vertical steep fall. The slope, first concavely then straight shaped, is covered with detritic bolders and fine material which is very much overgrown by bush vegetation. Between the scarp and the slope, another morphographic feature of resistant sandstone bank, forming a second low step, can be seen. This rocky step is frequently buried under erosion masses which have fallen down from the above scarp. Underneath the Bombouaka sandstone layers soft sand- and dark clay-stones follow and build up the foreland. The low water infiltration rate of the clayey layers are the reason for a wide swampy depression area in front of the escarpment (*Fosse aux Lions*).

Lateritic crusts can be observed again on top of the sandstone plateau in 440–460 m a.s.l. They obviously differ from the crusts found on the Dapaong-sandstone. Here crusts are less compact and more conglomerate-like and remind us of detritus. The high percentage of well subrounded sandstone fragments inside this crust leads to the assumption of a former fluvial rebedding and displacement of the material. Based on these results Poss & Rossi (1987: 30) recontructed fossile sections of a valley relief on the recent monoclinal ridge near *Pana Tièrou* (Fig. 1). It was dated from the Eocene or Miocene epoch. Therefore the formation of the lateritic crusts near Bombouaka took place before or during the initial formation of the monoclinal ridge. By this morphogenetic process the former drainage system coming from the sandstone plateau was separated from today's river net, and fossil remnants of fluvial deformation remained in top relief positions. Within the Bombouaka sandstone another planation level that obviously cut the layer surface can be observed in 340–360 m a.s.l. (Fig. 2c) but here lateritic crusts do not occur.

Cross section Mango (Fig. 2d): The small town Mango is located close to the river Oti at the lowest topographic position of northern Togo study area. Predominantly flat-laying calcareous clay-schists and clayey sandstone series form the geological underground. Only seldom do these rocks crop out at the surface or in river beds. The Mango area is furthermore characterized by thick layers of alluvial sandy deposits and also pebbles. The alluvial deposits frequently show an intensive ferralitic weathering, and sometimes they are in parts ferricrete-like consolidated. Around Mango lateritic crusts are visible on the surface in a wide expanse (120–150 m a.s.l.). The town Mango itself is situated on a laterite incrusted plateau in 120 m a.s.l.). This plateau is dissected by fluvial dynamics directed to the Oti. Adjoining to a low step formed by the crust is a slope of usually several 100 of meters in length. This slope is concavely to convexly shaped and is covered evenly by a loamy to sandy hillwash cover. VIELLEFON et al. (1965: 126) classified five erosion levels in the region around Mango. They are described as being fossilized by lateritic crusts over clay- and sand-stone as well as over former alluvial deposits. Such "planation-levels" in 220 m, 180 m and 160 m (VIELLEFON et al. 1965) cannot be confirmed by the present study.

Cross section Katchamba (Fig. 2e): The geological setting of the region where the village Katchamba (Fig. 1) is located is quite comparable to the Mango situation. Sandstone and claystone locally associated with schist-like rock structures occur. The layering varys between horizontal to graded. The genetic reason for this is due to so-called *microplissements*; e.g. foldings of the softer sediments which took place in connection with the south-easterly orientated pressure of the panafrican orogeny (LE COCQ 1986). The geomorphological appereance is identical to a gently undulated plain. Iron-comprised crusts of two different types have developed. On the one hand, in the river bed of the *Kara* (Fig. 1), small crust forms can be found that look like little tabular mountains. This type of crust consists of sand and gravel consolidated and hardened by ferric oxide. They can be interpreted as remnants of a formerly arid climate period when the evaporation of high concentrated iron-comprised "lakes" of residual water caused of the river sediments to consolidate in crust-like features. These crusts are not important for the present discussion. However, there are other lateritic crusts in the Katchamba region which are very interesting; they are marked on top positions in 200–220 m a.s.l. (Fig. 2e). These crusts form small and convex-buckled, fragmented boulder hills at around the 200 m contour line. These lateritic crusts seem to be strongly degraded already. There are hardly any plain-like distributions of laterites in the Katchamba area.

Cross section Défalé (Fig. 2f): Very resistant quartzites and micaceous schists of the structural unit *Atacora* form steep rising ridges and scarps. To the northwestern direction folded black schists of the *Kandé*-serie, and to the southeastern direction

Photo 1. Isolated rest of a planation surface in 500–550 m a.s.l. situated above the quartzitic scarps close to Défalé (Fig. 2f). The oblique dipping rock formations of the Atacora are cut by a nearly horizontal plain (02.02.1987).

Precambrian basement consisting of the orthogneiss of Niamtougou crops out. The quartzite banks are dipping with 50°–70° to E or SE (Fig. 2f). The Atacora-quartzite has an sugar-like fine grain and is very hard and compact. The grain's surfaces are covered with numerous muscovit micas. At the base of the outcropping quartzite layers debris covered slopes are noticeable. Also these slope sections are well overgrown with vegetation. The debris boulders are covered with an iron-coloured skin created by weathering processes. Planation surfaces without a cover of lateritic crusts are found at 500–550 m a.s.l. They cut the oblique dipping Atacora-structures nearly in a horizontal way (Photo 1). Lateritic crusts with a plain covering distribution exist on the Precambrian orthogneiss (age: 2064 M.A. République Togolaise 1984: 24) southeast of the scarp relief. Erosion processes have effected a great number of small tabular mountains with lateritic crusts in top relief positions. Within a geological window in between the Défalé scarp relief, a raft of Kara orthogneiss comes to the surface (Fig. 2f); it is interesting to note that this position is also covered by a lateritic crust. FAURE (1985: 17) dated these crust-formations from the early quaternary. That corresponds to the surface level of the *Haut Glacis* of MICHEL (1973, see Table 3). The surface remnants described decline gently to a SE-direction against the recent drainage and flow off in the direction of the Oti located in the NE. This causes one to assume that there was an earth history change of a formerly drainage direction leading to the river *Ouéme* in Bénin.

Cross section Kabou (Fig. 2g): From the geomorphological point of view, the region around Kabou is a typical landscape of inselbergs. But when we look at the inselberg massivs more closely, we see that they must be classified as monadnocks or severely weathered hogbacks (Fig. 2g). Here the *Buem* rocks consist mainly of quartzite, quartzitic sandstone, mica schist and metamorphic graywacke. As was the case with the cross section Défalé (Fig. 2f), the influence of the panafrican orogeny comes out in the layering and folding of the rocks. The inselbergs are surrounded by small footslope-like ledges at 300–320 m a.s.l. that carry a lateritic crust. Sometimes such lateritic crusts also form very small hills between the overtopping inselbergs. Le Cocq (1986) agrees with Michel (1973) by classifying these crusts as a planation surface level of the *Haut Glacis* (see Table 3).

3.2 *Geochemical analysis (%) of parent rocks and lateritic crusts*

The six cross sections introduced above show that lateritic crusts seem to exist above nearly all kinds of rocks and inside all geomorphological units and at different topographic levels (m a.s.l.). From the morphographic point of view the location and distribution of lateritic crusts as described above (see Fig. 2) does not allow for a clear assignment to a widely expanded planation surface. Nor could a causal connection between the formation of lateritic crusts and planation surfaces be proved. The following analysis of the geochemical composition of the parent rocks and their lateritic crusts will show whether the parent material plays a major role on the formation and the characteristics of in situ developed lateritic crusts. The theory of the climate-geomorphological point of view would expect hardly any dependency between the parent rock and the overlying crust, because of the very deep reaching weathering processes and the long span of time, going back, for example, to the tertiary, which led to the crust's formation. Table 1 now shows the chemical composition of the lateritic crusts described and their parent rocks. It is obvious that besides a sometime similar outward appearance of lateritic crusts, the chemical composition might be quite an individual one that differs from one crust to another enormously. The content of SiO_2 of the analyzed laterites varies between 21% (Bombouaka) and around 58% (Mango). The content of Al_2O_3 shows a span between 17.8 and 7.82% (Table 1). Very clear differences have also been measured with regard to the main feature of a lateritic crust: the content of ironoxide (Fe_2O_3).

The laterite sample from Katchamba (Fig. 2e) shows 52.5% Fe_2O_3, the sample from Mango only 24.4%. The lateritic crusts are obviously differentiated and individual weathering products of their respective locations. The combination of the elements of the parent rocks and lateritic crusts, examined under the premise of an autochthone development, shows considerable differences concerning the losses of quartz (=desilification) and the enrichment of Fe_2O_3 and Al_2O_3 in the weathering product. The example of Mango comes to an even higher SiO_2-content in the crust than in the parent rock. This is to be understood from secondary lateral and conglomeratic inclusions of quartz during the consolidation of the lateritic crust. Table 1 describes a very distinctive character of the lateritic crust that seem not compulsory to be a contemporaneous and uniform formation under the process of a planation surface genesis. A clear linear correlation between high primary contents

Table 1 Geochemical composition (%) of parent rocks and lateritic crusts in Northern Togo (RFA-Analysis).

Cross-section	Rock	SiO_2	TiO_2	Al_2O_3	Fe_2O_3	MnO	MgO	CaO	Na_2O	K_2O	P_2O_5	i.L.	Total
2b DAPAONG	Laterite Sandstone	35,81 94,43	0,58 0,07	12,62 1,21	39,91 2,50	0,05 0,01	0,04 0,04	0,03 0,02	0,03 0,18	0,14 0,07	0,13 0,02	10,05 1,05	99,40 99,65
2c BOMBOUAKA	Laterite Sandstone	21,04 90,77	0,71 0,13	17,80 4,23	49,19 0,55	0,03 0,01	0,00 0,06	0,04 0,06	0,21 0,20	0,08 2,14	0,20 0,05	10,15 1,18	99,47 99,37
2d MANGO	Laterite Clay-schist	58,49 48,45	0,46 0,42	7,82 9,56	25,42 3,84	0,13 0,69	0,08 1,28	0,12 16,48	0,00 2,61	0,22 1,29	0,12 0,49	6,73 14,56	99,59 99,68
2e KATCHAMBA	Laterite Arcose	21,03 79,98	0,52 0,38	12,45 8,72	52,50 3,10	0,07 0,11	0,13 0,62	0,13 0,29	0,06 2,31	0,30 1,88	0,30 0,11	11,83 1,81	99,33 99,31
2f DÉFALÉ	Laterite Orthogneiss	34,53 71,05	0,54 0,22	15,74 15,73	36,98 1,33	0,02 0,02	0,00 0,36	0,10 0,50	0,00 3,45	0,23 5,09	0,20 0,26	11,21 1,55	99,55 99,56
2g KABOU	Laterite Graywacke	47,16 79,39	0,62 0,50	10,62 8,63	30,84 3,60	0,54 0,10	0,16 0,77	0,06 0,70	0,00 2,02	0,59 1,80	0,13 0,10	8,66 2,10	99,40 99,71

i.L. = ignition losses

Table 2 Relationship between a primarily iron rich rock (haematite) and the corresponding lateritic crust (example of Manga, see Fig. 1).

Sample	Rock	SiO_2	Al_2O_3	Fe_2O_3
Manga 0°42'E/9°33'N	Laterite	31,44	13,03	53,12
230m a.s.l. (see fig. 1)	Haematite	38,59	03,19	43,43

of particular elements in the parent rock and likewise high contents of the same elements in the weathering product are not to discover. Such a correlation was proved by another rock/laterite crust-analysis from a location between Kabou and Katchamba. In the case of Manga (Fig. 1) a primarily high content of ferruginous silicate haematit in the underlying rocks effects an also high content of ironoxide in the lateritic crust. Table 2 shows the results of the main chemical components in %. A high content of ironoxide in the parent rocks seems to favour a high Fe_2O_3-content in the lateritic crust as well. In Table 1 again the data of the lateritic crusts of the examples Katchamba and Bombouaka are contradictory to these results. There the content of Fe_2O_3 in the parent rocks comes only to 3.1 and 0.55% respectively but in the lateritic crust they come up to 52.5 and 49.19% Fe_2O_3.

This led to the question as to whether the formation of lateritic crusts (for example in the course of the formation of planation surfaces) actually requires unique autochthoneous processes or whether multiphased polygenetic phases of laterization may also be a reason for the very high iron content of crusts on primarily iron-poor parent rocks.

3.3 *Autochthonous or allochthonous formation of lateritic crusts?*

Microsections and polished sections of lateritic crusts should give evidence if the crusts were developed *in situ* or if they were possibly redeposited; e.g. do the crusts show signs of a former morphodynamic combination of erosion processes?

Autochthonously formed lateritic crusts allow us to make certain extent conclusions concerning the climate-dependant exogene dynamic and the reconstruction of planation surfaces. Under the microscope the polished sections showed an undisturbed *in situ* formation of the lateritic crusts especially within the cross sections of Kabou, Défalé, Mango and obviously also in the case of the sandstone plateau close to Dapaong. The lateritic crusts of Katchamba and Bombouaka however show a disturbed and multiphased crust development; therefore we can presume that the older material of formerly existant lateritic crusts (Photo 3) was redeposited. The multiphase character of samples from Katchamba and Bombouaka can be confirmed by the high Fe_2O_3-content shown in Table 1. Photo 2 shows a polished section of the autochthonously developed laterite of the sandstone scarp at Dapaong. The quartz grains are clearly angular and partly mineralized, and they show no marks of assorting or another kind of morphodynamic stress. However, Photo 3 shows a section of an allochthonously developed lateritic crust taken from the cross section Katchamba. In the upper and lower thirds of the picture half circle-shaped structures can be seen. Partly subrounded, very mineralized, and especially small quartzitic grains can be observed inside these structures. Larger and less subrounded grains of quartz are situated between the spherical pisolites. The surrounding matrix between the quartz grains clearly contains less iron than those within the pisolites. So the two pisolites clearly belong to an older lateritic crust. This crust was probably eroded long time ago, and it was later on re-consolidated in a new lateritic crust (Photo 3).

4 *Results and conclusions*

The lateritic crusts of northern Togo can be understood as a result of intensive weathering under humid-tropical conditions. Contradictory to the viewpoint of Büdel (1981), they do not automatically mark wide and extended planation surfaces. The mineralogical and geochemical analysis of the lateritic crusts proved that, aside from a climatically induced lateritization, a petrographic predisposition accounts for the formation of lateritic crusts (see Table 2). The scientific literature contains numerous analysis on this topic emphasizing the close mineralogical relationships in the chemistry between lateritic crusts and their parent rocks (Ambrosi & Nahon 1986, Schellmann 1986). In spite of the many convergences in the chemical composition of the laterites (Table 1) that may occur, according to Büdel (1981), usually independant of the parent rock, the new analysis has shown a clear formative influence of the parent rocks. The "chemical heritage" of the basement rocks is still visible in the weathering products, even if the process of lateritization lasted for an extended period of time. A good example of this comes from Uganda (McFarlane 1976) where differences in the geological formations were correctly mapped by the indirect way of mapping differences in the outward appearance of existing surface lateritic crusts.

The common opinion on the importance of lateritic crusts for the development of planation surfaces is partly based on the fact that a lot of these studies were carried out exclusively in areas with a relatively homogeneous geologic underground. Differences of the laterite crusts cannot come out clearly as they would above a heterogeneously composed geology (compare Leprun 1979).

Further it has been shown that lateritic crusts can be autochthonous as well as allochthonous forms. This reduces their importance when there is a demand for reconstructing planation surfaces (Photos 2, 3). The development of planation surfaces in the part of northern Togo studied did not effect a widespread planation of the landscape during their genesis. The older relief complexes dating back to the panafrican orogeny obviously were not greatly effected by the planation process that took place more locally. The structural-impressed geomorphological forms as scarps, ridges and also inselbergs, mostly formed by weathering resistant rocks (for example quartzites) rather prove a tenacious ability of morphologic preservation against the climate-morphological and climate-controlled erosion processes. Planation surfaces were formed out but mostly above the already intensively saprolited geological underground. But this only happened locally, that means they were inserted into an already long existing relief of major structural forms (Fig. 2).

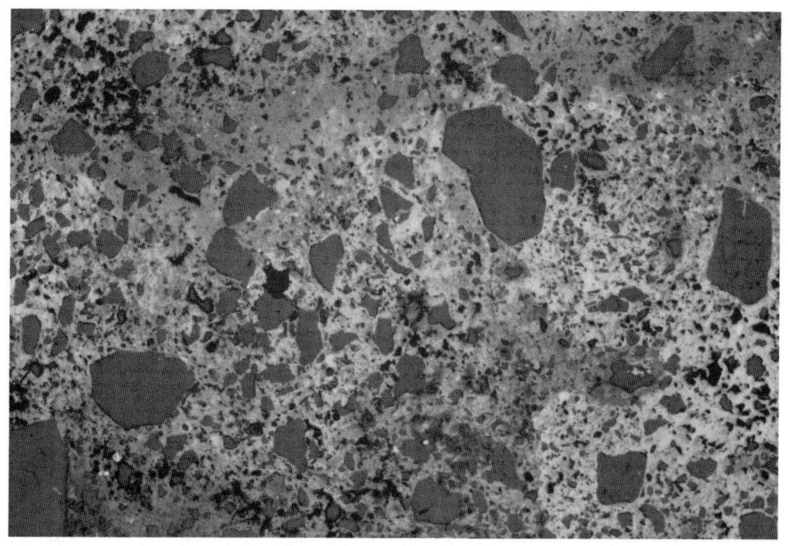

Photo 2. Polished section (3.2 mm large) of the autochthonously formed lateritic crust coming from the sandstone area of Dapaong (see Fig. 2b).

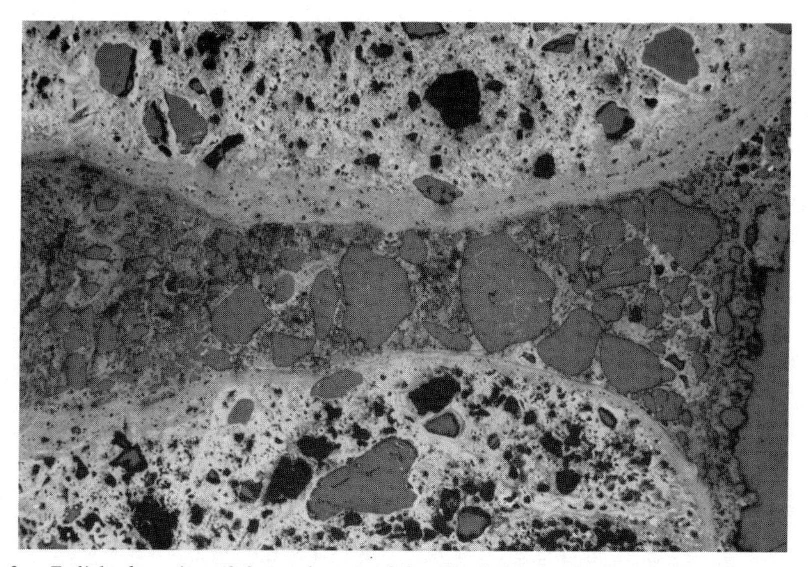

Photo 3. Polished section (3.2 mm large) of the allochthonously formed lateritic crust coming from the Katchamba cross section (see Fig. 2c).

Table 3 Planation surfaces chronology and lateritic crusts in Westafrica – a synoptic comparison from different authors and regions (compare Poss & Rossi 1987).

Locality / Authors — Period / Epoch	Senegal/Gambia (12°-16°N) (Michel 1973:292)	Burkina-Faso (12°-14°N) (Boulet 1970:245)	Northern Togo (10°30′-11°N) (Poss & Rossi 1987:41)	Northern Togo (9°-11°N) (Runge 1989, 1990b)
Late Jurassic	Top planation surface (*Labé*) in 1160-1200m a.s.l.	?	?	?
Cretaceous	*Dongol-Signon* planation surface in 850-950m a.s.l.	?	?	planation surface in 500-600m a.s.l. (*Défalé* escarpment, photo 1)
Early to Middle Eocene	*Fantofa*-planation surface (lateritic bauxite)	Lateritic weathering (bauxite)	?	
Eocene - Miocene	?	?	bauxite-free lateritic crust on planation surface in 500m a.s.l.	detritic lateritic crusts (440m a.s.l.) on the cuesta of *Bombouaka*
Pliocene	"transitional relief" Laterites of *Continental Terminal* (CT)	pisolitic lateritic crusts	planation formation in 300m a.s.l.	lateritic crusts within the major geologically controlled landforms (350-400m a.s.l.)
Lower Quaternary (2,0-0,55 M.A.)	*HAUT-GLACIS*	upper conglomeratic Glacis-surface	ferralitic weathering and formation of lateritic crusts on the Pliocene planation surface	secondary conglomeratic lateritic crusts (*GLACIS*-surfaces)
Middle Quaternary (0,55-0,11 M.A.)	*MOYEN-GLACIS*	lower conglomeratic Glacis-surface	polygenetic planation levels (*GLACIS*)	?
Late Quaternary since 110 000 B.P.	*BAS-GLACIS*		?	polycyclically and polygenetically composed lower slope section with local secondary crusts
Holocene since 10 000 B.P. Present	recent dissection of *BAS-GLACIS* levels	recent active erosion surface	recent incision	caused by frequent climatic changes a decay and degradation of older lateritic crusts started; alternation of flood plain deposition and dissection of sediments; advancing of surface extension by etchplanation processes

The chronological subdivision of the laterite incrusted levels adopted for Togo by numerous French authors from MICHEL (1973) who worked in Senegal (for example FAURE 1986, LE COCQ 1986), could not have been confirmed, nor was it conclusively disproved. The idea of reconstructing a wide expansed planation surface covering large parts of the West African landsurface with the help of lateritic crusts must be seen questionable on the base of the presented results. Nevertheless it was made an effort (see Table 3) to compare the new results from Togo with the existing results on planation surface chronology from West Africa. The highest planation-like rest on the Défalé escarpment at 500–550 m a.s.l. (Photo 1) was supposed to be dated from the upper Cretaceous to Paleocene considering the initial opening of the South Atlantic ocean and therefore interconnected tectonic updoming of the upper guinean ridge. The repeatedly redeposited laterites of the Bombouaka plateau were, according to POSS & ROSSI (1987), assigned to an intensive climate-morphological formation during the beginning and in the middle of the Tertiary. The tripartition of the quaternary planation relief of steps, according to MICHEL (1973), could be distinguished only locally in northern Togo (for example near Défalé, Fig. 2f) because of clear physiognomic similarities to the *Haut Glacis*. Because of the many gaps within this uncompleted planation step relief, it is probable that lower quaternary planation levels can be confused with each other and misinterpretated and therefore greater problems in understanding the planation chronology of these plain levels may occur.

Acknowledgements

The author wishes to thank Prof. Dr. K.-H. NITSCH and Prof. Dr. A. MÜCKE, Institut für Mineralogie, Universität Göttingen, and Dr. W. SCHELLMANN, Bundesanstalt für Geowissenschaften und Rohstoffe (BGR), Hannover, for carrying out RFA-Analysis and the making of thin sections as well as for the helpful discussions.

References

AFFATON, P. (1975): Etude géologique et structurale du Nord-Ouest Dahomey, du Nord-Togo et du Sud-Est de la Haute-Volta. – Trav. Lab. Sci. Terre, St. Jérome **10**: 1–201.

AMBROSI, J. P. & D. NAHON (1986): Petrological and Geochemical differentiation of lateritic iron crust profiles. – Chem. Geol. **57**: 371–393.

BARTH, H. K. (1970): Probleme der Schichtstufenlandschaften Westafrikas – am Beispiel der Bandiagara-, Gambaga- und Mampong-Stufenländer. – Tübinger Geogr. Stud. **38**: 1–215.

BESSOLES, B. & R. TROMPETTE (1980): Géologie de l'Afrique Vol. 2, La chaine panafricaine "zone mobile d'Afrique centrale (partie sud) et zone mobile soudanaise". – Mém. du B.R.G.M. **92**: 1–396.

BÜDEL, J. (1981): Klima-Geomorphologie. – 2. Aufl. 1–304; Berlin-Stuttgart.

CRENN, Y. (1957): Mesures gravimétriques et magnétiques dans la partie centrale de l'A.O.F. Interprétations géologiques. – ORSTOM Paris: 1–39.

FAURE, P. (1985): Les sols de la Kara, Nord-Est Togo. Relations avec l'environnement. – Carte Pédologique à 1:50 000. Trav. et Doc. de l'ORSTOM **183:** 1–281.

GOUDIE, A. (1973): Duricrusts in tropical and subtropical landscapes. – 1–174, Oxford.

LE COCQ, A. (1986): Les sols et leurs capacités agronomiques, Région de Bassar, Centre Ouest-Togo. Cartes 1:100 000. – ORSTOM – Notice explicative **102**: 1–103.

LEPRUN, J.-C. (1979): Les cuirasses ferrugineuses des pays cristallins de l'Afrique occidentale seche – Genèse, Transformation, Degradation. – Sci. Géol., Mém. **58**: 1–224.

MCFARLANE, M. (1976): Laterite and Landscape. – 1–151; London, New York.

MICHEL, P. (1973): Les bassins des fleuves Sénégal et Gambie: Etude Géomorphologique. – Mém. ORSTOM **63**: 1–752.

NAHON, D. & J. R. LAPPARTIENT (1977): Time factor and geochemistry in iron crusts genesis. – Catena **4**: 249–254.

OLLIER, C. (1981): Tectonics and landforms. – University of East Anglia, Geomorph. Texts **6**: 1–324.

POSS, R. & R. ROSSI (1987): Systèmes de versants et évolution morphopédologique au Nord-Togo. – Z. Geomorph. N. F. **31**: 21–43.

République Togolaise (1980): Etude pédologique de la vallée de l'Oti. – Ministere de l'Aménagement rural, Lomé, 1–140.

– (1984): Notice explicative de la carte géologique à 1:200 000, Feuille Kara, 1ère édition. – Mém. DGMG et BNRM **1**: 1–36.

– (1985): Notice explicative de la carte géologique à 1:200 000, Feuille Dapaong. – Mém. DGMG et BNRM **2**: 1–42.

RUNGE, J. (1989): Reliefentwicklung, Morphodynamik und Landnutzung in den wechselfeuchten Tropen Westafrikas – Beispiele aus Togo. – Diss. Math. Nat. Fachber. Univ. Göttingen: 1–181.

– (1990a): Geomorphological depressions (Bas-fonds) and present-day erosion processes on the planation surface of central Togo. – Erdkunde **45**: 52–65.

– (1990b): Morphogenese und Morphodynamik in Nord-Togo unter dem Einfluß spätquartären Klimawandels. – Gött. Geogr. Abh. **90**: 1–115.

SCHELLMANN, W. (1986): On the Geochemistry of Laterites. – Chem. Erde **45**: 39–52.

THOMAS, M. F. & M. B. THORP (1985): Environmental Change and episodic etchplanation in the humid tropics of Sierra Leone: The Kouidu Etchplain. – In: DOUGLAS, I. & T. SPENCER: Environmental Change and Tropical Geomorphology. – 239–267, London.

THORBECKE, F. (1927): Der Formenschatz im periodisch trockenen Tropenklima mit überwiegender Regenzeit. – Düsseldorfer Geogr. Vorträge, Teil III, Morphologie der Klimazonen: 10–17; Breslau.

VIELLEFON, P., O. COFFI & R. SANTANNA (1965): Etudes pédo-hydrologiques au Togo., Vol. II.: Les sols de la Région Maritime et de la Région des Savanes. – PNUD-FAO-ORSTOM 2: 1–248; Paris-Rome.

Address of the author: Dr. JÜRGEN RUNGE, Universität-GH Paderborn, FB 1: Physische Geographie, Postfach 1621, D-4790 Paderborn.

Z. Geomorph. N.F.	Suppl.-Bd. 92	217–230	Berlin · Stuttgart	Mai 1993

Sediment Yield and Sediment Retention in a Small Loess-Covered Catchment in SW-Germany

JUSSI BAADE, DIETRICH BARSCH, ROLAND MÄUSBACHER and GERD SCHUKRAFT, Heidelberg

with 4 figures and 3 tables

Summary. Soil erosion and sediment yield from catchments with similar land use patterns show a high spatial variability. Measurements of sediment yield from two catchments revealed differences of up to 700%. Ephemeral gully erosion in thalwegs is important. Against this background preventive measures including a grassed waterway and the establishment of a retention area were tested. Stabilizing a thalweg with a grassed waterway reduced ephemeral gully erosion by 50%. Measurements of sediment inflow and outflow at the retention area showed a sediment trap efficiency of 70% on average. This measure, using only 0.1% of the arable land in the catchment of the receiving stream diminished sediment load in the receiving stream by 10%. Considering the slow introduction of soil protective measures on the fields (in Europe) the described preventive measures might represent useful completions to other erosion and/or water pollution control strategies.

Zusammenfassung. Der durch Bodenerosion verursachte Feststoffaustrag aus Einzugsgebieten mit vergleichbarer Nutzung variiert stark. In den von uns untersuchten Teileinzugsgebieten ergaben sich Unterschiede von bis zu 700%. Lineare Erosion in den Tiefenlinien spielt dabei eine bedeutende Rolle. Vor diesem Hintergrund wurden im Rahmen eines Geländeexperiments selektive Erosionsschutzmaßnahmen untersucht. Sie umfassen eine Stillegung und Begrünung besonders gefährdeter Tiefenlinien und die Anlage von Retentionsflächen am Ausgang intensiv landwirtschaftlich genutzter Teileinzugsgebiete. Durch die Stabilisierung einer Tiefenlinie konnte hier die lineare Erosion um 50% vermindert werden. Die Eintrags- und Austragsmessungen an der Retentionsfläche zeigen, daß hier im Mittel 70% der Feststofffracht am Eintritt in den Vorfluter gehindert werden konnten. Mit dieser Maßnahme, die eine Stillegung von nur 0,1% der Ackerfläche im Einzugsgebiet des Vorfluters erfordert, konnte die Sedimentfracht im Vorfluter um 10% gesenkt werden. Da sich flächendeckende Erosionsschutzmaßnahmen nur langsam durchsetzen, bieten sich Retentionsflächen und die Stillegungen von Tiefenlinien als sinnvolle Ergänzung zu anderen Maßnahmen (z.B. Gewässerrandstreifen) an.

1 *Introduction*

For many years soil erosion on arable land was thought to be a problem only concerning farmers, reducing the productivity of their land in the longer term. The

0044-2798/93/0092-0217 $ 3.50

increased use of commercial fertilizers, genetic improvements, and other new technologies masked or even offset the productivity losses (NOWAK 1988). In recent years more attention has been given to various offsite damages (CLARK 1985, ARMSTRONG et al. 1990). Special attention has been drawn to the impact of soil erosion on water quality (CROSSON 1985). A recently published study (HAMM 1991) reveals that soil erosion accounts for 88% of the total phosphorus and 80% of the total nitrogen input from agricultural activities into the rivers and lakes in Germany. Approximately 70% of these nutrients are bound to the sediment.

Traditionally, soil erosion has been seen as a problem of sheet erosion and rill-interrill erosion. Therefore most attempts to control soil erosion on arable land and sediment yield to receiving streams promote preventive measures that are to be applied to the entire area of arable land, i.e. methods like a change of tillage system and/or land use. Frequently this adjustment requires a renewal and supply of the farmers' fleet of machines, which represent investments. But tradition and the increasing economic pressure on agricultural production are constraints to these necessary changes.

Therefore it seems worthwhile to consider alternative approaches to the reduction of on-site erosion and sediment yield. It is commonly accepted that soil erosion and sediment yield show a high degree of spatial variability: On the scale of a catchment both the variable source area and partial area models suggest that only a small proportion of the basin produces runoff and delivers sediment to the receiving stream (WALLING 1983, BURT 1989). From that it can be assumed that inside of a given catchment there is a restricted number of subcatchments showing higher runoff and/or sediment yield than other subcatchments or the entire catchment. On the scale of a field several recently published papers emphasize the role of thalwegs promoting ephemeral gully erosion (AUZET et al. 1990, DE PLOEY 1990, POESEN & GOVERS 1990). THORNE & ZEVENBERG (1990) estimate that the erosion due to ephemeral gullying may be comparable to rill and interrill erosion. Although most of these studies are based on observations in the European loess belt, it is also a worldwide phenomena (DE PLOEY 1990).

Against this background preventive measures which are little land consuming and cause only minor changes to the farmers' work have been tested to reduce sediment yield from arable land to receiving streams. These measures include a grassed waterway to cut down ephemeral gully erosion on the field and a retention area to diminish sediment yield from an intensively cultivated subcatchment to the receiving stream. Grassed waterways are used for terrace outlets in surface water disposal systems in the USA (HUDSON 1971, MOLDENHAUER & ONSTAD 1977) and are sometimes mentioned as preventive measures to control ephemeral gully erosion (EVANS & COOK 1986, POESEN 1989), but data on their efficiency are hardly available. Although the sediment trap efficiency of reservoirs is well-known (HEINEMANN 1984) and widely used for the estimation of soil erosion (e.g. FOSTER et al. 1990) there are only few examples where small basins were established in order to hold back the sediment (FERGUSON 1981, MIELKE 1985).

The aim of this paper is to present the results of our field experiments and to evaluate the efficiency of the applied preventive measures against the background of measured sediment yield. Special attention will be payed to the efficiency of the retention area.

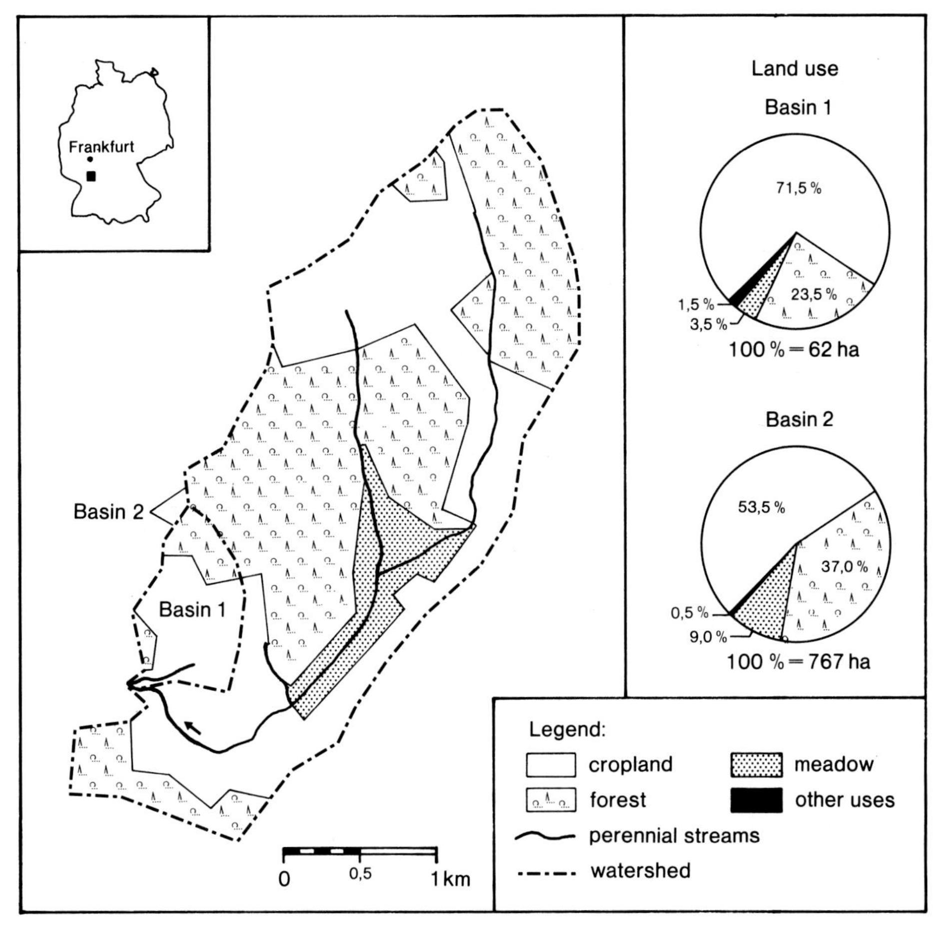

Fig. 1. Investigation area and land use.

2 *Investigation area*

The investigation takes place in the Kraichgau, an intensively cultivated loess-covered hilly region in SW-Germany, located approximately 100 km south of Frankfurt (cf. Fig. 1). The boundary of this region is marked by the Odenwald in the north, the SW-German scarpland to the east, the Black Forest in the south and the Rhine Rift Valley to the west (cf. AHNERT 1989, 2). Average longterm precipitation is 879 mm a^{-1} with the highest monthly precipitation during the summer months (June–August) and a secondary peak in December.

Eroded soil profiles, steps at the edge of forests, and sunken roads testify a long history of soil erosion in this region, which can be traced back to Neolithic times.

Estimates of the mean annual denudation rates on arable land vary from 0.5 mm a^{-1} to 3 mm a^{-1} or 7.5 t ha^{-1} a^{-1} to 45 t ha^{-1} a^{-1}, respectively. For individual events soil losses of several hundred metric tons per hectar are reported (EITEL 1989, 167). Ephemeral gully erosion is well-known to occur in the thalwegs of the dissected and undulating landscape.

The investigation area consists of two interlocked catchments (cf. Fig. 1): a first order basin of 62 ha (Basin 1) and a second order basin (Basin 2) of 767 ha or 7.67 km^2. Local relief is 55 m for Basin 1 and 140 m for Basin 2. The mean slope angle of the main drainage line is 5.7% and 2.8%, respectively. The loess-covered land is primarily used as cropland and forest; meadows and other uses are less important (cf. Fig. 1). On the arable land conventional tillage with cross slope farming is usually applied to grow maize, sugar beets, and winter wheat; a rotation typical for SW-Germany's loess regions. Based on the size of the catchments Basin 1 represents 8.1% of Basin 2, but a comparison of the amount of cropland in these two catchments reveals a considerable higher proportion (10.8%). This latter value is used to evaluate the contribution of Basin 1 to runoff and sediment yield of the receiving stream (Basin 2).

Mean discharge from Basin 2 is about 50 l s^{-1} and 2 l s^{-1} from Basin 1. The highest measured discharge in the period of January 1990 through December 1991 was 2650 l s^{-1} at the gauge station of Basin 2 and 330 l s^{-1} at the main outlet of Basin 1. The hydrological analysis shows that the subcatchment (Basin 1) plays a minor role in terms of total discharge from Basin 2 compared to its areal extent. It is responsible for only 3.0% of total annual discharge. In times of accelerated discharge during rainstorms its contribution increases to a mean of 5.6% and during individual events to a maximum of 12.9% of total discharge (cf. Table 1).

Table 1 Hydrological characteristics of individual events (Jan. 1990–Dec. 1991).

Date	Peak discharge (l s^{-1})		Total discharge (m^3)		Discharge ratio[1]
	GS-1A	GS-2	B1	B2	
15/02/90	130	2,200	4,580	90,125	5.1
27/02/90	150	2,650	3,190	79,285	4.0
30/06/90	330	500	860	6,690	12.9
22/09/90	130	400	800	9,600	8.3
18/11/90	45	280	805	18,165	4.4
20/11/90	80	875	2,450	42,790	5.7
26/02/91	20	180	490	5,200	10.1
20/03/91	30	800	1,050	30,721	2.9
19/12/91	50	675	970	16,130	5.6
20/12/91	52	560	890	15,170	5.4
21/12/91	310	2,280	4,500	63,350	6.0
22/12/91	95	1,300	3,820	57,100	5.7

[1] Discharge ratio = $B1/B2 * 100$.

3 *Applied prevention methods*

In order to reduce sediment yield from intensively cultivated catchments to the receiving streams the following conservation measures have been applied:

– Establishment of a grassed waterway in the thalweg of a zero-order basin to reduce ephemeral gully erosion.

Based on the mapping of ephemeral gullies in the investigation area in March 1990 the most affected thalweg in Basin 1 was selected for the survey. During two subsequent events in February 1990 it accounted for approx. 50% of the sediment load delivered from this catchment. In the valley bottom a 3 to 5 meters wide grassed waterway was established. In addition fascines made out of organic geotextiles were positioned across the thalweg. A total of 1800 m, which equals 4.5% of the arable land in this zero-order basin, was used for this measure. Comparative volumetric mapping of ephemeral gullies in this and another unstabilized thalweg before and after the construction of the grassed waterway indicate that ephemeral gully erosion could be reduced by approx. 50%. Details about the results are published elsewhere (BAADE et al. in print). The efficiency of this measure in terms of total sediment yield from Basin 1 is difficult to estimate because the long term contribution of ephemeral gully erosion in this thalweg to the sediment yield from the catchment is difficult to assess. One can, however, conclude that this grassed waterway will reduce sediment yield to the receiving stream by not more than 25%.

– Establishment of a sediment retention area at the outlet of an intensively cultivated first-order catchment to diminish sediment yield to the receiving stream.

At the outlet of the first-order catchment (Basin 1) an area of 4000 m^2 is used for the construction of a sediment retention area. A natural succession of plants covers this former field. In addition fascines, similar to the ones used in the thalweg, were set up across the slightly inclined slope (slope gradient 1.9°). Compared to the area of arable land of Basin 1 (44.5 ha) less than 1% is used for this measure. Details about the findings are given below.

4 *Measuring arrangement and methods*

Suspended sediment load from the investigation area is calculated from data received through continuous registrations of discharge and through event-triggered automatic water sampling. Fig. 2 shows the measuring arrangement at the outlet of the catchments and at the retention area. Runoff and suspended sediment leave Basin 1 mainly through the input gauge station at the retention area (GS-1A), but some water flows out in a spillway (GS-1B), which was constructed to avoid overtopping at the input gauge station. The gauge station of the receiving stream (GS-2) is located about 100 m upstream of the confluence with the channel from Basin 1. The input (GS-1A) and the output of the retention area (GS-1C) are equipped with a Parshall-flume with a capacity of 335 $l\ s^{-1}$; the spillway (GS-1B) is equipped with a Thomson-weir. In Basin 2 (GS-2) a rectangular overflow-weir with a capacity of 2900 $l\ s^{-1}$ is used. Total discharge and sediment load from Basin 1 are calculated adding up the data from the gauge stations GS-1A and GS-1B, and the data for Basin 2 is calculated adding up the results from the gauge stations GS-1A, GS-1B, and GS-2.

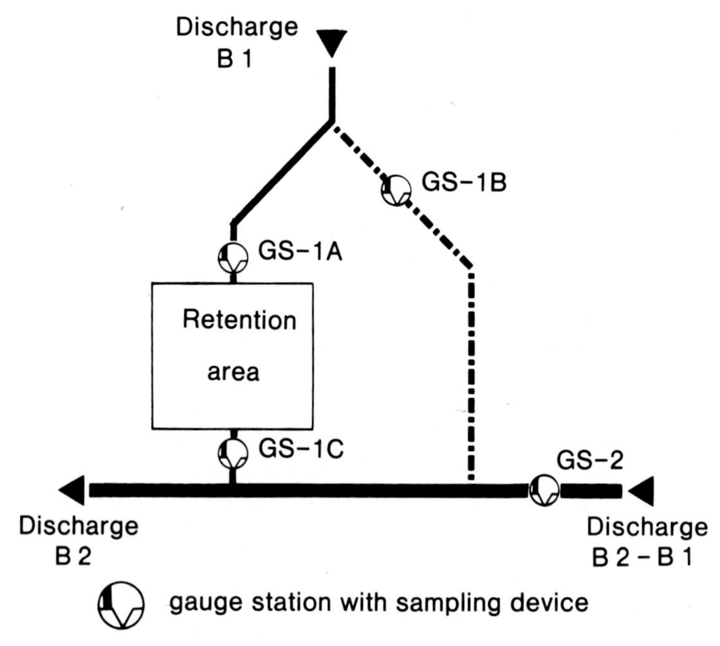

Fig. 2. Diagram of the measuring arrangement in the investigation area.

Water samples are taken in the stream line in a depth of 10 cm. In order to handle the fluctuations of the sediment concentrations through time the intervals for the sampling are 15 min to 30 min for Basin 1 and 30 min to 60 min for Basin 2. The higher temporal resolution is used during summer. Additional samples are taken by hand during most events to make sure that the selected intervals are sufficient. Sediment concentrations of the water samples are determined by filtration (0.2 μm).

Sediment load is primarily calculated on the base of measured sediment concentrations and discharge data. Rating curves for the concentrations/discharge relation were only used to bridge short time intervals with missing values for individual events. This restricted use of rating curves is explained by analysis of the concentration/discharge relation which showed the well-known poor correlation between these two variables (WALLING 1988). Even for individual events one has to distinguish between the rising and falling stage of the flood hydrograph to get a satisfying correlation.

To quantify the accuracy of the results the water level recording is assumed to be correct with a deviation of ± 1 cm. This rather conservative estimation is used, because the accuracy of the filtration process is difficult to quantify. It is therefore assumed that the sediment concentration data is free of errors. Using the discharge curves of the gauge stations the relative datum error for the discharge is calculated. It decreases with increasing discharge. Selected threshold values are the following: At the gauge station of Basin 2 (GS-2) the 10% error level is reached at a discharge of 95 $l\ s^{-1}$ and the 5% error level at a discharge of 255 $l\ s-^{1}$. For the Parshall-flume at the main gauge station of Basin 1 (GS-1A) and the output of the retention area (GS-1C) the corresponding threshold values are 31 l s^{-1} and 89 l s^{-1}, respectively. From this a mean datum error of 5% for the sediment load data from Basin 2 and of about 10% for the data from Basin 1 can be derived. We are aware of further possible inexactitudes, like the problems inherent in a close-to-surface one-point sampling for the determination of the suspended sediment concentration (WALLING 1988). But since the same sampling strategy is used at all sites, this should only influence absolute values of sediment load but not disturb proportions when several sampling sites are compared.

5 *Results*

5.1 *Sediment yield*

In order to establish a basis for the quantitative evaluation of the preventive measures described above, the sediment yield from Basin 1 and Basin 2 was measured. Another aim of these measurements was to confirm the assumption that Basin 1 is a subcatchment which delivers more sediment to the receiving stream than one would expect from its proportion of cropland. Whether this subcatchment always shows overproportional sediment yield or whether there are exceptions to that was another question under examination.

The results presented in this paper are based on the measurements from January 1990 to December 1991. During this period 12 events with quite different hydrological characteristics were sampled and analysed. Details about the peak discharge at the gauge stations GS-1A and GS-2 and the total discharge from Basin 1 and Basin 2 are given in Table 1. The comparison of the peak discharge reveals a wide range of different constellations with very high runoff from Basin 1 and moderate runoff from Basin 2 (June 30, 1990), with high runoff from Basin 1 and very high runoff from Basin 2 (Feb. 1990) or quite low runoff at both gauge stations (Feb. 26, 1991). This constrains the interpretation, but demonstrates the wide range of different events occurring in this rather small and homogeneous area.

Most events took place during winter and sediment yield and soil erosion, respectively, were highest at this time of year (cf. Table 2). This is unexpected because other papers (cf. DIKAU 1986, EITEL 1989, QUIST 1987) stress the dominant role of summer thunderstorms for soil erosion in this region. The unusually dry summer of 1991 and the short investigation period, which is far from being representative, might be an explanation for this uneven seasonal distribution. Nonetheless, this is advantageous for the interpretation of the results, too, because during winter several highly variable factors controlling soil erosion and sediment yield (like rainfall distribution, soil moisture and plant cover on arable land) are more evenly distributed in space than during summer.

In the two years' period the total suspended sediment load from Basin 2 amounted to 868.2 t. 114.4 t (13.2%) of this load stemmed from Basin 1. Comparing the mean contribution of Basin 1 to the sediment load from Basin 2 (13.2%) with its proportion of cropland (10.8%) shows already that Basin 1 can be considered an important source for the sediment. These findings are somewhat contradictionary to

Table 2 Suspended sediment load and sediment yield for individual events (Jan. 1990–Dec. 1991).

Date	Suspended sediment load (t)		Suspended sediment yield (t ha^{-1})		SYR (%)
	B1	B2	B1	B2	
15/02/90	11.5	153.0	0.639[1]	0.398[1]	161
27/02/90	8.9[2]	122.0	0.494[1]	0.318[1]	155
30/06/90	11.0	42.5	0.247	0.104	238
22/09/90	3.8	18.8	0.085	0.046	185
18/11/90	1.2	4.8	0.027	0.012	225
20/11/90	2.3	40.3	0.052	0.098	53
Sum 1990	38.7	381.4	1.544	0.930	
26/02/91	2.2	3.0	0.049	0.007	700
20/03/91	2.0	33.3	0.045	0.081	56
19/12/91	3.7	45.2	0.083	0.110	75
20/12/91	4.1	23.1	0.092	0.056	164
21/12/91	53.7[2]	316.8	1.207	0.772	156
22/12/91	10.0[2]	65.4	0.225	0.159	142
Sum 1991	75.7	486.8	1.701	1.185	

Sediment Yield Ratio (SYR) = Sediment yield (B1)/Sediment yield (B2) * 100.

[1] Because a large proportion of Basin 1 (26.5 ha) was covered with mustard (*Sinapis alba* L.), sediment yield is calculated for an area of 18 ha for Basin 1 and an area of 384 ha for Basin 2.

[2] Results is partly based on extrapolation.

the results of the hydrological analysis (cf. Table 1), because one would expect an important sediment contribution area to be an important storm runoff contribution area, too. Up to now the contribution of Basin 1 to storm runoff has only been calculated on the basis of total discharge. An analysis based on the separation of flow is underway and will probably reveal higher values for direct runoff.

Assuming that soil erosion occurred only on arable land the sediment yield (t ha^{-1}) and the sediment yield ratio (SYR) was calculated for every event (Table 2). The mean annual sediment yield amounted to 1.6 t ha^{-1} a^{-1} for Basin 1 and to 1 t ha^{-1} a^{-1} for Basin 2. These values are considerably lower than the soil loss estimations mentioned earlier. However, comments on that would be precipitate because of the short investigation period and the large influence of single events on the results.

During most events sediment yield from Basin 1 was 150% to 200% higher than from Basin 2. In one case even a value of 700% was calculated for the sediment yield ratio. This latter value has to be judged as an exception, because it occurred during a snowmelt event with an unequal distribution of the snow cover.

Beside of that three events are found with higher sediment yield from Basin 2 (i.e. the events of Nov. 20, 1990, March 20, 1991 and Dec. 19, 1991). The question arises, whether during these events soil erosion was really more intensive in the other parts of the catchment or whether these findings have to be explained in terms of

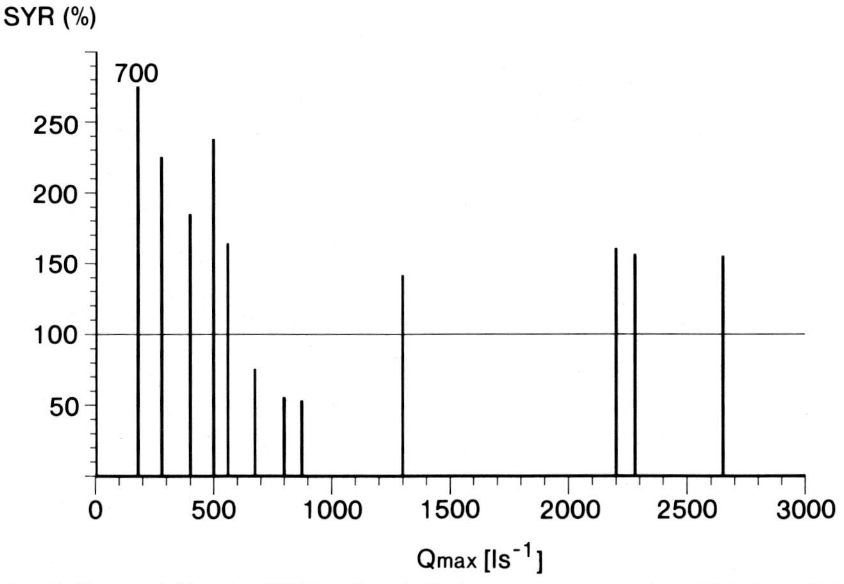

Fig. 3. Sediment yield ratio (SYR) and peak discharge at gauge station GS-2 for individual events (Jan. 1990–Dec. 1991).

Note: SYR > 100% indicates higher sediment yield from Basin 1, SYR = 100% indicates equal sediment yield from Basin 1 and Basin 2, SYR < 100% indicates higher sediment yield from Basin 2.

sediment delivery to and/or sediment transport in the receiving stream. All three events happened during winter and one can assume equal preconditions for soil erosion throughout the catchment. The two events in November 1990 and the events in December 1991 following each other immediately but revealing nonetheless different SYR values confirm this assumption. Therefore, we favor an explanation by in-stream processes. In Fig. 3 SYR values are graphed against the peak discharge at the gauge station of Basin 2 (GS-2). It is remarkable that all three events with SYR values < 100 are grouped around a maximum discharge between 650 l s^{-1} and 900 l s^{-1}. In the area with lower peak discharge SYR values are significantly higher and the same is true for the area with higher peak discharge. There is no similar pattern when SYR values are graphed against the maximum discharge at the outlet of Basin 1. This leads to the conclusion that the fluvial dynamics in the receiving stream might be responsible for this distribution. There are at least two possible explanations: erosion of stream bank, and/or the mobilisation of in-stream short time deposits (liquid mud). Stream bank erosion has been observed in the receiving stream and other channels with similar characteristics in this region (BARSCH et al. 1989), but it is generally assumed to occur rather during or shortly after bankfull floods (KNIGHTON 1984). On the other hand observations in the backwater area of gauge station GS-2 indicate a movement of liquid mud on the channel floor. Up to now it is impossible to decide which of these two processes is more likely to explain these observations. Direct measurements are planned to answer this question.

Despite of some events showing a higher sediment yield from the larger catchment, the first order catchment could be identified as an important sediment contribution area. During most events sediment yield from this catchment is 150% to 200% higher.

5.2 *Measures to increase sediment retention*

At the outlet of Basin 1 a retention area was established to reduce the sediment yield to the receiving stream. When runoff enters the retention area it spreads laterally; flow velocity is reduced and a large proportion of the suspended sediment is deposited. This process is further supported by the dense vegetation cover on the ground. Observations during several events indicate that the fascines have no visible effects. This is nonetheless advantageous, because a renunciation of fascines means a considerable reduction of the costs of a retention area.

Fig. 4 shows the flood hydrographs and the suspended sediment concentrations at the input and output of the retention area in the course of an individual event. The hydrographs of the input and output gauge stations being very similar reveal that the flood flow is not much influenced. But there is a visible reduction of suspended sediment concentrations in the runoff: The highest measured sediment concentration at the input was 11.1 g l^{-1}, whereas the concentration peak at the output was 2.7 g l^{-1}. Furthermore a delay of the sediment concentration peak of about half an hour

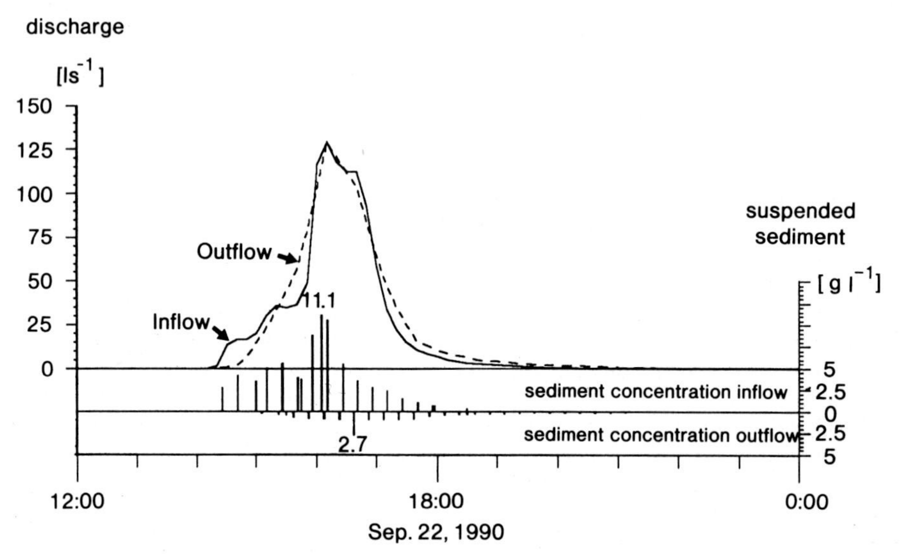

Fig. 4. Discharge and suspended sediment concentration at the input and output gauge stations of the retention area (Sep. 22, 1990).

has been observed frequently. This implies that the sediment is not transported uninterruptedly across the retention area by the flowing water but that sedimentation and remobilisation determine the transport process instead.

Peak discharge and peak suspended sediment concentrations at the input of the retention area, sediment inflow and outflow and the sediment trap efficiency (HEINEMANN 1984) of the retention area are presented in Table 3. Unfortunately, no data is available for the largest event (Dec. 21, 1991) because of a malfunction of the automatic sampling device at the outlet of the retention area. The mean efficiency of all events amounts to 65%. The comparison of the results for individual events reveals that the efficiency of the retention area is considerably higher and more stable in the second year after installation. Whereas the data shows considerable variation in 1990, the sediment trap efficiency stabilizes at a level of about 80% in 1991. This increasing efficiency is interpreted as a consequence of the development of the permanent vegetation cover. The high effectiveness during the spring events in 1991 is attributed to a high amount of organic litter on the ground after the winter. Investigations are continued to bolster the assumptions that the sediment trap efficiency remains on this level.

A comparison of the suspended sediment load from Basin 1 (cf. Table 2) and the sediment inflow to the retention area (Table 3) shows that only a small proportion of the sediment left this catchment through the spillway. It is assumed that even a slightly higher sediment input would not have influenced the sediment trap efficiency of the retention area. For the event of Dec. 21, 1991 a trap efficiency of 82% was assumed based on the findings of the neighboring events (cf. Table 3). Recalculation

Table 3 Sediment trap efficiency of the sediment retention area during individual events (Jan. 1990–Dec. 1991).

Date	Q_{max} [1] $(l\ s^{-1})$	Cs_{max} [2] $(g\ l^{-1})$	Sediment inflow (t)	Sediment outflow (t)	E [3]
15/02/90	133	6.5	8.45	4.88	42.2
27/02/90	150	>2.3	6.50 [4]	3.38	48.0
30/06/90	335	23.2	8.60	3.28 [4]	61.9
22/09/90	130	11.1	3.67	0.87	76.3
18/11/90	43	9.5	1.06	0.64	39.6
20/11/90	78	2.6	2.09	1.15	44.9
26/02/91	21	12.7	2.26	0.25	88.9
20/03/91	32	13.8	1.94	0.19	90.2
19/12/91	52	9.3	3.29	0.61	81.5
20/12/91	52	19.1	3.56	0.65	81.7
21/12/91	310	37.5	44.70 [4]	.	.
22/12/91	95	>7.4	8.10 [4]	1.40 [4]	82.7

[1] Q_{max} = peak discharge at input gauge station (GS-1A).
[2] Cs_{max} = maximum suspended sediment concentration at input gauge station (GS-1A).
[3] Sediment trap efficiency $E = (Sin - Sout)/Sin * 100$.
[4] Result is partly based on extrapolation.

of the sediment load from Basin 1 was undertaken, using the sediment trap efficiency for every individual event. So the sediment load from Basin 1 is reduced by 83 t or 72.5%, respectively. This means that the contribution of Basin 1 to the sediment delivered from Basin 2 is diminished to 4.0% as compared to the initial value of 13%. Related to the sediment load transported in the receiving stream a reduction by 9.6% could be attained. This reduction is achieved on an area which represents less than 1% of the arable land in Basin 1 and less than 0.1% of the arable land of Basin 2.

From the calculated amount of sediment deposited on the retention area (83 t) the life time and/or maintenance requirements of this preventive measure can be assessed. Assuming a uniform sedimentation a mean aggradation of 5 mm a^{-1} is calculated. After 20 years aggradation would reach a value of 10 cm and the trapped sediment should be returned to the contributing area or the enclosing structures should be built higher. Because the mean annual sediment yield from the catchment was quite low during the period under investigation compared to soil loss estimations reported in other papers, this estimation may be too high.

Evaluating the scrutinized preventive measures shows that the retention area is more efficient in reducing the sediment yield to receiving streams, that it uses less arable land and does not interfere with the farmers' work. Disadvantageous is only the fact that it does not diminish soil erosion itself.

6 *Conclusions*

The results of our field experiments show that grassed waterways and retention areas are effective measures to reduce ephemeral gully erosion and sediment yield to receiving streams, respectively. Further advantages are the very low proportion of arable land set aside for these measures and the low maintenance costs. Therefore they might represent useful completions to other erosion and/or water pollution control strategies, like vegetative filter strips. Investigations are continued to enhance the design of both preventive measures and to verify the present results. On the retention area special efforts will focus on an efficient flood control which most likely will further increase sediment retention.

Acknowledgement

The study was funded by the federal state Baden-Württemberg, FRG, as a part of the "Projekt Wasser–Abfall–Boden" (PW 88070) at the Kernforschungszentrum Karlsruhe GmbH (KfK). We are especially grateful to the landowner, Prinz zu Löwenstein, for his cooperation.

References

Ahnert, F. (1989): The major landform regions. – In: Ahnert, F. (ed.): Landforms and landform evolution in West Germany. – Catena Suppl. **15**: 1–9; Cremlingen-Destedt.

Armstrong, A. C., D. B. Davies & D. A. Castle (1990): Soil water management and the control of erosion on agricultural land. – In: Boardman, J. et al. (eds.): Soil erosion on agricultural land: 569–574; Chichester.

Auzet, A. V., J. Boiffin, F. Papy, J. Maucorps & J. F. Ouvry (1990): An approach to the assessment of erosion forms and erosion risk on agricultural land in the Northern Paris Basin, France. – In: Boardman, J. et al. (eds.): Soil erosion on agricultural land: 383–400; Chichester.

Baade, J., D. Barsch, R. Mäusbacher & G. Schukraft (in print): Field experiments on the reduction of sediment yield from arable land to receiving watercourses (N-Kraichgau, SW-Germany). – Submitted to 'Erosion des Terres agricoles en milieu tempéré des plaines', Proceedings of the Conference at St. Cloud, May 25–29, 1992.

Barsch, D., R. Mäusbacher, G. Schukraft & A. Schulte (1989): Beiträge zur aktuellen fluvialen Geomorphodynamik in einem Einzugsgebiet mittlerer Größe am Beispiel der Elsenz im Kraichgau. – In: Pörtge, K.-H. & J. Hagedorn (eds.): Beiträge zur aktuellen fluvialen Morphodynamik. – Gött. Geogr. Abh. **86**: 9–31.

Burt, T. P. (1989): Storm runoff gegeration in small catchments in relation to the flood response of large basins. – In: Beven, K. & P. Carling (eds.): Floods: Hydrological, sedimentological and geomorphological implications: 11–35; Chichester.

Clark, E. H. I. (1985): The off-site costs of soil erosion. – J. Soil Water Conserv. **40**: 19–22.

Crosson, P. (1985): Impact of erosion on land productivity and water quality in the United States. – In: El-Swaify, S. A. et al. (eds.): Soil Erosion and Conservation. (Soil Conservation Soc. of America, Internat. Conf. on Soil Erosion and Conservation, Jan. 16–22, 1983, Honolulu, Hawaii): 217–236. Ankeny, Iowa.

De Ploey, J. (1990): Threshold conditions for thalweg gullying with special reference to loess areas. – In: Bryan, R. B. (ed.): Soil erosion: experiments and models. – Catena Suppl. **17**: 147–151; Cremlingen-Destedt.

Dikau, R. (1986): Experimentelle Untersuchungen zu Oberflächenabfluß und Bodenabtrag von Meßparzellen und landwirtschaftlichen Nutzflächen. – Heidelb. Geogr. Arb. **81**; Heidelberg.

Eitel, B. (1989): Morphogenese im südlichen Kraichgau unter besonderer Berücksichtigung tertiärer und pleistozäner Decksedimente. Ein Beitrag zur Landschaftsgeschichte Südwestdeutschlands. – Stuttg. Geogr. Stud. **111**; Stuttgart.

Evans, R. & S. Cook (1986): Soil erosion in Britain. – SEESOIL **3**: 28–58.

Ferguson, B. K. (1981): Erosion and sedimentation control in regional and site planning. – J. Soil Water Conserv. **36**: 199–204.

Foster, I., R. Grew & J. Dearing (1990): Magnitude and frequency of sediment transport in agricultural catchments: A paired lake-catchment study in Midland England. – In: Boardman, J. et al. (ed.): Soil erosion on agricultural land: 153–171; Chichester.

Hamm, A. (ed.) (1991): Studie über Wirkungen und Qualitätsziele von Nährstoffen in Fließgewässern. – Herausgegeben vom Arbeitskreis 'Wirkungsstudie' im Hauptausschuß 'Phosphate und Gewässer' in der Gesellschaft Deutscher Chemiker durch den Obmann des Arbeitskreises A. Hamm; Sankt Augustin.

Heinemann, H. G. (1984): Reservoir trap efficiency. – In: Hadley, R. F. & D. E. Walling (ed.): Erosion and sediment yield: some methods of measurement and modelling: 201–218; Cambridge.

Hudson, N. (1971): Soil Conservation. – London.

Knighton, D. (1984): Fluvial Forms and Processes. – London.

Mielke, L. N. (1985): Performance of water and sediment control basins in northeastern Nebraska. – J. Soil Water Conserv. **40**: 524–528.

Moldenhauer, W. C. & C. A. Onstad (1977): Engineering practices to control erosion. – In: Greenland, D. J. & R. Lal (eds.): Soil conservation and management in the humid tropics: 87–92; Chichester.

NOWAK, P. J. (1988): The costs of excessive soil erosion. – J. Soil Water Conserv. **43**: 307–310.
POESEN, J. W. A. (1989): Conditions for gully formation in the Belgian loam belt and some ways to control them. – In: SCHWERTMANN et al. (eds.): Soil erosion protection measures in Europe. – Soil Technolog Series **1**: 39–52; Cremlingen-Destedt.
POESEN, J. W. A. & G. GOVERS (1990): Gully erosion in the Loam Belt of Belgium: Typology and control measures. – In: BOARDMAN, J. et al. (eds.): Soil erosion on agricultural land: 513–530; Chichester.
QUIST, D. (1987): Bodenerosion – Gefahr für die Landwirtschaft im Kraichgau? – Kraichgau **10**: 42–62.
THORNE, C. R. & L. W. ZEVENBERGEN (1990): Prediction of ephemeral gully erosion on cropland in the south-eastern United States. – In: BOARDMAN, J. et al. (eds.): Soil erosion on agricultural land: 447–460; Chichester.
WALLING, D. E. (1983): The sediment delivery problem. – J. Hydrol. **65**: 209–237.
– (1988): Measuring sediment yield from river basins. – In: LAL, R. (ed.): Soil erosion research methods: 39–73; Ankeny, Iowa.

Address of the authors: J. BAADE, D. BARSCH, R. MÄUSBACHER and G. SCHUKRAFT, Labor für Geomorphologie und Geoökologie, Geographisches Institut, Universität Heidelberg, Im Neuenheimer Feld 348, D-6900 Heidelberg.

Z. Geomorph. N.F.	Suppl.-Bd. 92	231–239	Berlin · Stuttgart	Mai 1993

Geographical Information Systems as Tools in Geomorphology

by

RICHARD DIKAU, Heidelberg

with 3 figures and 3 tables

Summary. Geographical information systems (GIS) are important tools for model development and application in geomorphology. Results are focused mainly on applying evaluative models to geoscientific map data to create derivatives useful for applied purposes and to predict natural hazards and risks. Much progress has been made in using digital elevation models (DEM) for morphometrical landform recognition and to improve morphometrical input data for geomorphological process models. The paper documents GIS applications with respect to different geomorphological processes and to specific analysis and modeling approaches.

Zusammenfassung. Geographische Informationssysteme (GIS) sind wichtige Werkzeuge für die Modellentwicklung und -anwendung in der Geomorphologie geworden. Ergebnisse konzentrieren sich vornehmlich auf die Anwendung von Bewertungsmodellen auf geowissenschaftliche Basisdaten, um zu Auswertungen zu gelangen, die für angewandte Zwecke und für die Abschätzung natürlicher Gefahren und Risiken nützlich sind. Wesentliche Fortschritte liegen im Einsatz digitaler Höhenmodelle, die für die Erkennung von geomorphographischen Reliefeinheiten und für die Verbesserung der geomorphographischen Eingabedaten in geomorphologischen Prozeßmodellen verwendet werden. Der Artikel dokumentiert GIS-Anwendungen im Hinblick auf unterschiedliche geomorphologische Prozeßtypen und auf spezielle Analyse- und Modellierungsansätze.

1 *Introduction*

The application of geographical information system (GIS) technologies in geomorphological research shows a high variety ranging from morphometrical landform recognition to groundwater hazard assessments. In many cases geomorphological informations are closely connected with theoretical assumptions and basic data from other geoscientific disciplines, e.g. soil science or geoecology. Therefore, the objective of this paper is to characterize some typical GIS applications, whereby geomorphological data and/or models are integral parts of the investigation, independent of the discipline involved.

The review is focused on GIS activities in Germany. Further examples refer to specific permafrost model developments in Switzerland. A first collection of articles concerning GIS activities in German Geography has been published by a GIS working group of the Zentralverband der Deutschen Geographen (GOSSMANN & SAURER 1991).

0044-2798/93/0092-0231 $ 2.25

Table 1 Examples of GIS applications with respect to different geomorphological processes.

Process/process factor	Author
Infiltration and filtering effects of soils	FRÄNZLE (1989, 1991) RASCHKE & MÜLLER (1991)
Landslides	DIKAU (1990) PÜSCHEL (1991) DIKAU & JÄGER (1993)
Permafrost	HOELZLE (1992) KELLER (1992)
Soil erosion	FLACKE et al. (1990) HENSEL (1991) ERDMANN & ROSCHER (1991) JÄGER (1993a, 1993b)
Runoff generation	FARRENKOPF (1993)

2 *GIS applications with respect to different geomorphological processes*

A classification of GIS applications in geomorphological research with respect to different geomorphological processes shows the 5 main topics summarized in Table 1. The importance of landform evolution and characteristics of Quaternary sediments and related soils to estimate spatially differentiated chemical pollution hazards of groundwater is shown by an integrated environmental monitoring project in the state of Schleswig-Holstein in northern Germany (FRÄNZLE 1989, 1991). Based on digitized maps (1:1 to 1:2 million scale) of soils, mean annual precipitation and evaporation, soil erodibility and land use, a groundwater sensivity map could be derived. This map has been used to determine chemical pollution hazard of groundwater on a regional scale. An approach to derive physical and chemical soil attributes and soil pollution assessments from soil maps at 1:25,000 scale and remote sensing data has been described by RASCHKE & MÜLLER (1991). They use static overlay models to estimate a hazard potential for leaching of cadmium depending on land use.

GIS tools applied to landslide hazard problems have been presented by DIKAU (1990). A model to estimate the relative landslide-susceptibility of a 12 km^2 study area in the Tertiary basin of Mainz (20 km south-west of Frankfurt) could be accomplished by using (1) geological maps, (2) landslide inventories of the state Geological Survey of Rheinland-Pfalz and (3) landform morphometry information delineated from digital elevation models (DEM). High correlations between landslide locations, geology and slope convergence could be found in the study area. The data resolution influences the hazard model significantly. PÜSCHEL (1991) examined the influence of DEMs with different resolutions and founds clear underestimations of high landslide-susceptibility classes using a 40 m-DEM compared with 20 m-grids. A temporal and spatial extension of these investigations is described by DIKAU

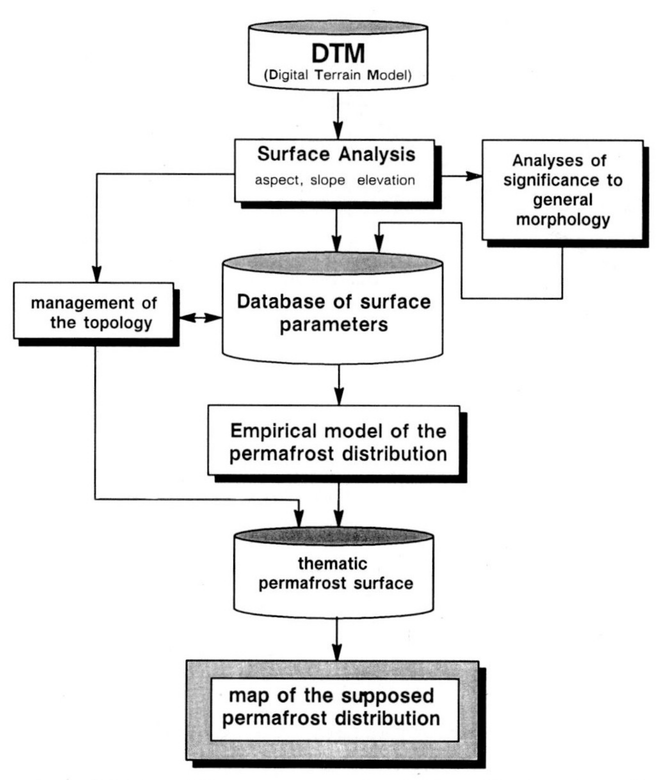

Fig. 1. Data and empirical model components of a GIS application to predict spatial permafrost distribution (KELLER 1992).

& JÄGER (1993) who use GIS tools to create regional landslide hazard models for the entire basin (1000 km²).

GIS applications to predict regional mountain permafrost distribution are presented by HOELZLE (1992) and KELLER (1992), Laboratory of Hydraulics, Hydrology and Glaciology, VAW-ETH Zürich, Switzerland. They develop two different types of models. The prediction model PERMAKART (KELLER 1992) is based on empirical knowledge about morphometrical parameters influencing ground temperatures and snow effects. A DEM approach is accomplished to derive the parameters and to create a spatial assessment of permafrost distribution in the Corvatsch area, Swiss Alps (Fig. 1). Based on highly significant relationships between potential direct solar radiation and the bottom temperature of the snow cover (BTS) an empirical model to predict mountain permafrost distribution has been developed by HOELZLE (1992)

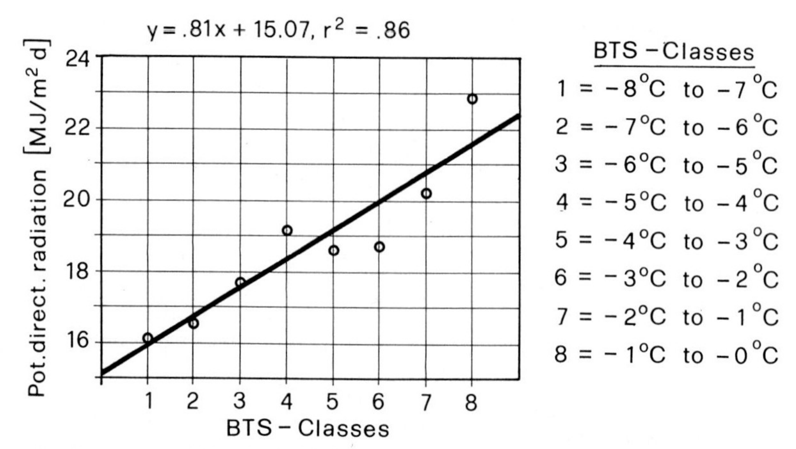

Fig. 2. Evaluation model to link field data of permafrost occurrence (expressed by bottom temperatures of the snow cover, BTS) and potential direct solar radiation derived from digital elevation models (HOELZLE 1992).

Table 2 Examples of analysis and modeling approaches using GIS tools.

Analysis and modeling approach	Author
Derivatives from detailed geoscientific maps	FRÄNZLE (1989, 1991) DIKAU (1990) ERDMANN & ROSCHER (1991) STÄBLEIN (1991) DIKAU & JÄGER (1993)
Regionalization, scale dependence of processes	FRÄNZLE (1989, 1991) DIKAU (1992) JÄGER (1993a)
Time analysis of process factors and process patterns	ERDMANN & ROSCHER (1991) DIKAU & JÄGER (1993)
Morphometrical landform modeling and recognition based on DEMs	FLACKE et al. (1990) JORDAN & LINDNER (1991) HENSEL (1991) KÖTHE & LEHMEIER (1991) DIKAU (1992) JÄGER (1993b) FARRENKOPF (1993)

(Fig. 2). Since radiation was derived from DEMs in 100 m-resolution and BTS reflects permafrost occurrence, the model could be used to create permafrost distribution maps.

In soil erosion research two topics of GIS applications are important, including (1) modifications of empirical models, such as the Universal Soil Loss Equation

(USLE) using improved morphometrical variables and (2) potential soil erosion assessments by derivatives from geoscientific maps. A modification of the slope length and slope angle factor (LS) of the USLE is described by BORK & HENSEL (1988), FLACKE et al. (1990) and HENSEL (1991). These approaches intend to improve limitations of the LS-factor in case of a complex slope morphometry. HENSEL (1991) developed a modified USLE (MUSLE87) to replace the slope length factor by the size of the catchment area calculated from gridded digital elevation data. A similar technique was developed by FLACKE et al. (1990) using digital elevation models in TIN mode (triangular irregular network) to calculate a differentiated slope length factor. This factor is based on partitioning of the slope into triangle-shape catchment units, which are characterized by a set of morphometrical attributes, e.g. the position of slope elements within the slope, area, mean slope length, flow direction etc.

The use of GIS techniques to integrate data from several state agencies of Baden-Württemberg (36,000 km^2) for assessments of regional soil erosion susceptibility shows JÄGER (1993a, 1993b). Soil loss is calculated by using the Universal Soil Loss Equation. The susceptibility model is based on a quantitative classification directed toward conservation policy. The data resolution of the final map collection is 2 · 2 km.

3 *Analysis and modeling approaches*

The GIS applications discussed above are based on different analysis and modeling procedures which are summarized in Table 2. One of the most frequent evaluative model is based on deriving secondary information from geoscientific maps. These models refer to theoretical and empirical knowledge about relationships between geoscientific data, e.g. slope material properties, and process behaviour, e.g. infiltration rates, which is highly suitable for research problems of geoecology or applied geomorphology. Examples show (1) FRÄNZLE (1989) who relates groundwater hazards to the average filtering capacity of soils in northern Germany, (2) STÄBLEIN (1989) who derives the natural potential for waste desposal from slope material information of a large scale geomorphological map, and (3) ERDMANN & ROSCHER (1991) who assess soil erodibility (K-factor of the USLE) based on soil map information.

An important problem of spatial modeling is emphasized by FRÄNZLE (1989) concerning scale constraints of the data and models included. FRÄNZLE (1989) aruges, that at the local scale, estimates of groundwater sensivity derived at regional scales has to include information about throughflow and groundwater hydraulics, which show high variability in space. Therefore, the regional approach presented cannot be adopted directly to assessments at a local scale. The necessity to develop a scale dependent morphometrical theory, adequate models, and GIS tools has been stressed by DIKAU (1992). He asserts, that morphometrical classification and portrayal of landforms requires the determination of concurrences of form and processes at different scales. From a morphometrical point of view this leads to a set of attributes assumed to be relevant for the different hierarchic levels of the classification system (Table 3). However, regarding GIS tools already available to transforming such hierarchic systems into adequate morphometrical models, it seems obvious

Table 3 Selection of scale dependent attributes of different hierarchic levels of landforms, which are a conceptual basis for advanced morphometrical GIS tools under development (DIKAU 1992).

Hierarchic level of relief unit/ characteristic dimension	A selection of primary attributes	
	Object attributes << inherent >>	Toposequences/patterns << context >>
Mesorelief B: Scale: 1:50,000 - 1:200,000 DEM: 100 - 600 m (approx.)	Attribute distribution, Attribute variability, Grain, Relief, Type of toposequences, Ridge/channel density	Type of low level unit, Position within macrorelief, Type of inhomogeneity, Process pattern
Mesorelief A: Scale: 1:5,000 - 1:50,000 DEM: 20 - 300 m (approx.)	Grain, Texture, Shape, Gradient, Curvature, Aspect	Toposequence position, Pattern of low-level units, Neighbours, Functional relationships
Microrelief A/B: Scale: 1:500 - 1:5000 DEM: 1 - 50 m (approx.)	Size, Shape, Gradient, Aspect, Curvature	Connectiveness, Density, Dispersion, Position within mesorelief

that we will need to develop in the future much more sophisticated software packages.

To include time into spatial modeling approaches is a largely underdeveloped task in geomorphological GIS applications. Here, time means change of process patterns and change of factors responsible for process occurrences. A recent example of this approach is found in the paper by ERDMANN & ROSCHER (1991), in which the authors reconstructed 4 land use pattern maps (1895, 1926, 1956 and 1990) to derive time dependent models for soil loss assessments on test areas near Bonn. Other basic data used in these investigations are soil maps and geomorphological maps at 1:25,000 scale. First approaches to include landslide process sequences for the last 100 years into a spatial hazard model has been presented by DIKAU & JÄGER (1993) who use landform morphometry and geology to regionalize local information.

A wide range of GIS applications is focused on morphometrical landform recognition and modeling based on digital elevation models. This includes calculations of input information for geomorphological process models previously discussed (BORK & HENSEL 1988, FLACKE et al. 1990, HENSEL 1991). JORDAN & LINDER (1991) investigated glacier ice budgets before and after the Nevado del Ruiz volcano eruption disaster in Columbia in 1985 by calculations based on gridded elevation data. The use of morphometrical landform analysis to predict spatial soil distributions within a project to construct large scale soil maps was presented by KÖTHE & LEHMEIER (1991). A set of morphometrical landform descriptors is created, e.g.

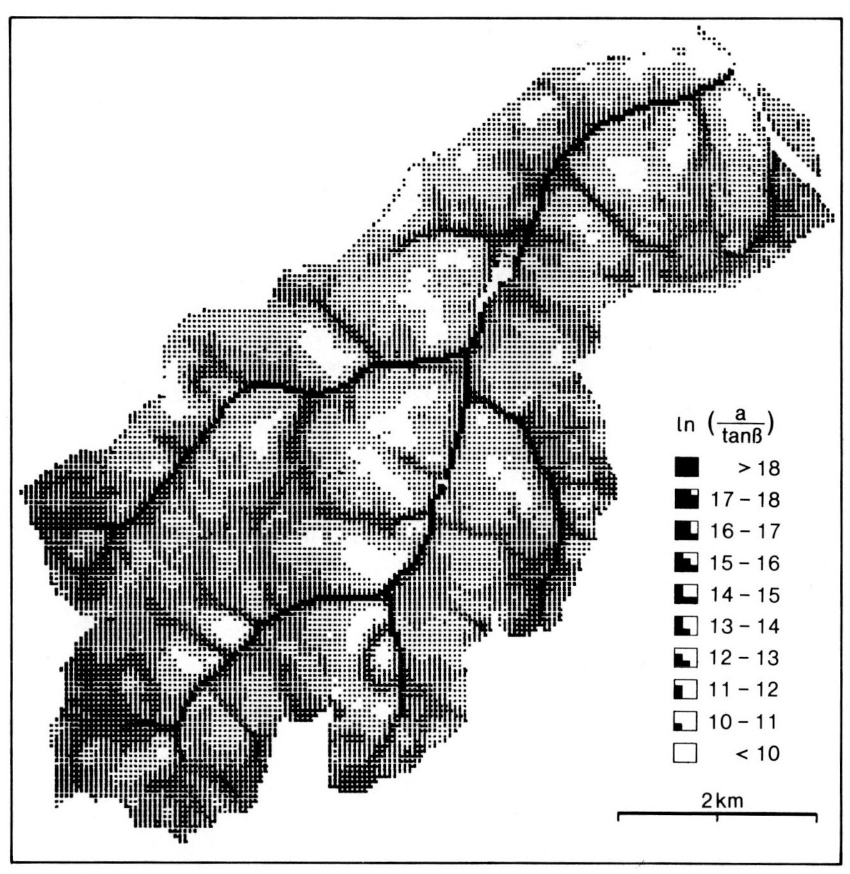

Fig. 3. Morphometrical index ln (a/tanβ) which is used to create variable source areas based on flow accumulation values (a) and slope angle (β) (FARRENKOPF 1993, calculated with computer models described by DIKAU 1992).

drainage lines and differently curved breaks of slopes, assumed to be adequate predictors for different soils. Within the prediction procedure, morphometrical data are used in combination with geological maps, detailed soil taxation data and historical land use maps. GÜNDRA (1992) developed a prediction model for soils of a loess area in northern Baden-Württemberg. He founds significant correlations between soil patterns and morphometrical attributes reflecting soil erosion and deposition processes in the test area, such as profile curvatures and slope convergence. These attributes could be derived from DEMs in 10 m resolution.

The development of morphometrical GIS tools applicable to geomorphological process studies and hazard prediction shows DIKAU (1992) and DIKAU & JÄGER (1992). On the basis of digital elevation models the authors derive sets of morphometrical attributes, e.g. profile curvatures and flow accumulation values. The objective of the project is to reconstruct scale dependent hierarchies of landform units and landform unit patterns, which can be used for morphometrical landform classifications and for improved models of spatial hazard assessments.

First modeling approaches to replace manually gathered data by deriving the topographical LS factors of the USLE on a regional scale from high resolution digital elevation models is described by JÄGER (1993b). He argues, that the highest accuracy should be observed for determining the LS factor, being the factor with the highest influence on the results. Therefore, topographical model variables will be derived from DEMs for the entire state in a forthcoming project.

The application of morphometrical analysis in modeling hydrological landform–runoff relationships is the task of a project in south-west Germany (FARRENKOPF 1993). Based on digital elevation models, the approach is concerned with linking spatially distributed morphometrical indices with runoff data of a 32 km^2 forested drainage basin. The indices are used to derive landform units assumed to have significant influence on runoff generation processes, e.g. variable source areas, which could be calculated by flow accumulation values, slope angle and slope curvatures (Fig. 3).

How empirical models can be linked with spatially distributed attributes of surface morphometry using GIS tools is presented in the previously cited paper by KELLER (1992) (Fig. 1). Here, digital elevation models are used to calculate sets of model variables, which are used to extrapolate the evaluative model over the entire area.

4 *Conclusions*

It seems obvious that geographical information systems appear promising tools for use in geomorphological process studies. The purpose of the paper is to report applications with typical data sources and model structures. This could be shown in a number of case studies from Germany and Switzerland which comprise:
- derivatives from different geoscientific maps,
- morphometrical derivatives from digital elevation models,
- groundwater and landslide hazard assessments,
- mountain permafrost and soil predictions,
- soil erosion and runoff generation models, and
- regionalized data and modeling approaches.

References

BORK, H.-R. & H. HENSEL (1988): Computer-aided construction of soil erosion and deposition maps. – Geol. Jb. **A104**: 357–371

DIKAU, R. (1990): Derivatives from detailed geoscientific maps using computer methods. – Z. Geomorph. N. F., Suppl. – Bd. **80**:45–55.

– (1992): Aspects of constructing a digital geomorphological base map. – Geol. Jb. **A122**: 357–370.

DIKAU, R. & S. JÄGER (1993): Landslide hazard modeling in New Mexico and Germany. (in prep.)

ERDMANN, K.-H. & S. ROSCHER (1991): Untersuchungen zur Bodenerosion im Bonner Raum unter Einsatz eines geographischen Informationssystems. – Arb. Rhein. Landeskunde **60**: 93–104.

FARRENKOPF, D. (1993): Die Steuerfunktion des Reliefs für den direkten Abfluß. – PhD Dissertation, Dept. of Geography, University of Heidelberg (in prep.).

FLACKE, W., K. AUERSWALD & L. NEUFANG (1990): Combining a modified Universal Soil Loss Equation with a digital terrain model for computing high resolution maps of soil loss resulting from rain wash. – Catena **17**: 383–397.

FRÄNZLE, O. (1989): Landform development and soil structure of the northern Federal Republic of Germany – Their role in groundwater resources management. – Catena Suppl. **15**: 95–106.

– (1991): Zukunftsorientierte Umweltforschung im Rahmen des Deutschen MAB-Programms. – In: RIEWENHERN, W. & H. LIETH (eds.): Verhandlungen der Gesellschaft für Ökologie, **19/3**: 545–562; Osnabrück

GOSSMANN, H. & H. SAURER (1991): GIS in der Geographie. – Freiburger Geogr. H. **34**: Freiburg.

GÜNDRA, H. (1992): Untersuchungen zu Relief und Bodenverbreitung im Einzugsgebiet des Biddersbach, Nord-Kraichgau. – Unpubl. MS Thesis, Dept. of Geography, University of Heidelberg.

HENSEL, H. (1991): Verfahren zur EDV-gestützten Abschätzung der Erosionsgefährdung von Hängen und Einzugsgebieten. – Bodenökologie u. Bodengenese **2**: Berlin.

HOELZLE, M. (1992): Permafrost occurrence from BTS measurements and climatic parameters in the eastern Swiss Alps. – Permafrost and Periglacial Processes **3**: 143–148.

JÄGER, S. (1993a): Modeling regional soil erosion of Baden-Württemberg using the Universal Soil Loss Equation and GIS tools. – Soil Technology (in print).

– (1993b): Bodenerosionsatlas Baden-Württemberg (in print).

JORDAN, E. & W. LINDER (1991): Einsatz und die Anwendung von digitalen Höhen-(Gelände-)Modellen am Universitätsstandort Vechta. – In: GOSSMANN, H. & H. SAURER (eds.): GIS in der Geographie. – Freiburger Geogr. H. **34**: 89–98; Freiburg.

KELLER, F. (1992): Automated mapping of mountain permafrost using the program PERMAKART within the geographical information system ARC/INFO. – Permafrost and Periglacial Processes **3**: 133–138.

KÖTHE, R. & F. LEHMEIER (1991): Digitale Reliefanalyse – Ein Projekt zur geomorphologischen Auswertung Digitaler Geländemodelle (DGM). – In: GOSSMANN H. & H. SAURER (eds.): GIS in der Geographie. – Freiburger Geogr. H. **34**: 99–109; Freiburg

PÜSCHEL, U. (1991): Computergestützte Untersuchungen zur Rutschungsempfindlichkeit an der nordwestlichen Randstufe des Rheinhessischen Plateaus. – Unpubl. MS Thesis, Dept. of Geography, University of Heidelberg.

RASCHKE, N. & K.-H. MÜLLER (1991): Auswertung von Bodenkarten mit Hilfe geographischer Informationssysteme unter Verwendung digitaler Fernerkundungsdaten. – Geoökodynamik **12**: 41–70.

Address of the author: Geographisches Institut der Universität Heidelberg, Labor für Geomorphologie und Geoökologie, Im Neuenheimer Feld 348, D-6900 Heidelberg.